# Gas Turbine Theory

By G. F. C. Rogers and Y. R. Mayhew
*Engineering Thermodynamics Work and Heat Transfer* (Longman)

*Thermodynamic and Transport Properties of Fluids:* SI (Blackwell)

*Diagram for Temperature Rise v. Fuel–Air Ratio for Combustion of a Gas Turbine Fuel* (Blackwell)

By G. F. C. Rogers
*The Nature of Engineering* (Macmillan)

THIRD EDITION

# GAS TURBINE THEORY

## H. Cohen
Fellow, Queens' College, University of Cambridge

## G. F. C. Rogers
Professor Emeritus, University of Bristol

## H. I. H. Saravanamuttoo
Professor, Department of Mechanical and Aeronautical Engineering,
Carleton University

Copublished in the United States with
John Wiley & Sons, Inc., New York

Longman Scientific & Technical
Longman Group UK Limited
Longman House, Burnt Mill, Harlow
Essex CM20 2JE, England
and Associated Companies throughout the world
Copublished in the United States with
John Wiley & Sons, Inc., 605 Third Avenue, New York, NY 10158

*First published* 1951
*Eighth impression* 1971
*Second edition published* 1972
*Tenth impression* 1985
*Third edition published under the Longman Scientific & Technical imprint* 1987

**Library of Congress Cataloging-in-Publication Data**

Cohen, Henry, 1921 Sept. 29–
  Gas turbine theory.

  Bibliography: p.
  Includes index.
  1. Gas-turbines.  I. Rogers, G. F. C. (Gordon
Frederick Crichton)  II. Saravanamuttoo, H. I. H.
III. Title.
TJ778.C6  1987    621.43'3    86-21478
ISBN 0-470-20705-1 (USA only)

**British Library Cataloguing in Publication Data**

Cohen, H.
  Gas turbine theory.—3rd ed.
  1. Gas-turbines
  I. Title  II. Rogers, G.F.C.
  III. Saravanamuttoo, H.I.H.
  621.43'3    TJ778

  ISBN 0-582-30539-X

Produced by Longman Singapore Publishers (Pte) Ltd.
Printed in Singapore.

# Contents

# *Preface*
# *to third edition*

The continued use of this book since the appearance of the first edition in 1951 suggests that the objectives of the original authors were sound. To the great regret of Professors Rogers and Saravanamuttoo, Dr Cohen was unable to join them in the preparation of this new edition. They would like to express their appreciation of his earlier contribution, however, and in particular for initiating the book in the first place.

Since the second edition was published in 1972 considerable advances have been made and the gas turbine has become well established in a variety of applications. This has required considerable updating, particularly of the Introduction. Sub-sections have been added to the chapter on shaft power cycles to include combined gas and steam cycles, cogeneration schemes and closed cycles, and many minor modifications have been made throughout the book to take account of recent developments. The chapter on axial flow compressors has been rewritten and enlarged, and describes in detail the design of a compressor to meet a particular aerodynamic specification which matches the worked example on turbine design. A section on radial flow turbines has been added to the turbine chapter.

In keeping with the introductory nature of the book, and to avoid losing sight of physical principles, no attempt has been made to introduce modern computational methods for predicting the flow in turbomachinery. Appropriate references to such matters have been added, however, to encourage the student to pursue these more advanced topics.

As a result of many requests from overseas readers, Professor Saravanamuttoo has agreed to produce a manual of solutions to the problems in Appendix B. The manual can be purchased by writing to: The Bookstore, Carleton University, Ottawa, Ontario K1S-5B6, Canada.

Finally, this edition would not have been possible without the generous support of Professor Saravanamuttoo's research by the Natural Sciences and Engineering Research Council of Canada. He would particularly like to express his thanks to Dr Bernard MacIsaac of GasTOPS and the many engineers from Pratt and Whitney Canada, Rolls-Royce, and the Royal Navy with whom he has collaborated for many years.

*March 1986*                                        G.F.C.R.
                                                   H.I.H.S.

# *Preface*
## *to second edition*

At the suggestion of the publishers the authors agreed to produce a new edition of Gas Turbine Theory in SI units. The continued use of the first edition encouraged the authors to believe that the general plan and scope of the book was basically correct for an introduction to the subject, and the object of the book remains as stated in the original Preface. So much development has taken place during the intervening twenty-one years, however, that the book has had to be rewritten completely, and even some changes in the general plan have been required. For example, because gas dynamics now forms part of most undergraduate courses in fluid mechanics a chapter devoted to this is out of place. Instead, the authors have provided a summary of the relevant aspects in an Appendix. Other changes of plan include the extension of the section on aircraft propulsion to a complete chapter, and the addition of a chapter on the prediction of performance of more complex jet engines and transient behaviour. Needless to say the nomenclature has been changed throughout to conform with international standards and current usage.

The original authors are glad to be joined by Dr Saravanamuttoo in the authorship of the new edition. He is actively engaged on aspects of the gas turbine with which they were not familiar and his contribution has been substantial. They are glad too that the publisher decided to use a wider page so that the third author's name could be added without abbreviation. Dr Saravanamuttoo in his turn would like to acknowledge his debt to the many members of staff of Rolls-Royce, the National Gas Turbine Establishment and Orenda Engines with whom he has been associated in the course of his work. The authors' sincere appreciation goes to Miss G. M. Davis for her expert typing of the manuscript.

*October 1971*

H.C.
G.F.C.R.
H.I.H.S.

*Preface*
*to first edition*

This book is confined to a presentation of the thermodynamic and aerodynamic theory forming the basis of gas turbine design, avoiding as far as possible the many controversial topics associated with this new form of prime-mover. While it is perhaps too early in the development of the gas turbine to say with complete assurance what are the basic principles involved, it is thought that some attempt must now be made if only to fill the gap between the necessarily scanty outline of a lecture course and the many articles which appear in the technical journals. Such articles are usually written by specialists for specialists and presuppose a general knowledge of the subject. Since this book is primarily intended for students, for the sake of clarity some parts of the work have been given a simpler treatment than the latest refinements of method would permit. Care has been taken to ensure that no fundamental principle has been misrepresented by this approach.

Practising engineers who have been engaged on the design and development of other types of power plant, but who now find themselves called upon to undertake work on gas turbines, may also find this book helpful. Although they will probably concentrate their efforts on one particular component, it is always more satisfactory if one's specialised knowledge is buttressed by a general understanding of the theory underlying the design of the complete unit.

Practical aspects, such as control systems and mechanical design features, have not been described as they do not form part of the basic theory and are subject to continuous development. Methods of stressing the various components have also been omitted since the basic principles are considered to have been adequately treated elsewhere. For a similar reason, the theory of heat-exchangers has not been included. Although heat-exchangers will undoubtedly be widely used in gas turbine plant, it is felt that at their present stage of development the appropriate theory may well be drawn from standard works on heat transmission.

Owing to the rate at which articles and papers on this subject now appear, and since the authors have been concerned purely with an attempt to define the basic structure of gas turbine theory, a comprehensive bibliography has not been included. A few references have been selected, however, and are appended to appropriate chapters. These are given not only as suggestions for further reading, but also as acknowledgements of sources of information. Some problems, with answers, are given at the end of the book. In most cases these have been chosen to illustrate various points which have not appeared in the worked examples in the text. Acknowledgements are due to the Universities of Cambridge, Bristol and Durham for permission to use problems which have appeared in their examination papers.

The authors have obtained such knowledge as they possess from contact with the work of a large number of people. It would therefore be invidious to acknowledge their debt by mention of one or two names which the authors associate in their minds with particular aspects of the work. They would, however, like to express their gratitude in a corporate way to their former colleagues in the research teams of what were once the Turbine Division of the Royal Aircraft Establishment and Power Jets (R. & D.) Ltd, now the National Gas Turbine Establishment.

In conclusion, the authors will welcome any criticisms both of detail and of the general scheme of the book. Only from such criticism can they hope to discover whether the right approach has been adopted in teaching the fundamentals of this new subject.

*April 1950*                                                          H.C.

                                                                     G.F.C.R.

# 1 *Introduction*

Of the various means of producing mechanical power the turbine is in many respects the most satisfactory. The absence of reciprocating and rubbing members means that balancing problems are few, that the lubricating oil consumption is exceptionally low, and that reliability can be high. The inherent advantages of the turbine were first realized using water as the working fluid, and hydro-electric power is still a significant contributor to the world's energy resources. Around the turn of the twentieth century the steam turbine began its career and, quite apart from its wide use as a marine power plant, it has become the most important prime mover for electricity generation. Steam turbine plant producing over 500 MW of shaft power with an efficiency of nearly 40 per cent are now being used. In spite of its successful development, the steam turbine does have an inherent disadvantage. It is that the production of high-pressure high-temperature steam involves the installation of bulky and expensive steam generating equipment, whether it be a conventional boiler or nuclear reactor. The significant feature is that the hot gases produced in the boiler furnace or reactor core never reach the turbine; they are merely used indirectly to produce an intermediate fluid, namely steam. Clearly a much more compact power plant results when the water to steam step is eliminated and the hot gases themselves are used to drive the turbine. Serious development of the gas turbine began not long before the Second World War with shaft power in mind, but attention was soon transferred to the turbojet engine for aircraft propulsion. The gas turbine began to compete successfully in other fields only in the mid nineteen fifties, but since then it has made a progressively greater impact in an increasing variety of applications.

In order to produce an expansion through a turbine a pressure ratio must be provided, and the first necessary step in the cycle of a gas turbine plant must therefore be compression of the working fluid. If after compression the working fluid were to be expanded directly in the turbine, and there were no losses in either component, the power developed by the turbine would just equal that absorbed by the compressor. Thus if the two were coupled together the combination would do no more than turn itself round. But the power developed by the turbine can be increased by the addition of energy to raise the temperature of the working fluid prior to expansion. When the working fluid is air a very suitable means of doing this is by combustion of fuel in the air which has been compressed. Expansion of the hot working fluid then produces a greater power output from the turbine, so that it is able to provide a useful output in addition to the power necessary to drive the compressor. This represents the gas turbine or internal-combustion turbine in its simplest form. The three main components are a compressor, combustion chamber and turbine, connected together as shown diagrammatically in Fig. 1.1.

**FIG. 1.1 Simple gas turbine system**

In practice, losses occur in both the compressor and turbine which increase the power absorbed by the compressor and decrease the power output of the turbine. A certain addition to the energy of the working fluid, and hence a certain fuel supply, will therefore be required before the one component can drive the other. This fuel produces no useful power, so that the component losses contribute to a lowering of the efficiency of the machine. Further addition of fuel will result in a useful power output, although for a given flow of air there is a limit to the rate at which fuel can be supplied and therefore to the net power output. The maximum fuel/air ratio that may be used is governed by the working temperature of the highly stressed turbine blades, which temperature must not be allowed to exceed a certain critical value. This value depends upon the creep strength of the materials used in the construction of the turbine and the working life required.

These then are the two main factors affecting the performance of gas turbines: component efficiencies and turbine working temperature. The higher they can be made, the better the all-round performance of the plant. It was, in fact, low efficiencies and poor turbine materials which brought about the failure of a number of early attempts to construct a gas turbine engine. For example, in 1904 two French engineers, Armengaud and Lemale, built a unit which did little more than turn itself over: the compressor efficiency was probably no more than 60 per cent and the maximum gas temperature that could be used was about 740 K.

It will be shown in Chapter 2 that the overall efficiency of the gas turbine cycle depends also upon the pressure ratio of the compressor. The difficulty of obtaining a sufficiently high pressure ratio with an adequate compressor efficiency was not resolved until the science of aerodynamics could be applied to the problem. The development of the gas turbine has gone hand in hand with the development of this science, and that of metallurgy, with the result that it is now possible to find advanced engines using pressure ratios of up to 35:1, component efficiencies of 85–90 per cent, and turbine inlet temperatures of up to 1650 K.

In the earliest days of the gas turbine, two possible systems of combustion were proposed: one at constant pressure, the other at constant volume. Theoretically, the thermal efficiency of the constant volume cycle is higher than that of the constant pressure cycle, but the mechanical difficulties are very much greater. With heat addition at constant volume, valves are necessary to isolate the combustion chamber from the compressor and turbine. Combustion is therefore intermittent, which impairs the smooth running of the machine. It is difficult to design a turbine to operate efficiently under such conditions and, although several fairly successful attempts were made in Germany during the period 1908–1930 to construct gas turbines operating on this system, the development of the constant volume type has been discontinued. In the constant pressure gas turbine, combustion is a continuous process in which valves are unnecessary and it was soon accepted that the constant pressure cycle had the greater possibilities for future development.

It is important to realize that in the gas turbine the processes of compression, combustion and expansion do not occur in a single component as they do in a reciprocating engine. They occur in components which are separate in the sense that they can be designed, tested and developed individually, and these components can be linked together to form a gas turbine unit in a variety of ways. The possible number of components is not limited to the three already mentioned. Other compressors and turbines can be added, with intercoolers between the compressors, and reheat combustion chambers between the turbines. A heat-exchanger which uses some of the energy in the turbine exhaust gas to preheat the air entering the combustion chamber may also be introduced. These refinements may be used to increase the power output and efficiency of the plant at the expense of added complexity, weight and cost. The way in which these components are linked together not only affects the maximum overall thermal efficiency, but also the variation of efficiency with power output and of output torque with rotational speed. One arrangement may be suitable for driving an alternator under varying load at constant speed, while another may be more suitable for driving a ship's propeller where the power varies as the cube of the speed.

Apart from variations of the simple cycle obtained by the addition of these other components, consideration must be given to two systems distinguished by the use of *open* and *closed* cycles. In the much more common open cycle gas turbine which we have been considering up to this point, fresh atmospheric air is drawn into the circuit continuously and energy is added by the combustion of fuel in the working fluid itself. In this case the products of combustion are expanded through the turbine and exhausted

to atmosphere. In the alternative closed cycle shown in Fig. 1.2 the same
working fluid, be it air or some other gas, is repeatedly circulated through
the machine. Clearly in this type of plant the fuel cannot be burnt in the
working fluid and the necessary energy must be added in a heater or 'gas-
boiler' wherein the fuel is burnt in a separate air stream supplied by an

FIG. 1.2 **Simple closed cycle**

auxiliary fan. The closed cycle is more akin to that of steam turbine
plant in that the combustion gases do not themselves pass through the
turbine. In the gas turbine the 'condenser' takes the form of a precooler for
cooling the gas before it re-enters the compressor. Numerous advantages
are claimed for the closed cycle and these will be put forward in section 1.3:
it is certainly the more suitable form of gas turbine cycle if a nuclear reactor
is to be used as the source of heat because it is dangerous to exhaust to
atmosphere a gas which has passed through the reactor core with the
possibility of radioactive contamination.

Finally, various *combined steam and gas cycles* have been proposed, with
the gas turbine exhaust supplying energy to the steam boiler. Figure 1.3

FIG. 1.3 **Combined steam and gas cycle**

shows such a system. It makes the best use of the comparatively low grade heat by employing a dual pressure steam cycle. This is similar to that used in nuclear power stations fuelled with natural uranium which also operate at a comparatively low temperature. Alternatively, because of the unused oxygen in the turbine's exhaust gas, it is possible to burn additional fuel in the steam boiler. This permits use of a single pressure steam cycle, but at the expense of the added complexity of a combustion system in the boiler. Although the characteristic compactness of the gas turbine is sacrificed in binary cycle plant, a significant number have been built for electricity generation. It should be noted that thermal efficiencies of up to 50 per cent can be achieved, which are considerably in excess of the values obtained by either steam turbine or diesel plant.

The gas turbine has proved itself to be an extremely adaptable source of power and has been used for a wide variety of functions, ranging from electric power generation and jet propulsion to the supply of compressed air and process heat, and the remainder of this Introduction is intended to emphasize this adaptability.† We shall commence, however, by discussing the various ways in which the components can be linked together when the object is the production of *shaft power*. In other words, we shall first have in mind gas turbines for electric power generation, pump drives for gas or liquid pipe lines, and land and sea transport. The vast majority of land-based gas turbines are in use for the first two of these, and applications to land and sea transport are still in their infancy, although the gas turbine is widely used in naval applications.

## 1.1  Open cycle single-shaft and twin-shaft arrangements

If the gas turbine is required to operate at a fixed speed and fixed load condition such as in peak load power generation schemes, the single-shaft arrangement shown in Fig. 1.1 may be suitable. Flexibility of operation, i.e. the rapidity with which the machine can accommodate itself to changes of load and rotational speed, and efficiency at part load, are in this case unimportant. Indeed the effectively high inertia due to the drag of the compressor is an advantage because it reduces the danger of overspeeding in the event of a loss of electrical load. A heat-exchanger might be added as in Fig. 1.4(a) to improve the thermal efficiency, although for a given size of plant the power output could be reduced by as much as 10 per cent due to frictional pressure losses in the heat-exchanger: the use of a heat-exchanger would be contemplated only if running hours were high.

Figure 1.4(b) shows a modified form proposed for use when the fuel, e.g. pulverized coal, is such that the products of combustion contain constituents

† Some of the remarks about the 'stability of operation' and 'part-load performance' will be more fully understood when the rest of the book, and Chapter 8 in particular, have been studied. It is suggested therefore that the remainder of the Introduction be given a second reading at that stage.

which corrode or erode the turbine blades. It is much less efficient than the normal cycle because the heat-exchanger, inevitably less than perfect, is transferring the whole of the energy input instead of merely a small part of it. Such a cycle would only be considered if a supply of 'dirty' fuel was available at very low cost. A serious effort was made to develop a coal burning gas turbine in the early nineteen fifties but with little success. More success has been achieved with residual oil, and provided that the maximum temperature is kept at a sufficiently low level the straightforward cycle can be used.

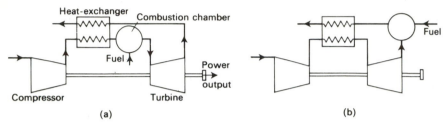

FIG. 1.4  **Single shaft open cycle gas turbines with heat-exchanger**

When flexibility in operation is of paramount importance, such as in road, rail and marine applications, the use of a mechanically independent (or *free*) *power turbine* is desirable. In this *twin-shaft* arrangement, Fig. 1.5,

FIG. 1.5  **Gas turbine with separate power turbine**

the high-pressure turbine drives the compressor and the combination acts as a *gas generator* for the low-pressure power turbine. Twin-shaft arrangements are also used for large scale electricity generating units, with the power turbine designed to run at the alternator speed without the need for an expensive reduction gearbox. A significant advantage is that the starter unit needs only be sized to turn over the gas generator. The starter may be a steam turbine, a diesel, or an expansion turbine operated from a supply of pipeline gas. A disadvantage of a separate power turbine, however, is that a shedding of electrical load can lead to rapid over-speeding of the turbine, and the control system must be designed to prevent this.

Variation of power for both single- and twin-shaft units is obtained by controlling the fuel flow supplied to the combustion chamber. Although they behave in rather different ways as will be explained in Chapter 8, in both cases the cycle pressure ratio and maximum temperature decrease as the power is reduced from the design value with the result that the thermal efficiency deteriorates considerably at part load.

The performance of a gas turbine may be improved substantially by reducing the work of compression and/or increasing the work of expansion. For any given compressor pressure ratio, the power required per unit quantity of working fluid is directly proportional to the inlet temperature. If therefore the compression process is carried out in two or more stages with intercooling, the work of compression will be reduced. Similarly, the turbine output can be increased by dividing the expansion into two or more stages, and reheating the gas to the maximum permissible temperature between the stages. Although the power output is improved the cost in additional fuel will be heavy unless a heat-exchanger is also employed. One arrangement of a plant incorporating intercooling, heat-exchange and reheat is shown in Fig. 1.6. Complex cycles of this type offer the possibility of varying the power output by controlling the fuel supply to the reheat chamber, leaving the gas generator operating closer to its optimum conditions. Much better economy at part load can therefore be achieved.

Complex cycles were proposed in the early days of gas turbines, when they were necessary to obtain a reasonable thermal efficiency at the low turbine

**FIG. 1.6 Complex plant with intercooling, heat exchange and reheat**

temperatures then possible. It can readily be seen, however, that the inherent simplicity and compactness of the gas turbine have been lost. In many applications small size and low capital cost are more important than high thermal efficiency, and it is significant that the gas turbine did not start to be widely used (apart from aircraft applications) until higher turbine inlet temperatures made the simple cycle economically viable. Unless the gas turbine begins to supplant the steam turbine in base load electricity generating stations, the preference for the simple cycle with or without heat-exchange is likely to remain.

## 1.2   Multi-spool arrangements

To obtain a high thermal efficiency without using a heat-exchanger, a high pressure ratio is essential. A certain difficulty then arises which follows from the nature of the compression process.

Because of the high air mass flow rates involved, non-positive displacement compressors are always used in gas turbines. Although the multi-stage centrifugal compressor is capable of producing a high pressure ratio for moderate powers, its efficiency is appreciably lower than that of the axial flow compressor. For this reason the axial compressor is normally preferred, particularly for large units. Unfortunately this type of compressor is more prone to instability when operating at conditions widely removed from its design operating point. When such a compressor operates at rotational speeds well below the design value, the air density in the last few stages is much too low, the axial flow velocity becomes excessive, and the blades stall. The unstable region, manifested by violent aerodynamic vibration, is likely to be encountered when a gas turbine is started up or operated at low power.

The problem is particularly severe if an attempt is made to obtain a pressure ratio of more than about 8:1 in one compressor. One way of overcoming this difficulty is to divide the compressor into two or more sections. In this context division means *mechanical* separation, permitting each section to run at a different rotational speed, unlike the intercooled compressor shown in Fig. 1.6. When the compressors are mechanically independent, each will require its own turbine, a suitable arrangement being shown in Fig. 1.7. The low-pressure compressor is driven by the low-pressure turbine

LP          HP                    HP        LP

**FIG. 1.7  Twin-spool engine**

and the high-pressure compressor by the high-pressure turbine. Power is normally taken either from the low-pressure turbine shaft, or from an additional free power turbine. The configuration shown in Fig. 1.7 is usually referred to as a *twin-spool* engine. It should be noted that although the two spools are mechanically independent, their speeds are related aerodynamically and this will be discussed further in Chapter 9.

The twin-spool layout is widely used both for shaft power units and for the aircraft engines discussed in section 1.4. In some cases, especially with engines of small air flow, the high pressure compressor is of the centrifugal type; this is because at the high pressures involved the volume flow rate is low and the blading required for an axial compressor would be too small for good efficiency. Twin-spool units were first introduced at a pressure ratio of about 10:1, and are suitable for cycle pressure ratios of at least 30:1. For very high pressure ratios a triple-spool arrangement may be used.

As an alternative to multi-spools, a high pressure ratio can be safely employed with a single compressor if several stages of variable stator blades are used. This approach has been pioneered by General Electric and pressure ratios of around 15:1 have been obtained in this manner. In advanced technology engines, combinations of multi-spool arrangements and variable stators are not uncommon.

## 1.3 Closed cycles

Outstanding among the many advantages claimed for the closed cycle is the possibility of using a high pressure (and hence a high gas density) throughout the cycle, which would result in a reduced size of turbomachinery for a given output and enable the power output to be altered by a change of pressure level in the circuit. This form of control means that a wide range of load can be accommodated without alteration of the maximum cycle temperature and hence with little variation of overall efficiency. The chief disadvantage of the closed cycle is the need for an external heating system, which involves the use of an auxiliary cycle and introduces a temperature difference between the combustion gases and the working fluid. The allowable working temperature of the surfaces in the heater will therefore impose an upper limit on the maximum temperature of the main cycle. A typical arrangement of a closed cycle gas turbine is shown in Fig. 1.8. The cycle includes a water cooled pre-

FIG. 1.8 Simple closed cycle gas turbine

cooler for the main cycle fluid between the heat-exchanger and compressor. In this particular arrangement the gas heater forms part of the cycle of an auxiliary gas turbine set, and power is controlled by means of a blow-off valve and an auxiliary supply of compressed gas as shown.

Besides the advantages of a smaller compressor and turbine, and efficient control, the closed cycle also avoids erosion of the turbine blades and other detrimental effects due to the products of combustion. Also, the need for filtration of the incoming air, which is a severe problem in the use of open cycle units operating in contaminated atmospheres, is eliminated. The high density of the working fluid improves heat transfer, so that more effective heat-exchange is possible. Finally, the closed circuit opens up the field for the use of gases other than air having more desirable thermal properties. As will be seen in the next chapter, the marked difference in the values of the specific heats for air and a monatomic gas such as helium does not affect the efficiency as much as might be supposed. But higher fluid velocities can be used with helium and optimum cycle pressure ratios are lower, so that in spite of the lower density the turbomachinery may not be much larger. On the credit side, the better heat transfer characteristics of helium mean that the size of the heat-exchanger and precooler can be about half that of units designed for use with air. The capital cost of the plant should therefore be less when helium is the working fluid.

At the time of writing, a dozen or so closed cycle plants of 2–20 MW output have been built, mostly from the Escher–Wyss stable, and all with air as the working fluid but burning various fuels such as coal, natural gas, blast-furnace gas and oil. Turbine inlet pressures of up to forty atmospheres have been used. A pilot plant of 25 MW using helium has been built in Germany, and with this working fluid large sets of up to 250 MW are thought to be possible. If so they may well be suitable for use in nuclear power plant. Considerable advantage accrues when the working fluid of the power cycle can be passed directly through the reactor core because the reactor coolant circulator pumps are no longer required and the unwanted temperature drop associated with an intermediate fluid (e.g. $CO_2$ temperature–steam temperature) is eliminated.

Finally, various small closed-cycle gas turbines (of 20–100 kW electrical output) have been considered for aerospace and underwater applications. In some the heat is derived from a radioactive isotope such as plutonium-238, and in others from the combustion of hydrogen.

## 1.4   Aircraft propulsion

Without any doubt the greatest impact of the gas turbine has been in the field of aircraft propulsion. The most important landmark in this development was the first experimental Whittle engine in 1937. Since then the gas turbine has completely supplanted the reciprocating engine, for all but light aircraft, because of its much higher power/weight ratio. The cycle for the simple turbojet engine is virtually that shown in Fig. 1.1 except that the turbine is designed to produce just sufficient power to drive the compressor. The exhaust gas is then expanded to atmospheric pressure in a propelling nozzle to produce a high velocity jet. Figure 1.9 shows a sectional view of a Rolls-Royce Olympus jet engine. This engine is of historical importance,

FIG. 1.9 Olympus turbojet engine [by courtesy of Rolls-Royce Ltd]

being the first two-spool engine in production; early versions were used in the Vulcan bomber and an advanced derivative similar to that shown is used to power the Concorde supersonic transport. (The Olympus has also been used widely as a gas generator to drive a power turbine, both for electricity generation and marine propulsion.)

For low-speed aircraft a combination of propeller and exhaust jet provides the best propulsive efficiency. Figure 1.10 shows a single-shaft *turboprop* engine (Rolls-Royce Dart) chosen to illustrate the use of a centrifugal compressor (two stage) and 'can' type combustion chambers. It is notable that this engine entered airline service around 1953 at a power of about 800 kW, and is still in production in 1985 with the latest version producing about 2500 kW with an improvement in specific fuel consumption of about 20 per cent. Turboprops are also designed with a free turbine driving the propeller or propeller plus LP compressor. The Pratt and Whitney of Canada PT-6, shown in Fig. 1.11, uses a free turbine; the use of a combined axial–centrifugal compressor and a reverse-flow annular combustor can also be seen. This engine is available in versions ranging from 450 to 1200 kW for aircraft ranging in size from single-engined trainers to four-engined STOL transport aircraft. Another variant using a free turbine is the turboshaft engine for use in helicopters; in this case the power turbine drives the helicopter and tail rotors through a complex gearbox and frequently two engines are coupled to a single rotor.

At high subsonic speeds a propulsive jet of smaller mass flow but higher velocity is required. This was originally provided by the turbojet engine, but these have now largely been superseded by *turbofan* (or *by-pass*) engines in which part of the air delivered by an LP compressor or fan by-passes the core of the engine (HP compressor, combustion chamber and turbines) to form an annular propulsive jet of cooler air surrounding the hot jet. This results in a jet of lower mean velocity which provides not only a better propulsive efficiency but also significantly reduces exhaust noise. Figure 1.12(a) is an example of a small turbofan engine, the Pratt and Whitney Canada JT-15D. This is an extremely simple mechanical design giving good performance, intended for small business aircraft where capital cost is critical. A twin-spool arrangement is used, again with a centrifugal HP compressor and a reverse flow annular combustor. The reverse flow combustor is well suited for use with the centrifugal compressor, where the flow must be diffused from a very high tangential velocity to a low axial velocity at entry to the combustor, and this configuration is widely used. Figure 1.12(b) shows an advanced turbofan designed for use with large civil aircraft, the V2500 engine designed by a consortium of five nations. In this application the fuel consumption is of paramount importance, requiring the use of both high by-pass ratio and high pressure ratio. It can be seen that all of the turbomachinery is of the axial type and a straight-through annular combustion chamber is used.

Heat-exchangers have as yet found no place in aircraft engines for reasons of bulk and weight, although they remain feasible for turboprop engines.

FIG. 1.10 Single-shaft turboprop engine [by courtesy of Rolls-Royce Ltd]

**FIG. 1.11  Turboprop engine [by courtesy of Pratt and Whitney Canada Ltd]**

FIG. 1.12(a) Small turbofan engine [by courtesy of Pratt and Whitney Canada Ltd]

FIG. 1.12(b) Large twin-spool turbofan engine [by courtesy of International Aero-
Engines]

This is because, with much of the net power output delivered to the propeller, the velocity of the gas leaving the turbine is relatively low and the frictional pressure loss need not be prohibitively high in a heat-exchanger of acceptable size. Around 1965 Allison developed a regenerative turboprop for the U.S. Navy, the object being to obtain an engine of exceptionally low specific fuel consumption for use on long endurance anti-submarine patrols. In this kind of application it is the total engine plus fuel weight which is critical, and it was thought that the extra weight of the heat-exchanger would be more than compensated by the low fuel consumption. It was proposed that the heat-exchanger should be by-passed on take-off to give maximum power. The engine did not go into production, but it is not impossible that regenerative turboprops will appear in the future, perhaps in the form of turboshaft engines for long endurance helicopters.

## 1.5   Industrial applications

Sometimes in this book we shall find it necessary to use the distinguishing terms 'aircraft gas turbine' and 'industrial gas turbine'. The first term is self-explanatory, while the second is intended to include all gas turbines not included in the first category. This broad distinction has to be made for three main reasons. Firstly, the life required of an industrial plant is of the order of 100 000 hours without major overhaul, whereas this is not expected of an aircraft gas turbine. Secondly, limitation of the size and weight of an aircraft power plant is much more important than in the case of most other applications of the gas turbine. Thirdly, the aircraft power plant can make use of the kinetic energy of the gases leaving the turbine, whereas it is wasted in other types and consequently must be kept as low as possible. These three differences in the requirements can have a considerable effect on design and, in spite of the fact that the fundamental theory applies to both categories, it will be necessary to make the distinction from time to time. Turbomachinery of gas turbines designed specifically for industrial purposes tends to look more like that of traditional steam turbines in mechanical appearance than the familiar lightweight constructions used in aircraft practice. Figure 1.13 shows the rugged construction employed in the Ruston Tornado designed for long life and to operate on either liquid or gaseous fuel.

Having said all this, it must be admitted that it has often been found more economical to employ a modified aircraft gas turbine for industrial applications than to design and develop a totally new power plant. This is simply because in this way a large part of the research and development cost is borne by a military budget rather than by the industrial user. The 'modification' will normally comprise a strengthening of the bearings, changes to the combustion system to enable it to burn cheaper fuel, the addition of a power turbine and reduction gear designed to suit the particular type of load (e.g. electric generator, marine propeller, gas or liquid pumps), a de-rating of the

FIG. 1.13 Industrial gas turbine with separate power turbine [by courtesy of Ruston Gas Turbines Ltd]

engine to give it a longer life, and possibly the addition of a heat-exchanger. Virtually all marine gas turbines have been developed from aircraft gas turbines in this way.

The widest applications of the gas turbine have been in pumping sets for gas and oil transmission pipelines, electricity generation and naval propulsion. In the case of natural gas pipelines, the turbines use the fluid being pumped as fuel and a typical pipeline might consume 7–10 per cent of the throughput for compression purposes. In recent years the value of gas has increased dramatically and this has led to a demand for high-efficiency pumping units. A major pipeline might have as much as 1000 MW of installed power and the fuel bills are comparable to those of a medium-sized airline. Pumping stations may be about 100 km apart and the gas turbines used range in power from 5 to 20 MW. Many compressor stations are in remote locations and aircraft derivative units of 15 to 20 MW are widely used. Other operators may prefer the use of industrial gas turbines and in recent years a considerable number of heat-exchangers have been retrofitted to simple cycle units. With oil pipelines the oil is often not suitable for burning in a gas turbine without expensive fuel treatment and it becomes necessary to bring a suitable liquid fuel in by road.

Gas turbines have been widely used for both peak load and base load electricity generation. In Great Britain and North America, gas turbines have been used almost exclusively for peak load purposes but in countries such as Saudi Arabia, where fuel is plentiful, they have been used for base load; a particular advantage of the gas turbine in desert locations is its freedom from any requirement for cooling water. In Great Britain the peak load market has been dominated by the aero-engine gas generator with a heavy-duty power turbine; one of the outstanding advantages of this type is its ability to produce full power from a standing start in two minutes, although this capability should be utilized only when needed because thermal shock will reduce life between overhauls. Following the major blackout of the eastern seaboard of the U.S.A. in the middle sixties many units were designed to be capable of a 'black' start, i.e. completely independent of the main electricity supply. By the end of 1985 about 3000 MW of emergency and peak load plant based on the Rolls-Royce Avon and Olympus gas turbines had been installed in Great Britain; these form a key part of the overall electricity generating system, but only run for a very small number of hours. Many similar units based on the Pratt and Whitney FT-4 have also been installed in the U.S.A. Figure 1.14(a) is a view of a generator set using four Olympus engines to produce about 80 MW, while Fig. 1.14(b) gives a clear idea of the difference in size between a conventional steam station of 128 MW and a peak load gas turbine of 160 MW shown ringed. The latter consist of two of the sets shown in Fig. 1.14(a). Aero-derivative units are capable of about 30 MW, and many electrical systems require much larger units. Heavy industrial gas turbines have been built in very large numbers and dominate the market on the basis of installed megawatts. The major manufacturers are General Electric, Westinghouse, Brown Boveri and KWU. Units designed for electricity

FIG. 1.14(a) Gas turbine generator [by courtesy of Rolls-Royce Ltd]

FIG. 1.14(b) Comparative size of steam and gas turbine power stations [by courtesy
        of Rolls-Royce Ltd]

generation use the single-shaft arrangement and are capable of delivering up to 130 MW per unit; the upper limit is fixed by considerations such as the size of disc forgings and the maximum width to permit shipping by rail. Unlike steam turbines, gas turbines are not often erected on the site and are usually delivered as complete packages ready to run. Single-shaft units running at 3000 and 3600 rpm respectively can drive 50 or 60 Hz generators directly without the need for an expensive gearbox. Compressor designs suitable for operation at these speeds result in 60 Hz machines of around 60–80 MW and 50 Hz machines of around 100–120 MW, with the power largely determined by the compressor airflow. Many heavy-duty units have run well in excess of 100 000 hours and some have exceeded 200 000 hours. It is perhaps worth noting that aero-derivative units have exceeded 50 000 hours before overhaul, in pipeline applications. Another major market for electricity generation is the provision of power for off-shore platforms, where gas turbines are used to provide base load power. Many Ruston units of 3–4 MW have been used, but for larger powers aero-derivatives such as the Rolls-Royce RB-211 and General Electric LM 2500 have been installed at ratings of about 20 MW; a big platform may require as much as 125 MW and both surface area and volume are at a premium.

Combined cycle installations offer thermal efficiencies in the range of 45–50 per cent and are now being installed as base load plant in some countries. It is interesting to note that Japan, a country with negligible fuel resources, is installing combined cycle stations of 2000 MW burning imported liquefied natural gas (LNG). This is because an expensive imported fuel must be utilized in the most efficient plant available, and the combined cycle is capable of higher thermal efficiency than either steam or diesel plant.

Gas turbines were used successfully in a few high-speed container ships, but the rapid increase in fuel prices in the mid seventies led to these ships being re-engined with diesels; the converted ships suffered a major loss in both speed and cargo capacity, but high speed could no longer be justified. The picture with respect to naval operations is quite different, however, and many navies (e.g. Britain, U.S.A., Canada, Netherlands) have now accumulated a large amount of gas turbine experience. A gas turbine was first used in a Motor Gun Boat in 1947, and aero-type engines (Rolls-Royce Proteus) were first used in fast patrol boats in 1958. The potential of aero-derivative engines for main propulsion of warships was soon realised and the Canadian DDH-280 class were the first all gas turbine powered warships in the western world, using a combination of Pratt and Whitney FT-4's for 'boost' power and FT-12's for 'cruise'. The Royal Navy selected the Olympus as boost engine and the Rolls-Royce Tyne for cruise duties; this machinery arrangement was also selected by the Royal Netherlands Navy. The Olympus and Tyne are the only naval gas turbines to have been proved in battle, operating with great success in the Falklands War. The U.S. Navy adopted the GE LM 2500, derived from the CF6 advanced turbofan, and this engine has been widely used around the world. With the increasing electrical needs of warships, and the absence of steam for use in turbogenerators, gas turbine driven generators also offer a very compact source of electricity.

A major disadvantage of the gas turbine in naval use is its poor specific fuel consumption at part load. If we consider a naval vessel having a maximum speed of, say, 36 knots and a cruise speed of 18 knots, with the power required proportional to the cube of the speed, the cruise power will be only one eighth of the maximum power; indeed, much time would be spent at speeds less than 18 knots. To overcome this problem, combined power plant consisting of gas turbines in conjunction with steam turbines, diesel engines and other gas turbines have been used. These go by names such as COSAG, CODOG, COGOG, COGAG, etc. CO stands for 'combined'; S, D, G for 'steam', 'diesel' and 'gas turbine'; the final A or O for 'and', 'or'. The 'or' requires explanation. The earliest arrangement used by the Royal Navy was COSAG where a ship's shaft was driven by both steam *and* gas turbines; the gearing was such that the ship could use either steam or gas turbine or both. The gas turbines were originally intended only for boost purposes, or for rapid starts, but in practice proved so versatile and popular with the operators that they became used for much longer periods. Another alternative is to combine a boost gas turbine with a cruise diesel; in this case the diesel power is so small relative to the gas turbine power that there is little advantage in adding the powers. The vessel therefore operates in either the diesel *or* gas turbine mode, i.e. as CODOG. For naval use the diesel engine has the advantage of very good cruise fuel consumption, with the disadvantage of large bulk for the power available and a loud underwater noise level. The COGOG arrangement, with a small cruise gas turbine and a large boost gas turbine, has been very widely used; the aim is to keep any gas turbine running near full power where its efficiency is a maximum. The small (4–5 MW) cruise gas turbines, however, are not competitive with diesels regarding specific fuel consumption and there appears to be a trend away from COGOG to CODOG. The COGAG arrangement, using gas turbines of the same size, has been used by the U.S. Navy, using four LM 2500 engines in large destroyers; this arrangement has also been used in the Royal Navy with four Olympus engines in the Invincible class of aircraft carriers.

For many years the gas turbine has been considered for road and rail transport, without making any real impact. Union Pacific successfully operated large freight trains with gas turbine power from about 1955 for 15–20 years, but these have now given way to diesels. Several high-speed passenger trains were built using helicopter-type gas turbines, but only with limited success; the most successful were those built by the French, but the very high-speed TGV (Train à Grand Vitesse) has been built with electric traction. A considerable amount of work was done on gas turbines for long haul trucks, with engines of 200–300 kW being developed. All used the low-pressure ratio cycle with a centrifugal compressor, free turbine and rotary heat exchanger. Similar efforts were expended on gas turbines for cars and, although these continue under U.S. government sponsorship, the automobile gas turbine still

**FIG. 1.15  Ruston cogeneration plant for a factory**

appears to be far over the horizon and may never appear. There is no doubt that the cost of gas turbines could be significantly reduced if they were built in numbers approaching those of piston engines. The major problem is still that of part load fuel consumption. The one breakthrough achieved by the gas turbine is its choice for propulsion of the M1 tank built for the US Army, but it has still not been proved that the gas turbine is superior to the diesel in this application.

Another concept of growing importance is *cogeneration* (or *total energy*) plant, where the gas turbine drives a generator and the exhaust gases are used as a source of low-grade heat. Heat at a relatively low temperature is required for heating buildings and operating air-conditioning systems. It is also required in many process industries—paper drying for example. Chemical industries often need large quantities of hot gas containing a high proportion of free oxygen at a pressure sufficient to overcome the pressure loss in chemical reactors. The temperature limitation in the gas turbine cycle means that high air/fuel ratios must be employed resulting in a large proportion of unused oxygen in the exhaust. The exhaust from a gas turbine is therefore often suitable. The unit can be designed to meet the hot gas requirement, with or without shaft power for other purposes, and sometimes to burn a fuel which is a by-product of the chemical process. Figure 1.15 illustrates an application of a cogeneration plant using Ruston gas turbines. It provides the whole of the electricity, process steam, heating steam, and chilled water required for a factory. The use of eight gas turbines feeding four auxiliary-fired waste-heat boilers enables the changing power and heat demands during the day to be met by running the necessary number of units at substantially full power and therefore at peak efficiency.

Finally, the gas turbine can be used as a compact air compressor suitable for supplying large quantities of air at moderate pressure. In this case the turbine produces just sufficient power to drive the compressor, and the net power output is in the form of compressed air bled from the compressor. Figure 1.16 illustrates a possible cycle for a blast furnace blower: here the blast furnace gas is used as fuel for the gas turbine.

**FIG. 1.16 Blast furnace power plant**

## 1.6   Some future possibilities

The increasing scarcity and therefore cost of high-grade fossil fuels will necessitate the wider use of poor-quality coal and heavy fuel oil with a high sulphur content. Such fuel can be burnt in steam power stations, but only with expensive boiler maintenance and the costly cleaning of stack gases to meet ever stringent anti-pollution regulations. Two other quite distinct approaches exist, both involving the use of gas turbines. The first makes use of the idea of *fluidized bed combustion*, and the second involves the transformation of the low-quality solid or liquid fuel into a clean gaseous fuel.

A fluidized bed combustor consists essentially of a refractory-brick lined cylinder containing sand-sized refractory particles kept in suspension by an upward flow of air. When used in conjunction with a gas turbine, the required air can be bled from the compressor. If coal is being burnt, the oxides of sulphur formed are trapped in the ash, and if oil is the fuel they can be trapped by particles of limestone or dolomite in the bed. Figure 1.17 shows one possible scheme. It makes use of the fact that heat is transferred between a fluidized bed and any solid surface immersed in it with very high heat transfer coefficients. In this scheme, most of the compressor air is heated in a tubular

**FIG. 1.17 Fluidized bed combustion**

heat-exchanger in the bed, and only the small amount of air bled for fluidization need be cleaned of dust in cyclone separators before being passed to the turbine. Corrosion and erosion problems are holding up development, but if they can be overcome the fluidized bed combustor opens up the possibility of burning coal mined by remote-controlled methods, or even the material in colliery spoil heaps. In the latter case not only would useful power be developed from hitherto unusable fuel, but valuable land would be reclaimed.

Before leaving the subject of fluidized bed combustion it is worth mentioning another possible application: the incineration of municipal waste. The waste is shredded and useful recoverable materials (steel, tin, aluminium etc.) are separated out using magnetic and flotation techniques, and vibrating screens. The remainder, about 85 per cent, is burnt in a fluidized bed combustor. Combustion of the waste maintains the temperature somewhere between 700°C and 800°C, which is high enough to consume the material without causing the bed to agglomerate. Supplementary oil burners are provided for starting the unit. The hot gases pass through several stages of cleaning to prevent erosion of the turbine and to satisfy air pollution standards. Income from the electricity generated and sale of recyclable materials is expected to reduce substantially the cost of waste disposal compared with the conventional land-fill method. It must be emphasized that the maximum temperature that can be used in a fluidized bed is unlikely to be very high so that the gas turbine efficiency will be low. Fluidized bed combustors are likely to be used only for burning cheap or otherwise unusable fuels.

The second approach to the problem of using poor-quality coal or heavy oil is its transformation into a clean gaseous fuel. Figure 1.18 illustrates a possible scheme in which a gasification plant is integrated with a COGAS power cycle.

FIG. 1.18 Gasification plant with COGAS cycle

The gasification process removes troublesome vanadium and sodium impurities which cause corrosion in the turbine, and also the sulphur which produces polluting oxides in the stack gases. Compressed air required for the process is bled from the gas turbine compressor. To overcome the pressure loss in the gasification plant the pressure is boosted in a separate compressor driven by a steam turbine. This would use some of the steam from the waste heat boiler, the major fraction supplying the power steam turbine. The gas produced by such a plant would have a very low calorific value—perhaps only $5000 \, kJ/m^3$ compared with about $39\,000 \, kJ/m^3$ for natural gas. This is because of dilution by the nitrogen in the air supplied by the gasifier. The low calorific value carries little penalty, however, because all gas turbines operate with a weak mixture to limit the turbine inlet temperature. It simply means that the nitrogen normally fed directly to the combustion chamber in the compressor delivery air, now enters as part of the fuel already at the pressure required.

One other possible future application of the gas turbine should be mentioned: its use as an energy storage device. The overall efficiency of a country's electricity generating system can be improved if sufficient energy storage capacity is provided to enable the most efficient base-load stations to run night and day under conditions yielding peak efficiency. This provision will become of particular importance as the contribution of the capital-intensive nuclear power stations increases. So far hydro-electric pumped storage plant have been built to meet the need, but suitable sites in Great Britain have virtually all been used. A possible alternative is illustrated in Fig. 1.19. Here a reversible motor/generator is coupled either to the compressor or turbine. During the night, off-peak power is used to drive the compressor which delivers air to an underground cavern via a 'pebble bed' regenerator.

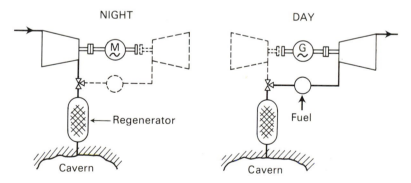

**FIG. 1.19 Energy storage scheme**

The regenerator stores the heat in pebbles of alumina or silica. During the day the compressed air is released through the regenerator, picking up stored energy on its way to the turbine. To satisfy peak demand it may prove desirable also to burn some fuel in a combustion chamber to make up for heat losses in the regenerator. If the cavern is to be sufficiently small to make such a scheme economic, the pressure must be high—perhaps as high as 100 bar.

This implies a high compressor delivery temperature of about 900°C. By cooling the air in the regenerator the volume is further reduced, and at the same time the walls of the cavern are protected from the high temperature. Salt caverns excavated by washing have been proposed, and disused mine workings are another possibility if economic means can be found of sealing them adequately.

The first air-storage gas turbine plant was built by Brown Boveri and commissioned in Germany in 1978. It has no regenerator, but it does incorporate a heat-exchanger and has two-stage compression with intercooling. An aftercooler protects the salt cavern walls from high temperature. The plant is for peak load generation and produces up to 290 MW for periods of 1–1.5 hours three times a day, using about 12 hours for pumping the reservoir up to pressure.

## 1.7   Gas turbine design procedure

It must be emphasized that this book provides an introduction to the theory of the gas turbine, and not to the design of gas turbines. To place the contents of the book in proper perspective, a schematic diagram representing a complete design procedure is shown in Fig. 1.20. This gives some idea of the inter-relationships between thermodynamic, aerodynamic, mechanical and control system design and emphasizes the need for feedback between the various specialists. The dotted lines enclose the areas covered in subsequent chapters: where they cut a block it indicates that the topic has received attention, but only in a cursory manner. Thus when dealing with the thermodynamic and aerodynamic theory which forms the core of this book, the reader will be reminded only of those mechanical aspects which interact directly with it.

The design process is shown as starting from a specification, resulting from either market research or a customer requirement. The development of a high-performance gas turbine is extremely costly, and so expensive that most large aero engines are developed by multi-national consortia. There are very few customers who are powerful enough to have an engine built to their requirement, and the specification usually results from market research. Successful engines are those which find a variety of applications, and their life-cycle from design to final service use may be in excess of 50 years. When the first edition of this book was written in 1950, the Rolls-Royce Dart was in the design stage and remained in production until 1986, and versions entering airline service can expect at least a 15-year life.

The specification is rarely a simple statement of required power and efficiency. Other factors of major importance, which vary with the application, include weight, cost, volume, life and noise, and many of these criteria act in opposition. For example, high efficiency inevitably incurs high capital cost, and a simple engine of lower efficiency may be perfectly acceptable if the running hours are low. An important decision facing the

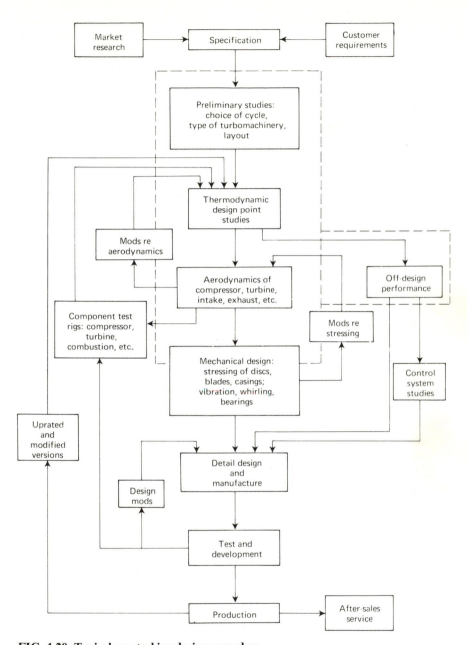

**FIG. 1.20  Typical gas turbine design procedure**

designer is the choice of cycle, and this aspect will be covered in Chapters 2 and 3. It is essential to consider at an early stage what type of turbomachinery to use, and this will in large measure depend on the size of the engine: turbomachinery and combustor design will be dealt with in Chapters 4–7. The layout of the engine must also be considered, for example whether a single or multi-shaft design should be used, and the behaviour of these different types of engine will be covered in Chapters 8 and 9.

The first major design step is to carry out *thermodynamic* design point studies. These are detailed calculations taking into account all-important factors such as expected component efficiencies, air-bleeds, variable fluid properties and pressure losses, and would be carried out over a reasonably restricted range of pressure ratio and turbine inlet temperature. A value for the specific output (i.e. power per unit mass of airflow) and specific fuel consumption will be determined over a range of values of the basic cycle parameters. Although in industry these calculations would be done on a digital computer, it should be clearly understood that there is *not* a mathematically defined optimum. For example, at a given turbine inlet temperature a large increase in pressure ratio may give a minimal improvement in thermal efficiency, and the resulting engine would be too complex and expensive to be practical. Once the designer has settled on a suitable choice of cycle parameters, he can make use of the specified output to determine the airflow required to give the specified power.

Knowing the airflow, pressure ratio and turbine inlet temperature, attention can be turned to the *aerodynamic* design of the turbomachinery. It is now possible to determine annulus dimensions, rotational speeds and number of stages. At this point it may well be found that difficulties arise which may cause the aerodynamacist to consult with the thermodynamacist to see if a change in the design point could be considered, perhaps a slight increase in temperature or decrease in pressure ratio.

The *mechanical* design can start only after aerodynamic and thermodynamic designs are well advanced. It will then likely be found that stress or vibration problems may lead to further changes, the requirements of the stress and aerodynamics groups often being in opposition. At the same time as these studies are proceeding, *off-design* performance and control system design must be considered; off-design operation will include the effects of varying ambient conditions, as well as reduced power operation. When designing a control system to ensure the safe and automatic operation of the engine, it is necessary to be able to predict temperature and pressure levels throughout the engine and to select some of these for use as control parameters.

Once the engine has entered service, there will be demands from customers for more powerful or more efficient versions, leading to the development of uprated engines. Indeed, such demands may often arise *before* the design process has been completed. When engines have to be uprated, the designer must consider such methods as increasing the mass flow, turbine inlet temperature, component efficiencies etc. while maintaining the same basic engine design. A successful engine may have its power tripled during a long development cycle. Eventually, however, the engine will become dated and no longer competitive. The timing of a decision to start a new design is of critical importance for the economic well-being of the manufacturer. References (1) and (2) describe the choice of design for an industrial gas turbine and turboprop respectively.

The foregoing should give the reader an overall, if superficial, view of the design process and may lead to the realization that the gas turbine industry can provide an exciting and rewarding technical career for a wide variety of highly skilled engineers.

# 2

# *Shaft power cycles*

Enough has been said in the foregoing chapter for the reader to realize how great is the number of possible varieties of gas turbine cycle when multi-stage compression and expansion, heat-exchange, reheat and intercooling are incorporated. A comprehensive study of the performance of all such cycles, allowing for the inefficiencies of the various components, would result in a very large number of performance curves. Reference (1) presents a study of this kind. We shall here concentrate mainly on describing methods of calculating cycle performance. For convenience the cycles are treated in two groups—shaft power cycles (this chapter) and aircraft propulsion cycles (Chapter 3). An important distinction between the two groups arises from the fact that the performance of aircraft propulsion cycles depends very significantly upon forward speed and altitude. These two variables do not enter into performance calculations for marine and land-based power plant to which this chapter is confined.

Before proceeding with the main task, it will be useful to review the performance of ideal gas turbine cycles in which perfection of the individual components is assumed. The specific work output and cycle efficiency then depend only on the pressure ratio and maximum cycle temperature. The limited number of performance curves so obtained enables the major effects of various additions to the simple cycle to be seen clearly. Such curves also show the upper limit of performance which can be expected of real cycles as the efficiency of gas turbine components is improved.

## 2.1  Ideal cycles

Analyses of ideal gas turbine cycles can be found in texts on engineering thermodynamics [e.g. Ref. (2)] and only a brief resumé will be given here. The assumption of ideal conditions will be taken to imply the following:

(*a*)  Compression and expansion processes are reversible and adiabatic, i.e. isentropic.

(*b*)  The change of kinetic energy of the working fluid between inlet and outlet of each component is negligible.

(c)   There are no pressure losses in the inlet ducting, combustion chambers, heat-exchangers, intercoolers, exhaust ducting, and ducts connecting the components.

(d)   The working fluid has the same composition throughout the cycle and is a perfect gas with constant specific heats.

(e)   The mass flow of gas is constant throughout the cycle.

(f)   Heat transfer in a heat-exchanger (assumed counterflow) is 'complete', so that in conjunction with (d) and (e) the temperature rise on the cold side is the maximum possible and exactly equal to the temperature drop on the hot side.

Assumptions (d) and (e) imply that the combustion chamber, in which fuel is introduced and burned, is considered as being replaced by a heater with an external heat source. For this reason, as far as the calculation of performance of ideal cycles is concerned, it makes no difference whether one is thinking of them as 'open' or 'closed' cycles. The diagrammatic sketches of plant will be drawn for the former, more usual, case.

## Simple gas turbine cycle

The ideal cycle for the simple gas turbine is the Joule (or Brayton) cycle, i.e. cycle 1234 in Fig. 2.1. The relevant steady flow energy equation is

$$Q = (h_2 - h_1) + \tfrac{1}{2}(C_2^2 - C_1^2) + W$$

where $Q$ and $W$ are the heat and work transfers per unit mass flow. Applying this to each component, bearing in mind assumption (b), we have

$$W_{12} = -(h_2 - h_1) = -c_p(T_2 - T_1)$$
$$Q_{23} = (h_3 - h_2) = c_p(T_3 - T_2)$$
$$W_{34} = (h_3 - h_4) = c_p(T_3 - T_4)$$

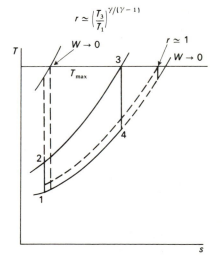

**FIG. 2.1 Simple cycle**

The cycle efficiency is

$$\eta = \frac{\text{net work output}}{\text{heat supplied}} = \frac{c_p(T_3 - T_4) - c_p(T_2 - T_1)}{c_p(T_3 - T_2)}$$

Making use of the isentropic $p$–$T$ relation, we have

$$T_2/T_1 = r^{(\gamma-1)/\gamma} = T_3/T_4$$

where $r$ is the pressure ratio $p_2/p_1 = r = p_3/p_4$. The cycle efficiency is then readily shown to be given by

$$\eta = 1 - \left(\frac{1}{r}\right)^{(\gamma-1)/\gamma} \tag{2.1}$$

The efficiency thus depends only on the pressure ratio and the nature of the gas. Figure 2.2(a) shows the relation between $\eta$ and $r$ when the working fluid is air ($\gamma = 1{\cdot}4$), or a monatomic gas such as argon ($\gamma = 1{\cdot}66$). For the

FIG. 2.2 **Efficiency and specific work output—simple cycle**

remaining curves in this section air will be assumed to be the working fluid. The effect of using helium instead of air in a closed cycle is studied in section 2.6, where it is shown that the theoretical advantage indicated in Fig. 2.2(a) is not realized when component losses are taken into account.

The specific work output $W$, upon which the size of plant for a given power depends, is found to be a function not only of pressure ratio but also of maximum cycle temperature $T_3$. Thus

$$W = c_p(T_3 - T_4) - c_p(T_2 - T_1)$$

which can be expressed as

$$\frac{W}{c_p T_1} = t\left(1 - \frac{1}{r^{(\gamma-1)/\gamma}}\right) - (r^{(\gamma-1)/\gamma} - 1) \tag{2.2}$$

where $t = T_3/T_1$; $T_1$ is normally atmospheric temperature and is not a major independent variable. It is therefore convenient to plot the specific work output in non-dimensional form $(W/c_pT_1)$ as a function of $r$ and $t$ as in Fig. 2.2(b). The value of $T_3$, and hence $t$, that can be used in practice, depends upon the maximum temperature which the highly stressed parts of the turbine can stand for the required working life: it is often called the 'metallurgical limit'. For a long-life industrial plant it might be between 3·5 and 4, whereas a value between 5 and 5·5 would be permissible for an aircraft engine with cooled turbine blades.

A glance at the $T-s$ diagram of Fig. 2.1 will show why a constant $t$ curve exhibits a maximum at a certain pressure ratio: $W$ is zero at $r = 1$ and also at the value of $r$ for which the compression and expansion processes coincide, namely $r = t^{\gamma/(\gamma-1)}$. For any given value of $t$ the optimum pressure ratio for maximum specific work output can be found by differentiating equation (2.2) with respect to $r^{(\gamma-1)/\gamma}$ and equating to zero: the result is

$$r_{\text{opt}}^{(\gamma-1)/\gamma} = \sqrt{t}$$

Since $r^{(\gamma-1)/\gamma} = T_2/T_1 = T_3/T_4$, this is equivalent to writing

$$\frac{T_2}{T_1} \times \frac{T_3}{T_4} = t$$

But $t = T_3/T_1$ and consequently it follows that $T_2 = T_4$. Thus the specific work output is a maximum when the pressure ratio is such that the compressor and turbine outlet temperatures are equal. For all values of $r$ between 1 and $t^{\gamma/2(\gamma-1)}$, $T_4$ will be greater than $T_2$ and a heat-exchanger can be incorporated to reduce the heat transfer from the external source and so increase the efficiency.

## Heat-exchange cycle

Using the nomenclature of Fig. 2.3, the cycle efficiency now becomes

$$\eta = \frac{c_p(T_3 - T_4) - c_p(T_2 - T_1)}{c_p(T_3 - T_5)}$$

FIG. 2.3 Simple cycle with heat-exchange

With ideal heat-exchange $T_5 = T_4$, and on substituting the isentropic $p - T$ relations the expression reduces to

$$\eta = 1 - \frac{r^{(\gamma - 1)/\gamma}}{t} \tag{2.3}$$

Thus the efficiency of the heat-exchange cycle is not independent of the maximum cycle temperature, and clearly it increases as $t$ is increased. Furthermore it is evident that, for a given value of $t$, the efficiency increases with decrease in pressure ratio and not with increase in pressure ratio as for the simple cycle. The full lines in Fig. 2.4 represent the equation, each constant

FIG. 2.4 **Efficiency—simple cycle with heat-exchange**

$t$ curve starting at $r = 1$ with a value of $\eta = 1 - 1/t$, i.e. the Carnot efficiency. This is to be expected because in this limiting case the Carnot requirement of complete external heat reception and rejection at the upper and lower cycle temperatures is satisfied. The curves fall with increasing pressure ratio until a value corresponding to $r^{(\gamma - 1)/\gamma} = \sqrt{t}$ is reached, and at this point equation (2.3) reduces to (2.1). This is the pressure ratio for which the specific work output curves of Fig. 2.2(b) reach a maximum and for which it was shown that $T_4 = T_2$. For higher values of $r$ a heat-exchanger would *cool* the air leaving the compressor and so reduce the efficiency, and therefore the constant $t$ lines have not been extended beyond the point where they meet the efficiency curve for the simple cycle which is shown dotted in Fig. 2.4.

The specific work output is unchanged by the addition of a heat-exchanger and the curves of Fig. 2.2(b) are still applicable. From these curves and those in Fig. 2.4 it can be concluded that to obtain an appreciable improvement in efficiency by heat-exchange, (a) a value of $r$ appreciably less than the optimum for maximum specific work output should be used and (b) it is not necessary to use a higher cycle pressure ratio as the maximum cycle temperature is increased. Later it will be shown that for real cycles conclusion (a) remains true but conclusion (b) requires modification.

## *Reheat cycle*

A substantial increase in specific work output can be obtained by splitting the expansion and reheating the gas between the high pressure and low pressure turbines. Figure 2.5(a) shows the relevant portion of the reheat cycle

(a)                                           (b)

**FIG. 2.5 Reheat cycle**

on the $T$–$s$ diagram. That the turbine work is increased is obvious when one remembers that the vertical distance between any pair of constant pressure lines increases as the entropy increases: thus $(T_3 - T_4)+(T_5 - T_6)>(T_3 - T_4')$.

Assuming that the gas is reheated to a temperature equal to $T_3$, differentiation of the expression for specific work output shows that the optimum point in the expansion at which to reheat is when the pressure ratios (and hence temperature drops and work transfers) for the HP and LP turbines are equal. With this optimum division, it is then possible to derive expressions for the specific output and efficiency in terms of $r$ and $t$ as before. Writing $c = r^{(\gamma - 1)/\gamma}$ they become

$$\frac{W}{c_p T_1} = 2t - c + 1 - \frac{2t}{\sqrt{c}} \tag{2.4}$$

$$\eta = \frac{2t - c + 1 - 2t/\sqrt{c}}{2t - c - t/\sqrt{c}} \tag{2.5}$$

Comparison of the $W/c_p T_1$ curves of Fig. 2.6 with those of Fig. 2.2(b) shows that reheat markedly increases the specific output. Figure 2.5(b), however, indicates that this is achieved at the expense of efficiency. This is to be expected because one is adding a less efficient cycle (4'456 in Fig. 2.5(a)) to the simple cycle—less efficient because it operates over a smaller temperature range. Note that the reduction in efficiency becomes less severe as the maximum cycle temperature is increased.

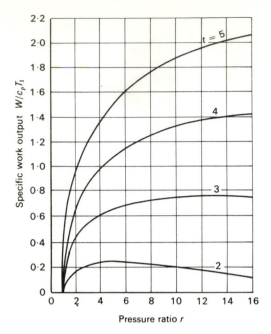

**FIG. 2.6 Work output—reheat cycle**

## *Cycle with reheat and heat-exchange*

The reduction in efficiency due to reheat can be overcome by adding heat-exchange as in Fig. 2.7. The higher exhaust gas temperature is now fully utilised in the heat-exchanger and the increase in work output is no longer offset by the increase in heat supplied. In fact, when a heat-exchanger is

**FIG. 2.7 Reheat cycle with heat-exchange**

employed, the efficiency is higher with reheat than without as shown by a comparison of Figs. 2.8 and 2.4. The family of constant $t$ lines exhibit the same features as those for the simple cycle with heat-exchange—each curve having the Carnot value at $r = 1$ and falling with increasing $r$ to meet the corresponding efficiency curve of the reheat cycle without heat-exchange at the value of $r$ corresponding to maximum specific work output.

FIG. 2.8  Efficiency—reheat cycle with heat-exchange

## Cycles with intercooled compression

A similar improvement in specific work output to that obtained by reheat can be achieved by splitting the compression and intercooling the gas between the LP and HP compressors; and, assuming that the air is intercooled to $T_1$, it can be shown that the specific work output is a maximum when the pressure ratios of the LP and HP compressors are equal. The use of intercoolers is seldom contemplated in practice because they are bulky and need large quantities of cooling water. The main advantage of the gas turbine, that it is compact and self-contained, is then lost. For this reason no performance curves for cycles with intercooling are included: suffice it to say that they are similar to Figs. 2.5(b) and 2.6, although the increase in specific work output and reduction in efficiency with respect to the simple cycle are not so marked. (A modification to the low temperature region of a cycle is normally less significant than a comparable modification to the high temperature region.) As with reheat, intercooling increases the cycle efficiency only when a heat-exchanger is also incorporated, and then curves almost identical to those of Fig. 2.8 are obtained.

This discussion of ideal cycles should be sufficient to indicate the main effects of additions to the simple gas turbine. We have seen that the choice of pressure ratio will depend on whether it is high efficiency or high specific work output (i.e. small size) which is required; and that in cycles without heat-exchange a higher pressure ratio must be used to take advantage of a higher permissible turbine inlet temperature. It will be evident from what follows that these conclusions are also broadly true of practical cycles in which component losses are taken into account.

## 2.2   Methods of accounting for component losses

The performance of real cycles differs from that of ideal cycles for the following reasons.

(a)  Because fluid velocities are high in turbomachinery the change in kinetic energy between inlet and outlet of each component cannot necessarily be ignored. A further consequence is that the compression and expansion processes are irreversible adiabatics and therefore involve an increase in entropy.

(b)  Fluid friction results in pressure losses in combustion chambers and heat-exchangers, and also in the inlet and exhaust ducts. (Losses in the ducts connecting components are usually included in the associated component losses.)

(c)  If a heat-exchanger is to be of economic size, terminal temperature differences are inevitable; i.e. the compressed air cannot be heated to the temperature of the gas leaving the turbine.

(d)  Slightly more work than that required for the compression process will be necessary to overcome bearing and 'windage' friction in the transmission between compressor and turbine, and to drive ancillary components such as fuel and oil pumps.

(e)  The values of $c_p$ and $\gamma$ of the working fluid vary throughout the cycle due to changes of temperature and, with internal combustion, due to changes in chemical composition.

(f)  The definition of the efficiency of an ideal cycle is unambiguous, but this is not the case for an open cycle with internal combustion. Knowing the compressor delivery temperature, composition of the fuel, and turbine inlet temperature required, a straightforward combustion calculation yields the fuel/air ratio necessary; and a combustion efficiency can also be included to allow for incomplete combustion. Thus it will be possible to express the cycle performance unambiguously in terms of fuel consumption per unit net work output, i.e. in terms of the *specific fuel consumption*. To convert this to a cycle efficiency it is necessary to adopt some convention for expressing the heating value of the fuel.

(g)  With internal combustion, the mass flow through the turbine might be thought to be greater than that through the compressor by virtue of the fuel added. In practice, about 1–2 per cent of the compressed air is bled off for cooling turbine discs and blade roots, and we shall see later that the fuel/air ratio employed is in the region 0·01–0·02. Thus for ordinary cycle calculations it is sufficiently accurate to assume that the fuel added merely compensates for this loss. In this book it will always be assumed that the mass flows through the compressor and turbine are equal. For modern aircraft engines, which operate with very high turbine inlet temperatures, greater quantities of cooling air are used (to pass through cooling passages in the turbine blading) and this approximation will not be adequate when estimating accurately the performance of a final design.

Methods of accounting for factors (*a*) to (*f*) must be discussed before giving examples of cycle performance calculations.

## Stagnation properties

The kinetic energy terms in the steady flow energy equation can be accounted for implicitly by making use of the concept of *stagnation* (or *total*) *enthalpy*. Physically, the stagnation enthalpy $h_0$ is the enthalpy which a gas stream of enthalpy $h$ and velocity $C$ would possess when brought to rest adiabatically and without work transfer. The energy equation then reduces to

$$(h_0 - h) + \tfrac{1}{2}(0 - C^2) = 0$$

and thus $h_0$ is defined by

$$h_0 = h + C^2/2 \tag{2.6}$$

When the fluid is a perfect gas, $c_p T$ can be substituted for $h$, and the corresponding concept of *stagnation* (or *total*) *temperature* $T_0$ is defined by

$$T_0 = T + C^2/2c_p \tag{2.7}$$

$C^2/2c_p$ is called the *dynamic temperature* and, when it is necessary to emphasize the distinction, $T$ is referred to as the *static temperature*. An idea of the order of magnitude of the difference between $T_0$ and $T$ is obtained by considering air at atmospheric temperature, for which $c_p = 1\cdot005$ kJ/kg K, flowing at 100 m/s. Then

$$T_0 - T = \frac{100^2}{2 \times 1\cdot005 \times 10^3} \simeq 5 \text{ K}$$

It follows from the energy equation that if there is no heat or work transfer $T_0$ will remain constant. If the duct is varying in cross-sectional area, or friction is degrading directed kinetic energy into random molecular energy, the static temperature will change—but $T_0$ will not. Applying the concept to an adiabatic compression, the energy equation becomes

$$W = -c_p(T_2 - T_1) - \tfrac{1}{2}(C_2^2 - C_1^2) = -c_p(T_{02} - T_{01})$$

Similarly for a heating process without work transfer,

$$Q = c_p(T_{02} - T_{01})$$

Thus if stagnation temperatures are employed there is no need to refer explicitly to the kinetic energy term. A practical advantage is that it is easier to measure the stagnation temperature of a high-velocity stream than the static temperature (see section 6.3).

When a gas is slowed down and the temperature rises there is a simultaneous rise in pressure. The *stagnation* (or *total*) *pressure* $p_0$ is defined in a similar way to $T_0$ but with the added restriction that the gas is imagined to be brought to rest not only adiabatically but also reversibly, i.e. isentropically. The stagnation pressure is thus defined by

$$\frac{p_0}{p} = \left(\frac{T_0}{T}\right)^{\gamma/(\gamma-1)} \tag{2.8}$$

Stagnation pressure, unlike stagnation temperature, is constant in a stream flowing without heat or work transfer only if friction is absent: the drop in stagnation pressure can be used as a measure of the fluid friction.

It is worth noting that $p_0$ is not identical with the usual pitot pressure $p_0^*$ defined for incompressible flow by

$$p_0^* = p + \rho C^2/2$$

Substituting equation (2.7) in (2.8), and making use of $c_p = \gamma R/(\gamma-1)$ and $p = \rho RT$, we have

$$p_0 = p\left(1 + \frac{\rho C^2}{2p} \times \frac{\gamma-1}{\gamma}\right)^{\gamma/(\gamma-1)}$$

$p_0^*$ is seen to be given by the first two terms of the binomial expansion. Thus $p_0$ approaches $p_0^*$ as the velocity is decreased and compressibility effects become negligible. As an example of the difference at high velocities, for air moving with sonic velocity (Mach number $M = 1$), $p_0/p = 1\cdot89$ whereas $p_0^*/p = 1\cdot7$. Thus by assuming the flow to be incompressible the stagnation pressure would be underestimated by about 11 per cent.

Applying equation (2.8) to an isentropic compression between inlet 1 and outlet 2 we have the stagnation pressure ratio given by:

$$\frac{p_{02}}{p_{01}} = \frac{p_{02}}{p_2} \times \frac{p_1}{p_{01}} \times \frac{p_2}{p_1} = \left(\frac{T_{02}}{T_2} \times \frac{T_1}{T_{01}} \times \frac{T_2}{T_1}\right)^{\gamma/(\gamma-1)} = \left(\frac{T_{02}}{T_{01}}\right)^{\gamma/(\gamma-1)}$$

Similarly, if required,

$$\frac{p_{02}}{p_1} = \left(\frac{T_{02}}{T_1}\right)^{\gamma/(\gamma-1)}$$

Thus $p_0$ and $T_0$ can be used in the same way as static values. Stagnation pressure and temperature are properties of the gas stream which can be used with static values to determine the combined thermodynamic and mechanical state of the stream. Such state points can be represented on the $T$–$s$ diagram as shown in Fig. 2.9 which depicts a compression process between 'static' states 1 and 2; the differences between the constant $p$ and $p_0$ lines have been exaggerated for clarity. The ideal stagnation state which would be reached after isentropic compression to the same, actual, outlet stagnation pressure is indicated by 02'. Primes will be attached to symbols to denote such ideal states throughout this book.

**FIG. 2.9 Stagnation states**

## Compressor and turbine efficiencies

The efficiency of any machine, the object of which is the absorption or production of work, is normally expressed in terms of the ratio of actual and ideal work transfers. Because turbomachines are essentially adiabatic, the ideal process is isentropic and the efficiency is called an *isentropic efficiency*. Making use of the concept of stagnation enthalpy or temperature to take account of any change in kinetic energy of the fluid between inlet and outlet we have, for the compressor,

$$\eta_c = \frac{W'}{W} = \frac{\Delta h_0'}{\Delta h_0}$$

For a perfect gas $\Delta h_0 = c_p \Delta T_0$ and this relation is usually sufficiently accurate for real gases under conditions encountered in gas turbines if a mean $c_p$ over the relevant range of temperature is used—see under heading 'Variation of specific heat' below (p. 41) for further discussion of this point. Furthermore, because the ideal and actual temperature changes are not very different, the mean $c_p$ can be assumed the same for both so that the isentropic efficiency is normally *defined* in terms of temperature as

$$\eta_c = \frac{T_{02}' - T_{01}}{T_{02} - T_{01}} \tag{2.9}$$

Similarly the turbine isentropic efficiency is defined as

$$\eta_t = \frac{W}{W'} = \frac{T_{03} - T_{04}}{T_{03} - T_{04}'} \tag{2.10}$$

When performing cycle calculations, values of $\eta_c$ and $\eta_t$ will be assumed and the temperature equivalents of the work transfers for a given pressure ratio are then found as follows:

$$T_{02} - T_{01} = \frac{1}{\eta_c}(T_{02}' - T_{01}) = \frac{T_{01}}{\eta_c}\left(\frac{T_{02}'}{T_{01}} - 1\right)$$

and finally

$$T_{02} - T_{01} = \frac{T_{01}}{\eta_c}\left[\left(\frac{p_{02}}{p_{01}}\right)^{(\gamma-1)/\gamma} - 1\right] \tag{2.11}$$

Similarly

$$T_{03} - T_{04} = \eta_t T_{03}\left[1 - \left(\frac{1}{p_{03}/p_{04}}\right)^{(\gamma-1)/\gamma}\right] \tag{2.12}$$

When the compressor is part of an industrial gas turbine, it usually has a short intake fairing which may be regarded as part of the compressor. $p_{01}$ and $T_{01}$ in equation (2.11) will then be equal to $p_a$ and $T_a$ respectively because the velocity of the ambient air is zero. This will be assumed to be the case throughout this chapter. When a long inlet duct and/or air filter is necessary, as in marine applications for example, the inlet pressure loss ($\Delta p_i$) must be deducted, i.e. $p_{01}$ will be $p_a - \Delta p_i$. The situation is rather different when the compressor is part of an aircraft propulsion unit because then there will be an intake duct of significant length in which ram compression takes place due to the forward speed of the aircraft. In this situation $p_{01}$ and $T_{01}$ would differ from $p_a$ and $T_a$ even if there were no friction losses, and it is always necessary to consider the intake and compressor as separate components. A discussion of how intake losses are then taken into account will be deferred until the next chapter.

In defining $\eta_t$ according to (2.10) and thus taking the ideal work as proportional to $(T_{03} - T'_{04})$, we are implying that the kinetic energy in the exhaust gas is going to be utilized, e.g. in a subsequent turbine or in the propelling nozzle of a jet engine. But if the turbine is part of an industrial plant exhausting directly to atmosphere this kinetic energy is wasted. The ideal quantity of turbine work would then seem to be taken more appropriately as that produced by an isentropic expansion from $p_{03}$ to the static outlet pressure $p_4$, with $p_4$ equal to the ambient pressure $p_a$. Thus $\eta_t$ would be defined by

$$\eta_t = \frac{T_{03} - T_{04}}{T_{03}\left[1 - \left(\dfrac{1}{p_{03}/p_a}\right)^{(\gamma-1)/\gamma}\right]} \tag{2.13}$$

In practice, even in such a case the kinetic energy of the gas immediately leaving the turbine is largely recovered in an exhaust diffuser which in effect increases the pressure ratio across the turbine: Fig. 2.10 indicates this for a diffuser which reduces the final velocity to a negligible value so that $p_{04} = p_4 = p_a$. The turbine pressure ratio is seen to be increased from $p_{03}/p_a$ to $p_{03}/p_x$. The temperature equivalent of the turbine work $(T_{03} - T_{0x})$ is still given by $(T_{03} - T_{04})$, because no work is done in the diffuser and $T_{0x} = T_{04}$, but $T_{04}$ is less than it would be if a diffuser were not fitted and $p_x$ was equal to $p_a$. For ordinary cycle calculations there is no need to consider the turbine expansion $3 \rightarrow x$ and diffusion process $x \rightarrow 4$ separately. We may put $p_{04} = p_a$ in equation (2.12) and regard $\eta_t$ as accounting also for the friction pressure loss in the diffuser $(p_{0x} - p_a)$. We then have equation (2.13), but must interpret it as applying to the turbine and exhaust diffuser combined rather than to the turbine alone. In this book, equation (2.12) with $p_{04}$ put equal to $p_a$ will be used for any turbine exhausting direct to atmosphere: for any turbine delivering gas to a propelling nozzle, or to a second turbine in series, equation (2.12) will be employed as it stands.

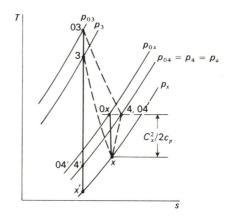

**FIG. 2.10 Turbine with exhaust diffuser**

So far we have been referring to overall efficiencies applied to the compressor or turbine as a whole. When performing cycle calculations covering a range of pressure ratio, say to determine the optimum pressure ratio for a particular application, the question arises as to whether it is reasonable to assume fixed typical values of $\eta_c$ and $\eta_t$. In fact it is found that $\eta_c$ tends to decrease and $\eta_t$ to increase as the pressure ratio for which the compressor and turbine are designed increases. The reason for this should be clear from the following argument based on Fig. 2.11. $p$ and $T$ are used instead of $p_0$ and $T_0$ to avoid a multiplicity of suffixes.

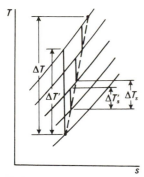

**FIG. 2.11**

Consider an axial flow compressor consisting of a number of successive stages. If the blade design is similar in successive blade rows it is reasonable to assume that the isentropic efficiency of a single stage, $\eta_s$, remains the same through the compressor. Then the overall temperature rise can be expressed by

$$\Delta T = \sum \frac{\Delta T_s'}{\eta_s} = \frac{1}{\eta_s} \Sigma \Delta T_s'$$

Also, $\Delta T = \Delta T'/\eta_c$ by definition of $\eta_c$, and thus

$$\frac{\eta_s}{\eta_c} = \frac{\Sigma \Delta T_s'}{\Delta T'}$$

But, because the vertical distance between a pair of constant pressure lines in the $T$–$s$ diagram increases as the entropy increases, it is clear from Fig. 2.11 that $\Sigma \Delta T_s' > \Delta T'$. It follows that $\eta_c < \eta_s$ and that the difference will increase with the number of stages, i.e. with increase of pressure ratio. A physical explanation is that the increase in temperature due to friction in one stage results in more work being required in the next stage: it might be termed the 'preheat' effect. A similar argument can be used to show that for a turbine $\eta_t > \eta_s$. In this case frictional 'reheating' in one stage is partially recovered as work in the next.

These considerations have led to the concept of *polytropic* (or *small-stage*) *efficiency* $\eta_\infty$, which is defined as the isentropic efficiency of an elemental stage in the process such that it is constant throughout the whole process. For a compression,

$$\eta_{\infty c} = \frac{\mathrm{d}T'}{\mathrm{d}T} = \text{constant}$$

But $T/p^{(\gamma-1)/\gamma} = $ constant for an isentropic process, which in differential form is

$$\frac{\mathrm{d}T'}{T} = \frac{\gamma-1}{\gamma} \frac{\mathrm{d}p}{p}$$

Substitution of $\mathrm{d}T'$ from the previous equation gives

$$\eta_{\infty c} \frac{\mathrm{d}T}{T} = \frac{\gamma-1}{\gamma} \frac{\mathrm{d}p}{p}$$

Integrating between inlet 1 and outlet 2, with $\eta_{\infty c}$ constant by definition, we have

$$\eta_{\infty c} = \frac{\ln(p_2/p_1)^{(\gamma-1)/\gamma}}{\ln(T_2/T_1)} \tag{2.14}$$

This equation enables $\eta_{\infty c}$ to be calculated from measured values of $p$ and $T$ at inlet and outlet of a compressor. Equation (2.14) can also be written in the form

$$\frac{T_2}{T_1} = \left(\frac{p_2}{p_1}\right)^{(\gamma-1)/\gamma \eta_{\infty c}} \tag{2.15}$$

Finally, the relation between $\eta_{\infty c}$ and $\eta_c$ is given by,

$$\eta_c = \frac{T_2'/T_1 - 1}{T_2/T_1 - 1} = \frac{(p_2/p_1)^{(\gamma-1)/\gamma} - 1}{(p_2/p_1)^{(\gamma-1)/\gamma\eta_{\infty c}} - 1} \qquad (2.16)$$

Note that if we write $(\gamma-1)/\gamma\eta_{\infty c}$ as $(n-1)/n$, equation (2.15) is the familiar relation between $p$ and $T$ for a polytropic process, and thus the definition of $\eta_\infty$ implies that the non-isentropic process is polytropic. This is the origin of the term polytropic efficiency.

Similarly, since $\eta_{\infty t}$ is $dT/dT'$, it can be shown that for an expansion between inlet 3 and outlet 4,

$$\frac{T_3}{T_4} = \left(\frac{p_3}{p_4}\right)^{\eta_{\infty t}(\gamma-1)/\gamma} \qquad (2.17)$$

and

$$\eta_t = \frac{1 - \left(\dfrac{1}{p_3/p_4}\right)^{\eta_{\infty t}(\gamma-1)/\gamma}}{1 - \left(\dfrac{1}{p_3/p_4}\right)^{(\gamma-1)/\gamma}} \qquad (2.18)$$

Making use of equations (2.16) and (2.18) with $\gamma = 1 \cdot 4$, Fig. 2.12 has been drawn to show how $\eta_c$ and $\eta_t$ vary with pressure ratio for a fixed value of polytropic efficiency of 85 per cent in each case.

FIG. 2.12  **Variation of turbine and compressor isentropic efficiency with pressure ratio for polytropic efficiency of 85%**

In practice, as with $\eta_c$ and $\eta_t$, it is normal to define the polytropic efficiencies in terms of stagnation temperatures and pressures. Furthermore, when employing them in cycle calculations, the most convenient equations to use will be shown to be those corresponding to (2.11) and (2.12), i.e. from equations (2.15) and (2.17),

$$T_{02} - T_{01} = T_{01}\left[\left(\frac{p_{02}}{p_{01}}\right)^{(n-1)/n} - 1\right] \qquad (2.19)$$

where $(n-1)/n = (\gamma-1)/\gamma\eta_{\infty c}$

$$T_{03} - T_{04} = T_{03}\left[1 - \left(\frac{1}{p_{03}/p_{04}}\right)^{(n-1)/n}\right] \qquad (2.20)$$

where $(n-1)/n = \eta_{\infty t}(\gamma - 1)/\gamma$.

And again, for a compressor of an industrial gas turbine we shall take $p_{01} = p_a$ and $T_{01} = T_a$, while for turbines exhausting to atmosphere $p_{04}$ will be put equal to $p_a$.

## Pressure losses

Pressure losses in the intake and exhaust ducting have been dealt with in the previous sub-section. In the combustion chamber a loss in stagnation pressure $(\Delta p_b)$ occurs due to the aerodynamic resistance of flame-stabilizing and mixing devices, and also due to momentum changes produced by the exothermic reaction. These sources of loss are referred to in detail in Chapter 6. When a heat-exchanger is included in the plant there will also be frictional pressure losses in the passages on the air-side $(\Delta p_{ha})$ and gas-side $(\Delta p_{hg})$. As shown in Fig. 2.13, the pressure losses have the effect of decreasing the turbine

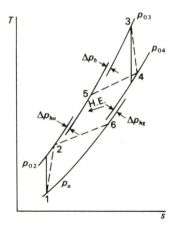

**FIG. 2.13  Pressure losses**

pressure ratio relative to the compressor pressure ratio and thus reduce the net work output from the plant. The gas turbine cycle is very sensitive to irreversibilities, because the net output is the difference of two large quantities (i.e. the 'work ratio' is low), so that the pressure losses have a significant effect on the cycle performance.

Fixed values of the losses can be fed into the cycle calculation directly. For example, for a simple cycle with heat-exchange we may determine the turbine pressure ratio $p_{03}/p_{04}$ from

$$p_{03} = p_{02} - \Delta p_b - \Delta p_{ha}, \text{ and } p_{04} = p_a + \Delta p_{hg}$$

But again the question arises as to whether it is reasonable to assume constant values for the pressure drops when cycles of different pressure ratio are being compared. The frictional pressure losses will be roughly proportional to the local dynamic head of the flow ($\frac{1}{2}\rho C^2$ for incompressible flow), as with ordinary pipe flow. It might therefore be expected that the pressure drops $\Delta p_{ha}$ and $\Delta p_b$ will increase with cycle pressure ratio because the density of the fluid in the air-side of the heat-exchanger and in the combustion chamber is increased. Even though $\rho$ is not proportional to $p$ because $T$ increases also, a better approximation might be to take $\Delta p_{ha}$ and $\Delta p_b$ as fixed proportions of the compressor delivery pressure. The turbine inlet pressure is then found from

$$p_{03} = p_{02}\left(1 - \frac{\Delta p_b}{p_{02}} - \frac{\Delta p_{ha}}{p_{02}}\right)$$

We shall express the pressure loss data in this way in subsequent numerical examples.

## Heat-exchanger effectiveness

Heat-exchangers for gas turbines can take many forms, including counter-flow and cross-flow recuperators (where the hot and cold streams exchange heat through a separating wall) or regenerators (where the streams are brought cyclically into contact with a matrix which alternately absorbs and rejects heat). In all cases, using the notation of Fig. 2.13, the fundamental process is that the turbine exhaust gases reject heat at the rate of $m_t c_{p46}$ $(T_{04} - T_{06})$ while the compressor delivery air receives heat at the rate of $m_c c_{p25}(T_{05} - T_{02})$. For conservation of energy, assuming that the mass flows $m_t$ and $m_c$ are equal,

$$c_{p46}(T_{04} - T_{06}) = c_{p25}(T_{05} - T_{02}) \tag{2.21}$$

But both $T_{05}$ and $T_{06}$ are unknown and a second equation is required for their evaluation. This is provided by the equation expressing the efficiency of the heat-exchanger.

Now the maximum possible value of $T_{05}$ is when the 'cold' air attains the temperature of the incoming hot gas $T_{04}$, and one possible measure of performance is the ratio of actual energy received by the cold air to the maximum possible value, i.e.

$$\frac{m_c c_{p25}(T_{05} - T_{02})}{m_c c_{p24}(T_{04} - T_{02})}$$

The mean specific heat of air will not be very different over the two temperature ranges, and it is usual to define the efficiency in terms of temperature alone and to call it the *effectiveness* (or *thermal-ratio*) of the heat-exchanger. Thus

$$\text{effectiveness} = \frac{T_{05} - T_{02}}{T_{04} - T_{02}} \tag{2.22}$$

When a value of effectiveness is specified, equation (2.22) enables the temperature at inlet to the combustion chamber $T_{05}$ to be determined. Equation (2.21) then yields $T_{06}$ if required. Note that the mean specific heats $c_{p46}$ and $c_{p25}$ are not approximately equal and cannot be cancelled, because the former is for turbine exhaust gas and the latter for air.

In general, the larger the volume of the heat-exchanger the higher can be the effectiveness, but with gas turbines for road or marine transport space is a vital limiting factor. Considerations of weight and space are so important in aircraft applications that heat-exchangers have not yet been employed to any advantage: recent developments with compact surface heat-exchangers of ceramic materials may change the picture in the future. Modern heat-exchangers have values of effectiveness around 0·90 and can withstand turbine exit temperatures of up to 900 K. Heat-exchangers are subjected to severe thermal stresses during start up and are not used where frequent starts are required, e.g. for peak load electricity generation. Pipeline applications are well suited to heat-exchangers, because the gas turbines may run for very extended periods at a base load condition. It should be noted, however, that many pipeline stations are located in remote areas where transportation and installation of a heat-exchanger present major problems; for this reason, some modern heat-exchangers are built up from a series of identical modules.

## Mechanical losses

In all gas turbines, the power necessary to drive the compressor is transmitted directly from the turbine without any intermediate gearing. Any loss that occurs is therefore due only to bearing friction and windage. This loss is very small and it is normal to assume that it amounts to about 1 per cent of the power necessary to drive the compressor. If the transmission efficiency is denoted by $\eta_m$, we have the work output required to drive the compressor given by

$$W = \frac{1}{\eta_m} c_{p12} (T_{02} - T_{01})$$

We shall take $\eta_m$ to be 99 per cent for all numerical examples.

Any power used to drive ancillary components such as fuel and oil pumps can often be accounted for simply by subtracting it from the net output of the unit. Power absorbed in any gearing between the gas turbine and the load can be dealt with similarly. Except to say that such losses can be significant, especially when small gas turbines of low power are under consideration as for road transport, we shall not consider them further. Their consideration would not require the method of cycle performance estimation to be modified except when the gas turbine has a separate power turbine. Power for fuel and oil pumps will then be taken from the compressor turbine (because under some operating conditions the power turbine is stationary) in which case the temperature at inlet to the power turbine is reduced.

## Variation of specific heat

The properties $c_p$ and $\gamma$ play an important part in the estimation of cycle performances, and it is necessary to take account of variations in values due to changing conditions through the cycle. In general, for real gases over normal working ranges of pressure and temperature, $c_p$ is a function of temperature alone. The same is true of $\gamma$ because it is related to $c_p$ by

$$\frac{\gamma-1}{\gamma} = \frac{R_0}{Mc_p} \qquad (2.23)$$

where $R_0$ is the universal gas constant and $M$ the molecular weight. The variation of $c_p$ and $\gamma$ with temperature for air is shown in Fig. 2.14 by the curves marked zero fuel/air ratio. Only the left-hand portion is of interest because even with a pressure ratio as high as 14 the compressor temperature rise will be no more than about 400 K.

In the turbine of an open cycle plant the working fluid will be a mixture of combustion gases. Most gas turbines run on kerosene which has a composition to which the formula $C_nH_{2n}$ is a good approximation. If some such composition is assumed, the products analysis can be calculated for various fuel/air ratios. Knowing the specific heats and molecular weights of the constituents it is then a simple matter to calculate the mean values of $c_p$ and $\gamma$ for the mixture. Figure 2.14 shows that $c_p$ increases and $\gamma$ decreases

**FIG 2.14** $c_p$ and $\gamma$ for air and typical combustion gases

with increase in fuel/air ratio. It is worth noting that the mean molecular weight of the combustion products from typical hydrocarbon fuels is little different from that of air, and therefore $c_p$ and $\gamma$ are closely related by equation (2.23) with $R_0/M = R_{air} = 0.287$ kJ/kg K.

The calculation of the products analyses is very lengthy when dissociation is taken into account and then, because pressure has a significant effect on the amount of dissociation, $c_p$ and $\gamma$ become a function of pressure as well as temperature. Accurate calculations of this kind have been made using a digital computer, and the results are tabulated in Ref. (3). Dissociation begins to have a significant effect on $c_p$ and $\gamma$ at a temperature of about 1500 K, and above this temperature the curves of Fig. 2.14 are strictly speaking applicable only to a pressure of 1 bar. In fact at 1800 K, both for air and products of combustion corresponding to low values of fuel/air ratio, a reduction of pressure to 0·01 bar increases $c_p$ by only about 4 per cent and an increase to 100 bar decreases $c_p$ by only about 1 per cent: the corresponding changes in $\gamma$ are even smaller. In this book we may safely ignore any effect of pressure because only the most sophisticated aircraft engines use a turbine inlet temperature as high as 1500 K, and a more typical operating temperature would be 1100 K.

Now compressor temperature rises and turbine temperature drops will be calculated using equations such as (2.11) and (2.12) or (2.19) and (2.20). For accurate calculations a method of successive approximation would be required, i.e. it would be necessary to guess a value of $\gamma$, calculate the temperature change, take a more accurate mean value of $\gamma$ and recalculate the temperature change. In fact, if this degree of accuracy is required it is better to use tables or curves of enthalpy and entropy as described, for example, in Ref. (1). For preliminary design calculations and comparative cycle calculations, however, it has been found to be sufficiently accurate to assume the following fixed values of $c_p$ and $\gamma$ for the compression and expansion processes respectively,

$$\text{air:} \quad c_{pa} = 1\cdot005 \text{ kJ/kg K}, \gamma_a = 1\cdot40 \text{ or } \left(\frac{\gamma}{\gamma-1}\right)_a = 3\cdot5$$

$$\text{combustion gases:} \quad c_{pg} = 1\cdot148 \text{ kJ/kg K}, \gamma_g = 1\cdot333 \text{ or } \left(\frac{\gamma}{\gamma-1}\right)_g = 4\cdot0$$

The reason why this does not lead to much inaccuracy is that $c_p$ and $\gamma$ vary in opposite senses with $T$. For cycle analysis we are interested in calculating compressor and turbine work from the product $c_p\Delta T$. Suppose that the temperature for which the above values of $c_p$ and $\gamma$ are the true values is lower than the actual mean temperature. $\gamma$ is then higher than it should be and $\Delta T$ will be overestimated. This will be compensated in the product $c_p\Delta T$ by the fact that $c_p$ will be lower than it should be. The actual temperatures at various points in the cycle will not be very accurate, however, and for the detailed design of the components it is necessary to know the exact conditions of the working fluid: the more accurate approaches mentioned above must then be employed.

## Fuel/air ratio, combustion efficiency and cycle efficiency

The performance of real cycles can be unambiguously expressed in terms of the specific fuel consumption, i.e. fuel mass flow per unit net power output. To obtain this the fuel/air ratio must be found. In the course of calculating the net output per unit mass flow of air the temperature at inlet to the combustion chamber ($T_{02}$) will have been obtained; and the temperature at outlet ($T_{03}$), which is the maximum cycle temperature, will normally be specified. The problem is therefore to calculate the fuel/air ratio $f$ required to transform unit mass of air at $T_{02}$ and $f$ kg of fuel at the fuel temperature $t_f$ to $(1 + f)$ kg of products at $T_{03}$.

Since the process is adiabatic with no work transfer, the energy equation is simply

$$\sum (m_i h_{io3}) - (h_{ao2} + f h_f) = 0$$

where $m_i$ is the mass of product $i$ per unit mass of air and $h_i$ its specific enthalpy. Making use of the enthalpy of reaction at a reference temperature of 25°C, $\Delta H_{25}$, the equation can be expanded in the usual way [see Ref. (2)] to become

$$(1 + f)c_{pg}(T_{03} - 298) + f\Delta H_{25} + c_{pa}(298 - T_{02}) + f c_{pf}(298 - T_f) = 0$$

where $c_{pg}$ is the mean specific heat of the products over the temperature range 298 K to $T_{03}$. $\Delta H_{25}$ should be the enthalpy of reaction per unit mass of fuel with the $H_2O$ in the products in the vapour phase, because $T_{03}$ is high and above the dew point. For common fuels $\Delta H_{25}$ may be taken from tables, or alternatively it may be evaluated from the enthalpies of formation of the reactants. It is usual to assume that the fuel temperature is the same as the reference temperature, so that the fourth term on the L.H.S. of the equation is zero. The term will certainly be small because $f$ is low ($<0.05$) and $c_{pf}$ for liquid hydrocarbon fuels is only about 2 kJ/kg K. Finally, we have already discussed in the previous section, 'Variation of specific heat', the calculation of the mean specific heat of the products $c_{pg}$ as a function of $f$ and $T$, so we are left with an equation from which $f$ can be obtained for any given values of $T_{02}$ and $T_{03}$.

Such calculations are too lengthy to be undertaken for every individual cycle calculation—particularly if dissociation is significant because then the $f\Delta H_{25}$ term must be modified to allow for the incompletely burnt carbon and hydrogen arising from the dissociated $CO_2$ and $H_2O$. It is usually sufficiently accurate to use tables or charts which have been compiled for a typical fuel composition. Figure 2.15 shows the combustion temperature rise ($T_{03} - T_{02}$) plotted against fuel/air ratio for various values of inlet temperature ($T_{02}$), and these curves will be used for all numerical examples in this book. It is a small-scale version of larger and more accurate graphs given in Ref. (4). The reference fuel for which the data has been calculated is a hypothetical liquid hydrocarbon containing 13·92 per cent H and 86·08 per cent C, for which the stoichiometric fuel/air ratio is 0·068 and $\Delta H_{25}$ is −43 100 kJ/kg. The curves are certainly adequate for any kerosene burnt in dry air. Methods are given in Ref. (4) whereby the data can be used for hydrocarbon fuels which differ widely in composition from the reference fuel, or where the fuel is burnt in a reheat chamber, i.e. not in air but in the products of combustion from a previous chamber in the cycle.

The data for Fig. 2.15 have been calculated on the assumption that the fuel is completely burnt, and thus the abscissa could be labelled 'theoretical fuel/air ratio'. The most convenient method of allowing for combustion loss is by introducing a *combustion efficiency* defined by

$$\eta_b = \frac{\text{theoretical } f \text{ for given } \Delta T}{\text{actual } f \text{ for given } \Delta T}$$

This is the definition used in this book. An alternative method is to regard the ordinate as the theoretical $\Delta T$ for a given $f$ and define the efficiency in

**FIG. 2.15 Combustion temperature rise v. fuel/air ratio**

**(A large-scale version of this diagram is published by Basil Blackwell)**

terms of the ratio: actual $\Delta T$/theoretical $\Delta T$. Neither definition is quite the same as the fundamental definition based on the ratio of actual energy released to the theoretical quantity obtainable. ($\eta_b$ differs from it because of the small additional heat capacity of the products arising from the increase in fuel needed to produce the given temperature.) But in practice combustion is so nearly complete—98–99 per cent—that the efficiency is difficult to measure accurately and the three definitions of efficiency yield virtually the same result.

Once the fuel/air ratio is known, the fuel consumption $m_f$ is simply $f \times m$ where $m$ is the air mass flow, and the specific fuel consumption can be found directly from

$$\text{s.f.c.} = \frac{f}{W_N}$$

Since the fuel consumption is normally measured in kg/h, while $W_N$ is in kW per kg/s of air flow, the s.f.c. in kg/kW h is given by the following numerical equation:†

$$\frac{\text{s.f.c.}}{[\text{kg/kW h}]} = \frac{f}{W_N/[\text{kW s/kg}]} \times \frac{[\text{s}]}{[\text{h}]} = \frac{3600\,f}{W_N/[\text{kW s/kg}]}$$

If the thermal efficiency of the cycle is required, for comparison with the ideal cycle efficiency, it must be defined in the form 'work output/heat supplied' even though the combustion process is adiabatic and in the thermodynamic sense no heat is supplied. We know that if the fuel is burnt under ideal conditions, such that the products and reactants are virtually at the same temperature (the reference temperature 25°C), the rate of energy release in the form of heat will be

$$m_f Q_{\text{gr},p} = f m Q_{\text{gr},p}$$

where $m_f$ is the fuel flow and $Q_{\text{gr},p}$ is the gross (or higher) calorific value at constant pressure. In the gas turbine it is not possible to utilize the latent heat of the $H_2O$ vapour in the products and the convention of using the net calorific value has been adopted in most countries. Thus the cycle efficiency may be defined as

$$\eta = \frac{W_N}{f Q_{\text{net},p}}$$

With the units used here, the equivalent numerical equation becomes

$$\eta = \frac{W_N/[\text{kW s/kg}]}{f \times Q_{\text{net},p}/[\text{kJ or kW s/kg}]}$$

† The fundamental principle employed here is that a physical quantity, or the symbol representing it, is equal to (pure number × unit) and *never* the number alone. Consequently an equation relating numbers only will, in addition to pure numbers and dimensional ratios, contain only quotients of symbols and units, e.g. $W_N/[\text{kW s/kg}]$.

or, using the above numerical equation for s.f.c.,

$$\eta = \frac{3600}{\text{s.f.c.}/[\text{kg/kW h}] \times Q_{\text{net},p}/[\text{kJ/kg}]}$$

$Q_{\text{net},p}$ is sensibly equal in magnitude but opposite in sign to the enthalpy of reaction $\Delta H_{25}$ referred to earlier, and the value of 43 100 kJ/kg will be used for all numerical examples.

When referring to the thermal efficiency of actual gas turbines, manufacturers often prefer to use the concept of *heat rate* rather than efficiency. The reason is that fuel prices are normally quoted in terms of pounds sterling (or dollars) per megajoule, and the heat rate can be used to evaluate fuel cost directly. Heat rate is defined as (s.f.c. $\times Q_{\text{net},p}$), and thus expresses the heat input required to produce a unit quantity of power. It is normally expressed in kJ/kW h, in which case the corresponding thermal efficiency can be found from 3600/(heat rate).

## 2.3   Design point performance calculations

Before assessing the effect of component losses on the general performance curves for the various cycles considered in section 2.1, it is necessary to outline the method of calculating the performance in any particular case for specified values of the design parameters. These parameters will include the compressor pressure ratio, turbine inlet temperature, component efficiencies and pressure losses.

For the first example we shall consider a unit operating on a simple heat-exchange cycle, with data chosen as typical of an industrial gas turbine in the 500–1000 kW range of power output. If it were to be used for road or marine transport it would certainly employ a separate power turbine but, assuming that the compressor turbine and power turbine have the same efficiency, this does not affect the design point performance calculation and the notation of Fig. 2.16 can be used. This example will illustrate the use of isentropic efficiencies, and polytropic efficiencies will be employed in the second example.

**FIG. 2.16 Heat-exchange cycle**

## EXAMPLE

Determination of specific work output, specific fuel consumption and cycle efficiency for a heat-exchange cycle having the following specification:

| | |
|---|---|
| Compressor pressure ratio | 4·0 |
| Turbine inlet temperature | 1100 K |
| Isentropic efficiency of compressor, $\eta_c$ | 0·85 |
| Isentropic efficiency of turbine, $\eta_t$ | 0·87 |
| Mechanical transmission efficiency $\eta_m$ | 0·99 |
| Combustion efficiency $\eta_b$ | 0·98 |
| Heat-exchanger effectiveness | 0·80 |
| Pressure losses— | |
|     Combustion chamber, $\Delta p_b$ | 2% comp. deliv. press. |
|     Heat-exchanger air-side, $\Delta p_{ha}$ | 3% ,, ,, ,, |
|     ,,    ,,    gas-side, $\Delta p_{hg}$ | 0·04 bar |
| Ambient conditions, $p_a, T_a$ | 1 bar, 288 K |

Since $T_{01} = T_a$ and $p_{01} = p_a$, and $\gamma = 1·4$, the temperature equivalent of the compressor work from equation (2.11) is

$$T_{02} - T_a = \frac{T_a}{\eta_c}\left[\left(\frac{p_{02}}{p_a}\right)^{(\gamma-1)/\gamma} - 1\right]$$

$$= \frac{288}{0·85}[4^{1/3·5} - 1] = 164·6 \text{ K}$$

Turbine work required to drive compressor per unit mass flow is

$$W_{tc} = \frac{c_{pa}(T_{02} - T_a)}{\eta_m} = \frac{1·005 \times 164·6}{0·99} = 167 \text{ kJ/kg}$$

$$p_{03} = p_{02}\left(1 - \frac{\Delta p_b}{p_{02}} - \frac{\Delta p_{ha}}{p_{02}}\right) = 4·0\,(1 - 0·02 - 0·03) = 3·8 \text{ bar}$$

$$p_{04} = p_a + \Delta p_{hg} = 1·04 \text{ bar, and hence } p_{03}/p_{04} = 3·65.$$

Since $\gamma = 1·333$ for the expanding gases, the temperature equivalent of the total turbine work from equation (2.13) is

$$T_{03} - T_{04} = \eta_t T_{03}\left[1 - \left(\frac{1}{p_{03}/p_{04}}\right)^{(\gamma-1)/\gamma}\right]$$

$$= 0·87 \times 1100\left[1 - \left(\frac{1}{3·65}\right)^{1/4}\right] = 265 \text{ K}$$

Total turbine work per unit mass flow is

$$W_t = c_{pg}(T_{03} - T_{04}) = 1·148 \times 265 = 304 \text{ kJ/kg}$$

Remembering that the mass flow is to be assumed the same throughout the unit, the specific work output is simply

$$W_t - W_{tc} = 304 - 167 = 137 \text{ kJ/kg (or kW s/kg)}†$$

† Note: 1 kW s/kg $\equiv$ 0·6083 hp s/lb.

(It follows that for a 1000 kW plant an air mass flow of 7·3 kg/s would be required.) To find the fuel/air ratio we must first calculate the combustion temperature rise $(T_{03} - T_{05})$.

From equation (2.22),

$$\text{heat-exchanger effectiveness} = 0{\cdot}80 = \frac{T_{05} - T_{02}}{T_{04} - T_{02}}$$

$$T_{02} = 164{\cdot}6 + 288 = 452{\cdot}6 \text{ K, and } T_{04} = 1100 - 265 = 835 \text{ K}$$

Hence

$$T_{05} = 0{\cdot}80 \times 382{\cdot}4 + 452{\cdot}6 = 758 \text{ K}$$

From Fig. 2.15, for a combustion chamber inlet air temperature of 758 K and a combustion temperature rise of $(1100 - 758) = 342$ K, the theoretical fuel/air ratio required is 0·0094 and thus

$$f = \frac{\text{theoretical } f}{\eta_b} = \frac{0{\cdot}0094}{0{\cdot}98} = 0{\cdot}0096$$

The specific fuel consumption is therefore

$$\text{s.f.c.} = \frac{f}{W_t - W_{tc}} = \frac{3600 \times 0{\cdot}0096}{137} = 0{\cdot}252 \text{ kg/kW h}†$$

Finally, the cycle efficiency is

$$\eta = \frac{3600}{\text{s.f.c.} \times Q_{net,p}} = \frac{3600}{0{\cdot}252 \times 43\,100} = 0{\cdot}33$$

This is a creditable performance, but it must be remembered that in many applications an intake air filter and/or exhaust silencer will be required and the additional pressure losses involved will reduce the specific work output and increase the s.f.c. The relative importance of such additional losses and of varying ambient conditions on the performance of gas turbines will be discussed in section 2.4.

The effect of adding a reheat chamber to the unit is examined in the second example. The polytropic efficiencies specified have been chosen so that the corresponding isentropic efficiencies are virtually the same as before. Fig. 2.17 shows the notation employed.

FIG. 2.17  **Cycle with heat-exchange and reheat**

† Note: 1 kg/kW h ≡ 1·644 lb/hp h.

## EXAMPLE

Data as in previous example, but assume that the compression and expansion efficiencies are quoted as

$$\eta_{\infty c} = 0 \cdot 878 \text{ and } \eta_{\infty t} = 0 \cdot 85 \text{ (for both turbines)}$$

In addition, the gas is reheated to 1100 K between the compressor turbine and power turbine. (From the previous example we can see that the desirable equal division of work between the two turbines will be approximately achieved.) The reheat chamber pressure loss $\Delta p_{br}$ is 2 per cent of the chamber inlet pressure, and the combustion efficiency is 0·98 as for the main chamber.

Since we shall be making use of equations (2.19) and (2.20) it is convenient to commence by evaluating $(n-1)/n$ for the polytropic compression and expansion:

$$\text{for compression, } \frac{n-1}{n} = \frac{1}{\eta_{\infty c}} \left( \frac{\gamma - 1}{\gamma} \right)_a = \frac{1}{0 \cdot 878 \times 3 \cdot 5} = 0 \cdot 3253$$

$$\text{for expansion, } \frac{n-1}{n} = \eta_{\infty t} \left( \frac{\gamma - 1}{\gamma} \right)_g = \frac{0 \cdot 85}{4} = 0 \cdot 2125$$

From equation (2.19), with $p_{01} = p_a$ and $T_{01} = T_a$

$$T_{02} - T_a = T_a \left[ \left( \frac{p_{02}}{p_a} \right)^{(n-1)/n} - 1 \right]$$

$$= 288 \left[ 4^{0 \cdot 3253} - 1 \right] = 164 \cdot 3 \text{ K}$$

And thus, as for the previous example, $W_{tc} = 167$ kJ/kg.

The intermediate pressure between the two turbines is unknown, but it can be determined from the fact that the compressor turbine produces just sufficient work to drive the compressor. The temperature equivalent of the compressor turbine work is therefore

$$T_{03} - T_{04} = \frac{W_{tc}}{c_{pg}} = \frac{167}{1 \cdot 148} = 145 \cdot 5 \text{ K}$$

The corresponding pressure ratio $p_{03}/p_{04}$ can be found using equation (2.20).

$$T_{03} - T_{04} = T_{03} \left[ 1 - \left( \frac{1}{p_{03}/p_{04}} \right)^{(n-1)/n} \right]$$

$$145 \cdot 5 = 1100 \left[ 1 - \left( \frac{1}{p_{03}/p_{04}} \right)^{0 \cdot 2125} \right]$$

$$\frac{p_{03}}{p_{04}} = 1 \cdot 95$$

And since $p_{03} = 3{\cdot}8$ bar as in the previous example, we have

$$p_{04} = 1{\cdot}948 \text{ bar}$$

$$p_{05} = p_{04}\left(1 - \frac{\Delta p_{br}}{p_{04}}\right) = 1{\cdot}948\,(1 - 0{\cdot}02) = 1{\cdot}91 \text{ bar}$$

$$p_{06} = p_a + \Delta p_{hg} = 1{\cdot}04 \text{ bar, and hence } p_{05}/p_{06} = 1{\cdot}836.$$

Having found the power turbine pressure ratio, the temperature equivalent of the work output becomes

$$T_{05} - T_{06} = T_{05}\left[1 - \left(\frac{1}{p_{05}/p_{06}}\right)^{(n-1)/n}\right]$$

$$= 1100\left[1 - \left(\frac{1}{1{\cdot}836}\right)^{0{\cdot}2125}\right] = 133{\cdot}7 \text{ K}$$

and the specific work output, i.e. power turbine work per unit air mass flow, is

$$W_{tp} = c_{pg}(T_{05} - T_{06}) = 1{\cdot}148 \times 133{\cdot}7 = 153 \text{ kJ/kg (or kW s/kg)}$$

To find the fuel/air ratio in the two combustion chambers we may proceed as follows.

$$\text{Heat-exchanger effectiveness} = 0{\cdot}80 = \frac{T_{07} - T_{02}}{T_{06} - T_{02}}$$

$$T_{02} = 452{\cdot}3 \text{ K}, \quad T_{06} = 1100 - 133{\cdot}7 = 966{\cdot}3 \text{ K}$$

Hence inlet temperature to main combustion chamber is

$$T_{07} = 0{\cdot}80 \times 514 + 452{\cdot}3 = 864 \text{ K}$$

and combustion temperature rise is

$$T_{03} - T_{07} = 1100 - 864 = 236 \text{ K}$$

From Fig. 2.15, theoretical $f = 0{\cdot}006\,55$.
For the reheat chamber,

$$T_{04} = 1100 - 145{\cdot}5 = 954{\cdot}5, \text{ and } T_{05} - T_{04} = 1100 - 954{\cdot}5 = 145{\cdot}5 \text{ K}$$

From Fig. 2.15, theoretical $f = 0{\cdot}004\,05$.
Then the total actual fuel/air ratio required is

$$f = \frac{0{\cdot}006\,55 + 0{\cdot}004\,05}{0{\cdot}98} = 0{\cdot}0108$$

$$\text{s.f.c.} = \frac{f}{W_{tp}} = \frac{3600 \times 0{\cdot}0108}{153} = 0{\cdot}254 \text{ kg/kW h}$$

$$\text{cycle efficiency, } \eta = \frac{3600}{0{\cdot}254 \times 43\,100} = 0{\cdot}329$$

Thus the addition of reheat has increased the specific work output by $(153 - 137)/137 = 11\cdot7$ per cent, but at the expense of a slight increase in specific fuel consumption of $(0\cdot254 - 0\cdot252)/0\cdot252 = 0\cdot8$ per cent. These changes are small, as might be expected for a cycle of low pressure ratio. The implications of this comparison will be discussed further in section 2.4. For the moment it is sufficient to point out that when small changes of this order of magnitude are being considered it is necessary to take account of such finer points as the following:

(a) More air must be bled from the compressor to cool the hotter power turbine, and the effective mass flow through the compressor turbine will be reduced: the mass flow through the power turbine might be unchanged because of the additional fuel supplied in the reheat chamber.

(b) The reheat fuel/air ratio would have to be calculated accurately remembering that the fuel is burnt not in air but in the combustion gases from the main combustion chamber.

## 2.4 Comparative performance of practical cycles

The large number of variables involved make it impracticable to derive algebraic expressions for the specific output and efficiency of real cycles. On the other hand, the type of step-by-step calculation illustrated in the previous section is ideally suited for computer programming, each of the design parameters being given a set of values in turn to elicit their effect upon the performance.

Some performance curves will now be presented to show the main differences between practical and ideal cycles, and the relative importance of some of the parameters. The curves are definitely *not* a comprehensive set from which designers can make a choice of cycle for a particular application. To emphasize that too much importance should not be attached to the values of specific output and efficiency the full specification of the parameters has not been given: it is sufficient to note that those parameters which are not specified on the curves are kept constant. All the curves use compressor pressure ratio $r_c$ as abscissa, the turbine pressure ratio being less than $r_c$ by virtue of the pressure losses. The cycle efficiency has been evaluated to facilitate comparison with the ideal curves of section 2.1. In practice it is usual to quote s.f.c. rather than efficiency, not only because its definition is unambiguous, but also because it provides both a direct indication of fuel consumption and a measure of cycle efficiency to which it is inversely proportional.

### Simple gas turbine cycle

When component losses are taken into account the efficiency of the simple cycle becomes dependent upon the maximum cycle temperature $T_{03}$ as well as pressure ratio, Fig. 2.18. Furthermore, for each temperature the efficiency

**FIG. 2.18  Cycle efficiency and specific output of simple gas turbine**

has a peak value at a particular pressure ratio. The fall in efficiency at higher pressure ratios is due to the fact that the reduction in fuel supply to give the fixed turbine inlet temperature, resulting from the higher compressor delivery temperature, is outweighed by the increased work necessary to drive the compressor. Although the optimum pressure ratio for maximum efficiency differs from that for maximum specific output, the curves are fairly flat near the peak and a pressure ratio between the two optima can be used without much loss in efficiency. It is perhaps worth pointing out that the lowest pressure ratio which will give an acceptable performance is always chosen: it might even be slightly lower than either optimum value. Mechanical design considerations beyond the scope of this book may affect the choice: such considerations include the number of compressor and turbine stages required, the avoidance of excessively small blades at the high pressure end of the compressor, and whirling speed and bearing problems associated with the length of the compressor–turbine combination.

The advantage of using as high a value of $T_{03}$ as possible, and the need to use a higher pressure ratio to take advantage of a higher permissible temperature, is evident from the curves. The efficiency increases with $T_{03}$ because the component losses become relatively less important as the ratio of positive turbine work to negative compressor work increases, although the gain in efficiency becomes marginal as $T_{03}$ is increased beyond 1200 K (particularly if a higher temperature requires a complex turbine blade cooling system which incurs additional losses). There is nothing marginal, however, about the gain in specific work output with increase in $T_{03}$. The consequent reduction in size of plant for a given power is very marked, and this is particularly important for aircraft gas turbines as will be emphasized in the next chapter.

The following figures illustrate the relative importance of some of the other parameters. Changes in efficiency are quoted as simple *differences* in percentages. With a $T_{03}$ of 1500 K, and a pressure ratio near the optimum value, an increase of 5 per cent in the polytropic efficiency of either the compressor or turbine would increase the cycle efficiency by about 4 per cent and the specific output by about 65 kW s/kg. (If isentropic efficiencies had been used, the turbine loss would have been seen to be more important than the compressor loss but the use of polytropic efficiencies obscures this fact.) A reduction in combustion chamber pressure loss from 5 per cent of the compressor delivery pressure to zero would increase the cycle efficiency by about 1·5 per cent and the specific output by about 12 kW s/kg. The remaining parameter of importance is the ambient temperature, to which the performance of gas turbines is particularly sensitive.

The ambient temperature affects both the compressor work (proportional to $T_a$) and the fuel consumption (a function of $T_{03} - T_{02}$). An increase in $T_a$ reduces both specific output and cycle efficiency, although the latter is less affected than the former because for a given $T_{03}$ the combustion temperature rise is reduced. Considering again the case of $T_{03} = 1500$ K and a pressure ratio near the optimum, an increase in $T_a$ from 15 to 40°C reduces the efficiency by about 2·5 per cent and the specific output by about 62 kW s/kg. The latter is nearly 20 per cent of the output, which emphasizes the importance of designing a gas turbine to give the required power output at the highest ambient temperature likely to be encountered.

## Heat-exchange (or regenerative) cycle

As far as the specific work output is concerned, the addition of a heat-exchanger merely causes a slight reduction due to the additional pressure losses: the curves retain essentially the same form as those in Fig. 2.18. The efficiency curves are very different, however, as shown in Fig. 2.19. Heat-exchange increases the efficiency substantially and markedly reduces the optimum pressure ratio for maximum efficiency. Unlike the corresponding curves for the ideal cycle, they do not rise to the Carnot value at $r_c = 1$ but

FIG. 2.19 Heat-exchange cycle

fall to zero at the pressure ratio at which the turbine provides just sufficient work to drive the compressor: at this point there will be positive heat input with zero net work output. The spacing of the constant $T_{03}$ curves in Fig. 2.19 indicate that when a heat-exchanger is used the gain in efficiency is no longer merely marginal as $T_{03}$ is raised above 1200 K. This is a most important feature of the heat-exchange cycle, because progress in materials science and blade cooling techniques has enabled the permissible temperature to be increased at an average rate of about 10 K per year and there is every hope that this will continue. It should also be noted that the optimum pressure ratio for maximum efficiency increases as $T_{03}$ is increased. Our study of the ideal heat-exchange cycle suggested that no such increase in pressure ratio was required: this was the conclusion under the heading 'Heat-exchange cycle' (p. 27) in section 2.1 which it was stated would have to be modified. Nevertheless, it remains true to say that no very high pressure ratio is ever required for a heat-exchange cycle, and the increase in weight and cost due to a heat-exchanger is partially offset by the reduction in size of the compressor.

The dotted curves have been added to show the effect of heat-exchanger effectiveness. Not only does an increase in effectiveness raise the cycle efficiency appreciably, but it also reduces the value of the optimum pressure ratio still further. Since the optimum $r_c$ for maximum efficiency is below that for maximum specific output, it is inevitable that a plant designed for high efficiency will suffer a space and weight penalty. For example, with a $T_{03}$ of 1500 K and an effectiveness of 0.75, the optimum $r_c$ for maximum efficiency is about 10 at which (from Fig. 2.18) the specific output is not far short of the peak value. But with the effectiveness increased to 0.875 the optimum $r_c$ is reduced to about 6, at which the specific output is only about 90 per cent of the peak value. The position is similar at the lower (and at present more realistic) values of $T_{03}$.

An alternative method of presenting performance characteristics is to plot the variation of specific fuel consumption and specific output on a single figure for a range of values of pressure ratio and turbine inlet temperature. Examples for realistic values of polytropic efficiency, pressure losses and heat-exchanger effectiveness are shown in Figs. 2.20(a) and (b) for simple and heat-exchange cycles respectively. The marked effect of increasing $T_{03}$ on specific output in both cases is clearly evident. Such plots also contrast clearly the small effect of $T_{03}$ on s.f.c. for the simple cycle with the much greater effect when a heat-exchanger is used.

## Heat-exchange cycle with reheat or intercooling

Our study of ideal cycles showed that there is no virtue in employing reheat without heat-exchange because of the deleterious effect upon efficiency: consequently this case is not considered here. With heat-exchange, addition of reheat improves the specific output considerably without loss of efficiency (cf. Figs. 2.21 and 2.19). The curves of Fig. 2.21 are based on the assumption

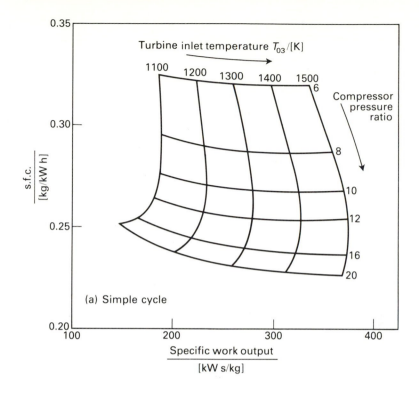

(a) Simple cycle

Specific work output

[kW s/kg]

(b) Heat-exchange cycle

Specific work output

[kW s/kg]

FIG. 2.20 Cycle performance curves

FIG. 2.21 Cycle with heat-exchange and reheat

that the gas is reheated to the maximum cycle temperature at the point in the expansion giving equal pressure ratios for the two turbines. The gain in efficiency due to reheat obtained with the ideal cycle is not realized in practice, partly because of the additional pressure loss in the reheat chamber and the inefficiency of the expansion process, but primarily because the effectiveness of the heat-exchanger is well short of unity and the additional energy in the exhaust gas is not wholly recovered. It is important to use a pressure ratio not less than the optimum value for maximum efficiency, because at lower pressure ratios the addition of reheat can actually reduce the efficiency as indicated by the curves and by the example in section 2.3.

Reheat has not been widely used in practice because the additional combustion chamber, and the associated control problems, can offset the advantage gained from the decrease in size of the main components consequent upon the increase in specific output. With the possible exception of the application mentioned at the end of this section, reheat would certainly be considered only (a) if the expansion had to be split between two turbines for other reasons and (b) if the additional flexibility of control provided by the reheat fuel supply was thought to be desirable. With regard to (a), it must be noted that the natural division of expansion between a compressor turbine and power turbine may not be the optimum point at which to reheat and if so the full advantage of reheat will not be realized. Finally, readers familiar with steam turbine design will understand that reheat also introduces additional mechanical problems arising from the decrease in gas density, and hence the need for longer blading, in the low-pressure stages.

Intercooling, which has a similar effect upon the performance of the ideal heat-exchange cycle as reheat, does not suffer from the same defects. When incorporated in a practical cycle, even allowing for the additional pressure loss there is, in addition to the marked increase in specific output, a worthwhile improvement in efficiency. Nevertheless, as pointed out in section 2.1, intercoolers tend to be bulky, and if they require cooling water the self-contained nature of the gas turbine is lost. The cycle is attractive for naval applications, however, because cooling water is readily available—the ocean. Not only does the intercooled cycle with heat-exchange yield a thermal efficiency in excess of 40 per cent, but it also has a good part-load efficiency. The latter feature means that it might be possible to use a single engine rather than a COGOG arrangement, thus offsetting the greater bulk and cost of the engine.

Complex cycles using intercooling, heat-exchange and reheat are unlikely to be used unless gas turbines become competitive with steam plant for large power stations. This may happen if the present trend towards higher permissible core temperatures in nuclear reactors continues. The gas turbine would probably be used in the form of a closed cycle plant with the working fluid acting also as the reactor coolant: separate coolant circulating compressors would then be unnecessary. Closed cycles are considered in section 2.6.

We may conclude this section by emphasising that in practice most gas turbines utilise a high-pressure ratio simple cycle. The only other widely used cycle is the low-pressure ratio heat-exchange variety, but even so the number of engines built with heat-exchangers is only a small fraction of the output of the gas turbine industry. The other modifications mentioned do not normally show sufficient advantage to offset the increased complexity and cost.

## 2.5   COGAS cycles and cogeneration schemes

In the gas turbine, practically all the energy not converted to shaft power is available in the exhaust gases for other uses. The only limitation is that the final exhaust temperature, i.e. the stack temperature, should not be reduced much below 170°C to avoid dewpoint corrosion problems arising from sulphur in the fuel. The exhaust heat may be used in a variety of ways. If it is wholly used to produce steam in a waste heat boiler for a steam turbine, with the object of augmenting the shaft power produced, the system is referred to as a *combined gas/steam cycle power plant* or *COGAS plant* (Fig. 1.3). Alternatively, the exhaust heat may be used to produce hot water or steam for district or factory heating, hot gas or steam for some chemical process, hot gas for distillation plant, or steam for operating an absorption refrigerator in water chilling or air-conditioning plant. The shaft power will normally be used to produce electricity. In such circumstances the system is referred to as a *cogeneration* or *total energy* plant (Fig. 1.15). We shall consider briefly the main characteristics of these two types of system in turn.

## COGAS cycles

Methods of determining the optimum design conditions for the gas and steam cycles in a COGAS plant are too complex to be discussed in detail here. The more important considerations, however, can be stated quite simply. For any given turbine inlet temperature of the gas turbine, an increase in compressor pressure ratio will lead to a reduction in exhaust temperature (Fig. 2.1) and hence in the heat available to the steam cycle. Now, as shown in Fig. 2.22(a), the enthalpy rise between feed water inlet and steam outlet must equal the enthalpy drop of the exhaust gases in the boiler, and the pinch-point and

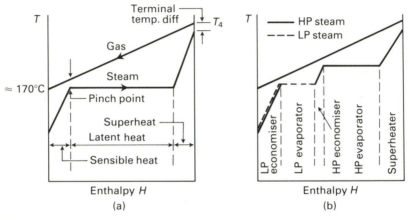

FIG. 2.22  **T–H diagrams for single and dual-pressure COGAS schemes**

terminal temperature differences cannot be less than about 20°C if the boiler is to be of economic size. It follows that a reduction in gas turbine exhaust temperature will lead to a reduction in the steam pressure that can be used for the steam cycle. In the combined plant, therefore, selection of a higher compressor ratio to improve the gas turbine efficiency may lead to a fall in steam cycle efficiency and no net gain in overall thermal efficiency. In practice, however, a higher pressure ratio is accompanied by a higher turbine inlet temperature and the most advanced combined cycles use high-pressure ratio gas turbines.

The foregoing assumes that the plant designer has a free choice of cycle parameters. In fact, most COGAS plants are produced by adding a suitable exhaust-heated Rankine cycle to an existing gas turbine. It is then a question of choosing the Rankine cycle conditions which will best match the gas turbine. The Rankine plant may of course use a working fluid other than water, and at least one plant has been built which employs ammonia.

Another point of importance is that COGAS plant are used for large base-load generating stations and the overall thermal efficiency, while very important, is not the ultimate criterion. The cost of a unit of electricity sent out is what matters, and this depends also on the capital cost of the plant. For example, if a dual-pressure steam cycle was used, as in the plant depicted in Fig. 1.3, a higher efficiency would be obtained because the average temperature at which heat is transferred to the steam is increased. Figure 2.22(b) will make this clear. The additional complication would certainly increase the cost of the boiler and steam turbine, and only a detailed study could show whether it would lead to a reduction in the cost of the electricity produced. The difficulty of optimizing design conditions is further increased by the fact that in a large power station more than one gas turbine may be associated with each waste heat boiler and steam turbine, and also there is always the option of burning additional fuel in the boiler. The latter choice would certainly involve a boiler of increased capital cost, but offers the advantage that cheap heavy oil or coal can be used because the gases arising from the auxiliary firing do not pass through a turbine. Normally, however, this option is ruled out by logistic difficulties in supplying a large plant with more than one fuel.

COGAS plants that have been built so far have efficiencies ranging from 43 to 50 per cent.

## Cogeneration plant

Figure 2.23 is a diagrammatic sketch of a flexible cogeneration plant suitable for applications in which the required ratio of heat output to electrical output might vary over a wide range. When only power is required, the waste heat boiler would be bypassed completely, and when the maximum heat/power ratio is required the heat-exchanger would be bypassed and supplementary fuel burnt in the boiler.

The overall efficiency of a cogeneration plant may be defined as the sum of the net work and useful heat output per unit air mass flow, divided by the product of the fuel/air ratio and $Q_{net,p}$ of the fuel. For the purpose of cycle

**FIG. 2.23 Cogeneration plant**

calculations, it is sufficiently accurate to evaluate the useful heat output per unit mass flow as $c_{pg}[T_{in} - 443]$ bearing in mind the need to keep the stack temperature $T_7$ from falling much below 170°C. $T_{in}$ will be $T_6$, $T_4$ or some intermediate temperature, depending on whether or not turbine exhaust gas is being passed through the heat-exchanger on its way to the boiler.

Figure 2.24, drawn for a fixed turbine inlet temperature and typical values of component efficiency and pressure loss, shows that at high values of heat/power ratio the pressure ratio $r_c$ has little effect on overall efficiency. The

**FIG. 2.24  Efficiency of cogeneration plant**

pressure ratio would normally be chosen to give maximum specific work output and hence minimum capital cost: it would be about 6 for the conditions considered here. Furthermore, it can be seen that the best efficiency is achieved with the heat-exchanger by-passed. In fact, the heat-exchanger is useful only for heat/power ratios of less than 1·5 and for most cogeneration plants the expense of a heat-exchanger would probably not be worthwhile. Further details can be found in Reference (5): the paper also discusses the part-load performance of such systems which will be of interest after the reader has studied Chapter 8.

## 2.6   Closed-cycle gas turbines

The main features of the closed-cycle gas turbine have been described in section 1.3. Although a few plants have been built, they are unlikely to be widely used unless the high temperature nuclear reactor (HTR) comes into service. The HTR is the ideal application, in which all the advantages of the closed cycle can be realised to the full. Such reactors will probably use helium as the coolant because it has the required nuclear characteristics and, as suggested in section 1.3, helium has advantages over air as the working fluid in a gas turbine. To quantify these advantages, the performance of a particular closed cycle will be evaluated for air and helium.

Figure 2.25 is a sketch of the plant, annotated with some of the assumed operating conditions, viz. LP compressor inlet temperature and pressure, HP compressor inlet temperature, and turbine inlet temperature. The compressor

**FIG. 2.25  Closed-cycle plant for nuclear reactor**

inlet temperatures are fixed by the cooling water temperature and the required temperature difference between water and gas. A turbine inlet temperature of 1100 K might be achievable, but it is limited by the permissible temperature of the fuel in the nuclear reactor. The higher we can make the pressure in the circuit the smaller the plant becomes, and we shall assume that a minimum pressure of 20 bar is practicable. Typical values of component efficiencies and pressure loss are: $\eta_{\infty c} = 0.89$, $\eta_{\infty t} = 0.88$, heat exchanger effectiveness 0.7; pressure loss as percentage of component inlet pressures: in precooler and intercooler 1.0 per cent each, in hot and cold sides of heat exchanger 2.5 per cent each, in nuclear reactor 3 per cent.

The performance will be evaluated over a range of compressor pressure ratio $p_{04}/p_{01}$. It will be assumed that the compression is split between LP and HP compressors such that $p_{02}/p_{01} = (p_{04}/p_{01})^{0.5}$ which leads to an approximately equal division of work input. The usual values of $\gamma = 1.4$ and $c_p = 1.005\ \text{kJ/kg K}$ are assumed to hold throughout the cycle when air is the working fluid, and for helium $\gamma = 1.666$ and $c_p = 5.193\ \text{kJ/kg K}$. The cycle efficiency is given by:

$$\eta = \frac{[(T_{06} - T_{07}) - (T_{04} - T_{03}) - (T_{02} - T_{01})]}{(T_{06} - T_{05})}$$

When comparing the specific work output of cycles using fluids of different density, it is more useful to express it in terms of output per unit volume flow than per unit mass flow, because the size of plant is determined by the former. Here we shall evaluate the specific output from the product of the output per unit mass flow and the density at inlet to the compressor where $T = 300$ K and $p = 20$ bar. The density of air at this state is $23 \cdot 23$ kg/m$^3$ whereas for helium (molar mass 4 kg/kmol) it is only $3 \cdot 207$ kg/m$^3$.

Figure 2.26 shows the results of the calculations and we will consider first the efficiency curves. They suggest that the helium cycle has a slightly lower

FIG. 2.26 **Comparison of air and helium closed-cycle performance**

efficiency. In fact more accurate calculations, allowing for the variation of $c_p$ and $\gamma$ with temperature in the case of air, lead to almost identical maximum efficiencies. (With helium, $c_p$ and $\gamma$ do not vary significantly with temperature over the range of interest.) As will be shown below, however, the heat transfer characteristics of helium are better than those of air so that it is probable that a higher heat-exchanger effectiveness can be used with helium without making the heat-exchanger excessively large. The portion of a dotted curve in Fig. 2.26 indicates the benefit obtained by raising the effectiveness from $0 \cdot 7$ to $0 \cdot 8$ for the helium cycle: the efficiency is then 39·5 per cent compared with about 38 per cent for the air cycle.

That the heat transfer characteristics are better for helium can be deduced quite simply from the accepted correlation for heat transfer in turbulent flow in tubes [Ref. (2)], viz.

$$\mathrm{Nu} = 0 \cdot 023 \, \mathrm{Re}^{0 \cdot 8} \, \mathrm{Pr}^{0 \cdot 4}$$

where Nu is the Nusselt number ($hd/k$). The Prandtl number ($c_p \mu/k$) is approximately the same for air and helium, so that the heat transfer coefficients will be in the ratio

$$\frac{h_h}{h_a} = \left(\frac{Re_h}{Re_a}\right)^{0\cdot 8} \left(\frac{k_h}{k_a}\right) \tag{2.24}$$

Suffixes $h$ and $a$ refer to helium and air respectively. From tables of properties [e.g. Ref. (1) of Chapter 3], it may be seen that the thermal conductivities vary with temperature, but the ratio $k_h/k_a$ is of the order of 5 over the relevant range of temperature. The Reynolds number ($\rho C d/\mu$) is a function of velocity as well as properties of the fluid, and the flow velocity in the heat-exchanger (and coolers) is determined by the pressure drop entailed. As may be found in any text on fluid mechanics, or Ref. (2), the pressure loss for turbulent flow in tubes is given by

$$\Delta p = \left(\frac{4L}{d}\right)\tau = \left(\frac{4L}{d}\right)f\left(\frac{\rho C^2}{2}\right)$$

where $\tau$ is the wall shear stress and $f$ the friction factor given by the Blasius law

$$f = \frac{0\cdot 0791}{Re^{0\cdot 25}} = \frac{0\cdot 0791}{(\rho C d/\mu)^{0\cdot 25}}$$

It follows that for equal pressure losses,

$$(\rho^{0\cdot 75}C^{1\cdot 75}\mu^{0.25})_h = (\rho^{0\cdot 75}C^{1\cdot 75}\mu^{0\cdot 25})_a$$

$\rho_h/\rho_a$ is $0\cdot 138$ and $\mu_h/\mu_a$ is approximately $1\cdot 10$, so that

$$\frac{C_h}{C_a} = \left(\frac{1}{0\cdot 138^{0\cdot 75} \times 1\cdot 10^{0\cdot 25}}\right)^{1/1\cdot 75} = 2\cdot 3$$

Thus for similar pressure losses the flow velocity of helium can be double that of air.

Using this result we have

$$\frac{Re_h}{Re_a} = \frac{\rho_h C_h \mu_h}{\rho_a C_a \mu_a} = \frac{0\cdot 138 \times 2\cdot 3}{1\cdot 10} = 0\cdot 29$$

Finally, from equation (2.24)

$$\frac{h_h}{h_a} = 0\cdot 29^{0\cdot 8} \times 5 = 1\cdot 86$$

and the heat transfer coefficient for helium is therefore almost twice that for air. This implies that the heat-exchanger need have only half the surface area of tubing for the same temperature difference, or that a higher effectiveness can be used economically.

Turning now to the specific work output curves of Fig. 2.26, from which the comparative size of plant can be deduced, it will be appropriate to make the comparison at the pressure ratio yielding maximum efficiency in each case, i.e.

around 4 for helium and 7 for air. Evidently about 45 per cent greater volume flow is required with helium than with air for a given power output. The reason is that under conditions of maximum efficiency, in each case the compressor temperature rises and turbine temperature drops are similar for helium and air, so that the specific outputs on a *mass* flow basis are in the ratio $c_{ph}/c_{pa} \approx 5$. The density of helium is only 0·138 times that of air, however, so that on a volume flow basis the specific output with helium is about 0·7 of that with air.

The increase in size of the heat-exchanger and coolers due to this higher volume flow will be more than offset by the reduction arising from the better heat transfer coefficients and the higher flow velocities that can be used with helium. Higher velocities will also reduce the diameter of the ducts connecting the components of the plant. A 45 per cent increase in volume flow with a doubling of the velocity would imply a 15 per cent reduction in diameter.

The effect of using helium instead of air on the size of the turbomachinery will only be fully appreciated after the chapters on compressors and turbines have been studied. We have already noted that the work done per unit mass flow in these components is about five times as large with helium as with air. One might expect that five times the number of stages of blading will be required. This is not so, however, because the Mach number limitation on flow velocity and peripheral speed (which determine the work that can be done in each stage) is virtually removed when helium is the working fluid. The reason is that the sonic velocity, $(\gamma RT)^{0·5}$, is much higher in helium. The gas constant is inversely proportional to the molar mass so that the ratio of sonic velocities at any given temperature becomes

$$\frac{a_h}{a_a} = \left(\frac{\gamma_h R_h}{\gamma_a R_a}\right)^{0·5} = \left(\frac{1·666 \times 29}{1·40 \times 4}\right)^{0·5} = 2·94$$

The work per stage of blading can probably be quadrupled with helium, so that the number of stages is only increased in the ratio 5/4. On top of this we have a reduction in annulus area, and therefore height of blading, consequent upon the higher flow velocities through the turbomachinery. Only a detailed study would show whether the turbomachinery will in fact be much larger for the helium cycle.

Because helium is a relatively scarce resource, the decision to use it in large power plant will probably rest not so much on thermodynamic considerations as the satisfactory solution of the practical problem of sealing the system at the high pressures required. Experience with sealing high-pressure carbon dioxide in gas-cooled reactors has shown that leakage is substantial, and helium, being a ligher gas, will be even more difficult to contain.

## NOMENCLATURE

The most widely used symbols in the book are introduced here and they will not be repeated in the lists at the end of other chapters.

| | |
|---|---|
| $C$ | velocity |
| $c_p$ | specific heat at constant pressure |
| $f$ | fuel/air ratio by weight |
| $h$ | specific enthalpy |
| $\Delta H$ | enthalpy of reaction |
| $m$ | mass flow |
| $M$ | molecular weight, Mach number |
| $n$ | polytropic index |
| $p$ | absolute pressure |
| $Q$ | heat transfer per unit mass flow |
| $Q_{net,p}$ | net calorific value at constant $p$ |
| $R, R_0$ | specific, universal, gas constant |
| $r$ | pressure ratio |
| $s$ | specific entropy |
| $T$ | absolute temperature |
| $t$ | temperature ratio |
| $W$ | specific work output |
| $\gamma$ | ratio of specific heats |
| $\eta$ | efficiency |
| $\rho$ | density |

*Suffixes*

| | |
|---|---|
| 0 | stagnation value |
| 1, 2, 3, etc. | reference planes |
| $\infty$ | polytropic |
| $a$ | ambient, air |
| $b$ | combustion chamber |
| $c$ | compressor |
| $f$ | fuel |
| $g$ | gas |
| $h$ | heat-exchanger |
| $i$ | intake, mixture constituent |
| $m$ | mechanical |
| $N$ | net |
| $s$ | stage |
| $t$ | turbine |

# 3

# Gas turbine cycles for aircraft propulsion

Aircraft gas turbine cycles differ from shaft power cycles in that the useful power output of the former is produced wholly or in part as a result of expansion in a propelling nozzle: wholly in turbojet and turbofan engines, and partly in turboprop engines. A second distinguishing feature is the need to consider the effect of forward speed and altitude on the performance of aircraft engines. The beneficial effect of these parameters, together with an inherently high power/weight ratio, are what enabled the gas turbine to so rapidly and completely supplant the reciprocating engine for aircraft propulsion except for the very lowest power levels.

The chapter opens with a discussion of the criteria appropriate for evaluating the performance of jet propulsion cycles, and of the additional parameters required to allow the losses in the intake and propelling nozzle to be taken into account. The cycle performances of the turbojet, turbofan and turboprop engines are then discussed in turn. For the analysis of other forms of jet power plant, such as ramjets and rockets, the reader must turn to specialized texts on aircraft propulsion.

## 3.1 Criteria of performance

Consider the schematic diagram of a propulsive duct shown in Fig. 3.1. Relative to the engine, the air enters the intake with a velocity $C_a$ equal and opposite to the forward speed of the aircraft, and the power unit accelerates the air so that it leaves with the jet velocity $C_j$. The 'power unit' may consist of a gas turbine in which the turbine merely drives the compressor, one in which part of the expansion is carried out in a power turbine driving a propeller, or simply a combustion chamber as in ramjet engines. For simplicity we shall assume here that the mass flow $m$ is constant (i.e. that the fuel flow is negligible), and thus the *net thrust* $F$ due to the rate of change of momentum is

$$F = m(C_j - C_a) \tag{3.1}$$

$mC_j$ is called the *gross momentum thrust* and $mC_a$ the *intake momentum drag*.

When the exhaust gases are not expanded completely to $p_a$ in the propulsive duct, the pressure $p_j$ in the plane of the exit will be greater than $p_a$ and there will be an additional pressure thrust exerted over the jet exit area $A_j$ equal to $A_j(p_j-p_a)$ as indicated in Fig. 3.1. The net thrust is then the sum of the *momentum thrust* and the *pressure thrust*, namely

$$F = m(C_j-C_a)+A_j(p_j-p_a) \qquad (3.2)$$

When the aircraft is flying at a uniform speed $C_a$ in level flight the thrust must be equal and opposite to the drag of the aircraft at that speed.

**FIG. 3.1  Propulsive duct**

In what follows we shall assume there is complete expansion to $p_a$ in the propelling nozzle and therefore that equation (3.1) is applicable. From this equation it is clear that the required thrust can be obtained by designing the engine to produce either a high velocity jet of small mass flow or a low velocity jet of high mass flow. The question arises as to what is the most efficient combination of these two variables and a qualitative answer is provided by the following simple analysis.

The *propulsive efficiency* $\eta_p$ can be defined as the ratio of the useful propulsive energy or thrust power ($FC_a$) to the sum of that energy and the unused kinetic energy of the jet. The latter is the kinetic energy of the jet *relative to the earth*, namely $m(C_j-C_a)^2/2$. Thus

$$\eta_p = \frac{mC_a(C_j-C_a)}{m[C_a(C_j-C_a)+(C_j-C_a)^2/2]} = \frac{2}{1+(C_j/C_a)} \qquad (3.3)$$

$\eta_p$ is often called the *Froude efficiency*. Note that it is in no sense an overall power plant efficiency, because the unused enthalpy in the jet is ignored. From equations (3.1) and (3.3) it is evident that

(a)  $F$ is a maximum when $C_a = 0$, i.e. under static conditions, but $\eta_p$ is then zero;

(b)  $\eta_p$ is a maximum when $C_j/C_a = 1$ but then the thrust is zero.

We may conclude that although $C_j$ must be greater than $C_a$ the difference should not be too great. This is the reason for the development of the family of propulsion units shown in Fig. 3.2. Taken in the order shown they provide propulsive jets of decreasing mass flow and increasing jet velocity, and therefore in that order they will be suitable for aircraft of increasing design cruising speed. In practice the choice of power plant can be made only when the specification of the aircraft is known: it will depend not only on the

(a) Piston engine          (b) Turboprop engine

(c) Turbofan engine

(d) Turbojet engine          (e) Ramjet engine

**FIG. 3.2  Propulsion engines**

required cruising speed but also on such factors as the desired range of the aircraft and maximum rate of climb. Because the thrust and fuel consumption of a jet propulsion unit vary with both cruising speed and altitude (air density), the latter is also an important parameter. Figure 3.3 indicates the

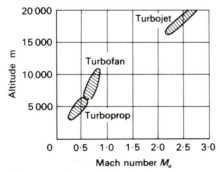

**FIG. 3.3   Flight regimes**

flight regimes found to be suitable for the broad categories of power plant installed in civil aircraft.

The propulsion efficiency is a measure of the effectiveness with which the propulsive duct is being used for propelling the aircraft, and it is not the *efficiency of energy conversion* within the power plant itself which we may symbolize by $\eta_e$. Now the rate of energy supplied in the fuel may be regarded as $m_f Q_{net,p}$ where $m_f$ is the fuel mass flow. This is converted into potentially useful kinetic energy for propulsion $m(C_j^2 - C_a^2)/2$, together with unusable enthalpy in the jet $mc_p(T_j - T_a)$. $\eta_e$ is therefore defined by

$$\eta_e = \frac{m(C_j^2 - C_a^2)/2}{m_f Q_{net,p}} \tag{3.4}$$

The *overall efficiency* $\eta_o$ is the ratio of the useful work done in overcoming drag to the energy in the fuel supplied, i.e.

$$\eta_o = \frac{mC_a(C_j - C_a)}{m_f Q_{net,p}} = \frac{FC_a}{m_f Q_{net,p}} \tag{3.5}$$

It is easy to see that the denominator of equation (3.3), namely $m[C_a(C_j - C_a) + (C_j - C_a)^2/2]$ is equal to the numerator of equation (3.4), and hence that

$$\eta_o = \eta_p \eta_e \tag{3.6}$$

The object of the simple analysis leading to equation (3.6) is to make the point that the efficiency of an aircraft power plant is inextricably linked to the aircraft speed. A crude comparison of different engines can be made, however, if the engine performance is quoted at two operating conditions: the sea-level static performance at maximum power (i.e. at maximum turbine inlet temperature) which must meet the aircraft take-off requirements, and the cruising performance at the optimum cruising speed and altitude of the aircraft for which it is intended. The ambiguous concept of efficiency is discarded in favour of the *specific fuel consumption* which, for aircraft engines, is usually defined as the fuel consumption *per unit thrust* (e.g. kg/h N). The overall efficiency given by equation (3.5) can be written in the form

$$\eta_o = \frac{C_a}{\text{s.f.c.}} \times \frac{1}{Q_{net,p}} \tag{3.7}$$

With a given fuel the value of $Q_{net,p}$ will be constant and it can be seen that the overall efficiency is proportional to $C_a/\text{s.f.c.}$, rather than $1/\text{s.f.c.}$ as for shaft power units.

Another important performance parameter is the *specific thrust* $F_s$, namely the thrust *per unit mass flow of air* (e.g. N s/kg). This provides an indication of the relative size of engines producing the same thrust because the dimensions of the engine are primarily determined by the air flow requirements. Size is important because of its association not only with weight but also with frontal area and the consequent drag. Note that the s.f.c. and specific thrust are related by

$$\text{s.f.c.} = \frac{f}{F_s} \tag{3.8}$$

where $f$ is the fuel/air ratio.

When estimating the cycle performance at altitude we shall need to know the way in which ambient pressure and temperature vary with height above sea level. The variation depends to some extent upon the season and latitude, but it is usual to work with an average or 'standard' atmosphere. The *International Standard Atmosphere* corresponds to average values at middling latitudes and yields a temperature decreasing by about 3·2 K per 500 m of altitude up to 11 000 m after which it is constant at 216·7 K until 20 000 m. Above this height the temperature starts to increase again slowly. Once the temperature is fixed, the variation of pressure follows according to the laws of hydrostatics. An abridged tabulated version of the I.S.A. is given in Ref. (1) and this will be used for subsequent numerical examples; it should be realized, however, that actual ambient conditions can vary widely from I.S.A. values both at sea level and high altitude. For high-subsonic and supersonic aircraft it is more appropriate to use Mach number rather than m/s for aircraft speed, because the drag is more a function of the former. It must be remembered that for a given speed in m/s the Mach number will rise with altitude up to 11 000 m because the temperature is falling. Figure 3.4 shows a plot of $M_a$ versus $C_a$ for sea level and 11 000 m obtained from $M_a = C_a/(\gamma R T_a)^{\frac{1}{2}}$ and I.S.A. data.

FIG. 3.4

## 3.2  Intake and propelling nozzle efficiencies

The nomenclature to be adopted is indicated in Fig. 3.5, which illustrates a simple turbojet engine and the ideal cycle upon which it operates. The turbine produces just sufficient work to drive the compressor, and the remaining part of the expansion is carried out in the propelling nozzle. Because of the significant effect of forward speed, the intake must be considered as a separate component: it cannot be regarded as part of the compressor as it was in Chapter 2. Before proceeding to discuss the performance of aircraft propulsion cycles, it is necessary to describe how the losses in the two additional components—intake and propelling nozzle—are to be taken into account.

**FIG. 3.5 Simple turbojet engine and ideal cycle**

## Intakes

Although the designer of supersonic intakes would not agree with the description, the intake is simply an adiabatic duct. It is indeed only 'simple' for subsonic aircraft. Since there is no heat or work transfer, the stagnation temperature is constant although there will be a loss of stagnation pressure due to friction and due to shock waves at supersonic flight speeds. Under static conditions or at very low forward speeds the intake acts as a nozzle in which the air accelerates from zero velocity or low $C_a$ to $C_1$ at the compressor inlet. At normal forward speeds, however, the intake performs as a diffuser with the air decelerating from $C_a$ to $C_1$ and the static pressure rising from $p_a$ to $p_1$. Since it is the stagnation pressure at the compressor inlet which is required for cycle calculations, it is the pressure rise $(p_{01} - p_a)$ which is of interest and which is referred to as the *ram pressure rise*. At supersonic speeds it will comprise the pressure rise across a system of shock waves at the inlet (see Appendix A.7) followed by that due to subsonic diffusion in the remainder of the duct.

The intake efficiency can be expressed in a variety of ways, but the two most commonly used are the *isentropic efficiency* $\eta_i$ (defined in terms of temperature rises) and the *ram efficiency* $\eta_r$ (defined in terms of pressure rises). Referring to Fig. 3.6, we have

$$T_{01} = T_{0a} = T_a + \frac{C_a^2}{2c_p}$$

**FIG. 3.6 Intake loss**

and

$$\frac{p_{01}}{p_a} = \left(\frac{T'_{01}}{T_a}\right)^{\gamma/(\gamma-1)}$$

where $T'_{01}$ is the temperature which would have been reached after an isentropic ram compression to $p_{01}$. $T'_{01}$ can be related to $T_{01}$ by introducing an isentropic efficiency $\eta_i$ defined by

$$\eta_i = \frac{T'_{01} - T_a}{T_{01} - T_a} \tag{3.9}$$

It follows that

$$T'_{01} - T_a = \eta_i \frac{C_a^2}{2c_p}$$

so that $\eta_i$ can be regarded as the fraction of the inlet dynamic temperature which is made available for isentropic compression in the intake. The intake pressure ratio can then be found from

$$\frac{p_{01}}{p_a} = \left[1 + \frac{T'_{01} - T_a}{T_a}\right]^{\gamma/(\gamma-1)} = \left[1 + \eta_i \frac{C_a^2}{2c_p T_a}\right]^{\gamma/(\gamma-1)} \tag{3.10a}$$

Remembering that $M = C/(\gamma R T)^{\frac{1}{2}}$ and $\gamma R = c_p(\gamma - 1)$, this equation can be written as

$$\frac{p_{01}}{p_a} = \left[1 + \eta_i \frac{\gamma - 1}{2} M_a^2\right]^{\gamma/(\gamma-1)} \tag{3.10b}$$

The stagnation temperature can also be expressed in terms of $M_a$ as

$$\frac{T_{01}}{T_a} = \left[1 + \frac{\gamma - 1}{2} M_a^2\right] \tag{3.11}$$

The ram efficiency $\eta_r$ is defined by the ratio of the ram pressure rise to the inlet dynamic head, namely

$$\eta_r = \frac{p_{01} - p_a}{p_{0a} - p_a}$$

$\eta_r$ can be shown to be almost identical in magnitude to $\eta_i$ and the two quantities are interchangeable: curves relating them can be found in Ref. (2). Apart from the fact that $\eta_r$ is easier to measure experimentally, it has no advantage over $\eta_i$ and we shall use the latter in this book. For subsonic intakes, both $\eta_i$ and $\eta_r$ are found to be independent of inlet Mach number up to a value of about 0·8 and thence their suitability for cycle calculations. They both suffer equally from the drawback of implying zero stagnation pressure loss when $C_a$ is zero, because then $p_{01}/p_a = 1$ and $p_a = p_{0a}$. This is not serious because under these conditions the average velocity in the intake is low, and the flow is accelerating, so that the effect of friction is very small.

The intake efficiency will depend upon the location of the engine in the aircraft (in wing, pod or fuselage), but we shall assume a value of 0·93 for $\eta_i$ in the numerical examples relating to subsonic aircraft which follow. It would be less than this for supersonic intakes, the value decreasing with increase in inlet Mach number. In practice, neither $\eta_i$ nor $\eta_r$ is used for supersonic intakes and it is more usual to quote values of stagnation pressure ratio $p_{01}/p_{0a}$ as a function of Mach number. $p_{01}/p_{0a}$ is called the *pressure recovery factor* of the intake. Knowing the pressure recovery factor, the pressure ratio $p_{01}/p_a$ can be found from

$$\frac{p_{01}}{p_a} = \frac{p_{01}}{p_{0a}} \times \frac{p_{0a}}{p_a}$$

where $p_{0a}/p_a$ is given in terms of $M_a$ by the isentropic relation (8) of Appendix A, namely

$$\frac{p_{0a}}{p_a} = \left[1 + \frac{\gamma-1}{2} M_a^2\right]^{\gamma/(\gamma-1)}$$

Some data on the performance of supersonic intakes can be found in Ref. (3); their design is the province of highly specialized aerodynamicists and much of the information is classified. A rough working rule adopted by the American Department of Defense for the pressure recovery factor *relating to the shock system itself* is

$$\left(\frac{p_{01}}{p_{0a}}\right)_{shock} = 1·0 - 0·075\,(M_a - 1)^{1·35}$$

which is valid when $1 < M_a < 5$. To obtain the overall pressure recovery factor, $(p_{01}/p_{0a})_{shock}$ must be multiplied by the pressure recovery factor for the subsonic part of the intake.

## Propelling nozzles

We shall here use the term 'propelling nozzle' to refer to the remaining part of the engine aft of the last turbine stage. Depending on the location of the engine in the aircraft, and on whether reheat is to be incorporated for thrust boosting, the 'propelling nozzle' will comprise some or all of the items shown in Fig. 3.7. In the transition from turbine annulus to circular jet pipe some increase in area is provided to reduce the velocity, and hence friction loss, in the jet pipe.

The question immediately arises as to whether a simple convergent nozzle is adequate or whether a convergent-divergent nozzle should be employed. As we shall see from subsequent cycle calculations, even with moderate cycle pressure ratios the pressure ratio $p_{04}/p_a$ will be greater than the critical pressure ratio† over at least part of the operational range of forward speed

---

† An estimate of the value of the critical pressure ratio can be obtained by assuming isentropic flow, i.e. by putting $\gamma = 1·333$ in equation (12) of Appendix A,

$$\frac{p_{04}}{p_c} = \left(\frac{\gamma+1}{2}\right)^{\gamma/(\gamma-1)} = 1·853$$

and altitude. Although a convergent-divergent nozzle might therefore appear to be necessary, it must be remembered that it is thrust which is required and not maximum possible jet velocity. Certainly it can be shown that for an isentropic expansion the thrust produced is a maximum when complete expansion to $p_a$ occurs in the nozzle: the pressure thrust $A_5(p_5 - p_a)$ arising from incomplete expansion does not entirely compensate for the loss of momentum thrust due to a smaller jet velocity. But this is no longer true when friction is taken into account because the theoretical jet velocity is not achieved.

Reheat fuel

Diffuser      Jet pipe      Nozzle

**FIG. 3.7 'Propelling nozzle' system**

For values of $p_{04}/p_a$ up to about 3 there is no doubt, from experiments described in Ref. (4), that the thrust produced by a convergent nozzle is as large as that from a convergent-divergent nozzle even when the latter has an exit/throat area ratio suited precisely to the pressure ratio. At operating pressure ratios less than the design value, a convergent-divergent nozzle of fixed proportions would certainly be *less* efficient because of the loss incurred by the formation of a shock wave in the divergent portion. For these reasons most aircraft gas turbines employ a convergent propelling nozzle. A secondary advantage of this type is the relative ease with which the following desirable features can be incorporated:

(a) *Variable area* to improve starting and part-load performance, the importance of which will be discussed in Chapter 8. Variable area is also essential with reheat systems. Figure 3.8(a) illustrates the 'iris' and 'central plug' methods of achieving variation of propelling nozzle area.

(b) *Thrust spoiler* and reverser to reduce the length of runway required for landing.

(c) *Noise suppressor.* Most of the jet noise is due to the mixing of the high velocity hot stream with the cold atmosphere, and the intensity decreases with decrease in jet velocity. For this reason the jet noise of the turbofan engine is less than that of the simple turbojet. In any given case, the noise level can be reduced by accelerating the mixing process and this is normally achieved by increasing the surface area of the jet stream as shown in Fig. 3.8(b).

It should not be thought that convergent-divergent nozzles are never used: they are necessary for the high pressure ratio engines used to power supersonic aircraft. The value of $p_{04}/p_a$ is then many times larger than the critical

(a)                                    (b)

**FIG. 3.8 Variable area, thrust reversal and noise suppression**

pressure ratio and may be as high as 10–20 for flight Mach numbers in the range 2–3. Variable exit/throat area ratio is essential to avoid shock losses over as much of the operating range as possible, and the additional mechanical complexity has to be accepted. The main limitations on the design are:

(a) the exit diameter must be within the overall diameter of the engine—otherwise the additional thrust is offset by increased external drag;

(b) in spite of the weight penalty, the included angle of divergence must be kept below about 30° because the loss in thrust associated with divergence (non-axiality) of the jet increases sharply at greater angles.

After these opening remarks, we may now turn our attention to methods of allowing for propelling nozzle loss in cycle calculations. We shall restrict our attention to the normal case of a convergent nozzle. Two approaches are commonly used: one via an isentropic efficiency $\eta_j$ and the other via a specific thrust coefficient $K_F$. The latter is defined as the ratio of the actual specific gross thrust, namely $[mC_5 + A_5(p_5 - p_a)]/m$, to that which would have resulted from isentropic flow. When the expansion to $p_a$ is completed in the nozzle, i.e. when $p_{04}/p_a < p_{04}/p_c$, $K_F$ becomes simply the ratio of actual to isentropic jet velocity which is the 'velocity coefficient' often used by steam turbine designers. Under these conditions it will be easy to see from the following that $\eta_j = K_F^2$. Although $K_F$ is the easier to measure on nozzle test rigs, it is not so useful for our present purpose as $\eta_j$.

Figure 3.9 illustrates the real and isentropic processes on the $T$–$s$ diagram; and $\eta_j$ is defined by

$$\eta_j = \frac{T_{04} - T_5}{T_{04} - T_5'}$$

It follows that for given inlet conditions ($p_{04}$, $T_{04}$) and as assumed value of $\eta_j$, $T_5$ is given by

$$T_{04} - T_5 = \eta_j T_{04}\left[1 - \left(\frac{1}{p_{04}/p_5}\right)^{(\gamma-1)/\gamma}\right]$$  (3.12)

This is also the temperature equivalent of the jet velocity ($C_5^2/2c_p$) because $T_{05} = T_{04}$. For pressure ratios up to the critical value, $p_5$ will be put equal to $p_a$ in equation (3.12), and the pressure thrust is zero. Above the critical pressure ratio the nozzle is choked, $p_5$ remains at the critical value $p_c$, and $C_5$ remains at the sonic value $(\gamma R T_5)^{\frac{1}{2}}$. The outstanding question is how the critical pressure should be evaluated for non-isentropic flow.

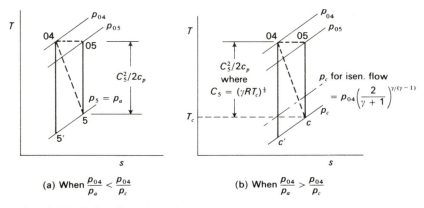

FIG. 3.9 Nozzle loss for unchoked and choked flows

The critical pressure ratio $p_{04}/p_c$ is the pressure ratio $p_{04}/p_5$ which yields $M_5 = 1$. Consider the corresponding critical temperature ratio $T_{04}/T_c$: it is the same for both isentropic and irreversible adiabatic flow. This follows from the fact that $T_{04} = T_{05}$ in all cases of adiabatic flow with no work transfer, and thus

$$\frac{T_{04}}{T_5} = \frac{T_{05}}{T_5} = 1 + \frac{C_5^2}{2c_p T_5} = 1 + \frac{\gamma-1}{2} M_5^2$$

Putting $M_5 = 1$ we have the familiar expression

$$\frac{T_{04}}{T_c} = \frac{\gamma+1}{2}$$  (3.13)

Having found $T_c$ from equation (3.13), we see from Fig. 3.9(b) that the value of $\eta_j$ yields $T_c'$ which is the temperature reached after an isentropic expansion to the real critical pressure $p_c$, namely

$$T_c' = T_{04} - \frac{1}{\eta_j}(T_{04} - T_c)$$

$p_c$ is then given by

$$p_c = p_{04}\left(\frac{T_c'}{T_{04}}\right)^{\gamma/(\gamma-1)} = p_{04}\left[1 - \frac{1}{\eta_j}\left(1 - \frac{T_c}{T_{04}}\right)\right]^{\gamma/(\gamma-1)}$$

On substituting for $T_c/T_{04}$ from equation (3.13), we have the critical pressure ratio in the convenient form

$$\frac{p_{04}}{p_c} = \frac{1}{\left[1 - \frac{1}{\eta_j}\left(\frac{\gamma-1}{\gamma+1}\right)\right]^{\gamma/(\gamma-1)}} \qquad (3.14)$$

This method of using $\eta_j$ to determine the critical pressure ratio yields results consistent with a more detailed analysis involving the momentum equation given in Ref. (5). In particular, it shows that the effect of friction is to increase the pressure drop required to achieve a Mach number of unity.

The remaining quantity necessary for evaluating the pressure thrust $A_5(p_c - p_a)$, is the nozzle area $A_5$. For a given mass flow $m$, this is given *approximately* by

$$A_5 = \frac{m}{\rho_c C_c} \qquad (3.15)$$

with $\rho_c$ obtained from $p_c/RT_c$ and $C_c$ from $[2c_p(T_{04} - T_c)]^{\frac{1}{2}}$ or $(\gamma R T_c)^{\frac{1}{2}}$. It is only an approximate value of the exit area because allowance must be made for the thickness of the boundary layer in the real flow. Furthermore, for conical nozzles, the more complete analysis in Ref. (5) shows that the condition $M = 1$ is reached just downstream of the plane of the exit when the flow is irreversible. In practice, the exit area necessary to give the required engine operating conditions is found by trial and error during development tests. Furthermore, it is found that individual engines of the same type will require slightly different nozzle areas, due to tolerance build-ups and small changes in component efficiencies. Minor changes to the nozzle area are made using *nozzle trimmers*, which are small tabs used to block a portion of the nozzle area.

The value of $\eta_j$ is obviously dependent on a wide variety of factors, such as the length of the jet pipe, and whether the various auxiliary features mentioned earlier are incorporated because they inevitably introduce additional frictional losses. Another factor is the amount of swirl in the gases leaving the turbine, which should be as low as possible (see Chapter 7). We shall assume a value of 0·95 for $\eta_j$ in the following cycle calculations.

## 3.3 Simple turbojet cycle

Figure 3.10 shows the real turbojet cycle on the $T$–$s$ diagram for comparison with the ideal cycle of Fig. 3.5. The method of calculating the design point performance at any given aircraft speed and altitude is illustrated in the following example.

**FIG. 3.10 Turbojet cycle with losses**

## EXAMPLE

Determination of the specific thrust and s.f.c. for a simple turbojet engine, having the following component performance at the design point at which the cruising speed and altitude are 270 m/s and 5000 m.

| | |
|---|---|
| Compressor pressure ratio | 8·0 |
| Turbine inlet temperature | 1200 K |
| Isentropic efficiency: | |
| of compressor, $\eta_c$ | 0·87 |
| of turbine, $\eta_t$ | 0·90 |
| of intake, $\eta_i$ | 0·93 |
| of propelling nozzle, $\eta_j$ | 0·95 |
| Mechanical transmission efficiency $\eta_m$ | 0·99 |
| Combustion efficiency $\eta_b$ | 0·98 |
| Combustion pressure loss $\Delta p_b$ | 4% comp. deliv. press. |

From the I.S.A. table, at 5000 m

$$p_a = 0\cdot5405 \text{ bar and } T_a = 255\cdot7 \text{ K}$$

The stagnation conditions after the intake may be obtained as follows:

$$\frac{C_a^2}{2c_p} = \frac{270^2}{2 \times 1\cdot005 \times 1000} = 36\cdot3 \text{ K}$$

$$T_{01} = T_a + \frac{C_a^2}{2c_p} = 255\cdot7 + 36\cdot3 = 292 \text{ K}$$

$$p_{01} = p_a \left[ 1 + \eta_i \frac{C_a^2}{2c_p T_a} \right]^{\gamma/(\gamma-1)} = 0\cdot5405 \left[ 1 + \frac{0\cdot93 \times 36\cdot3}{255\cdot7} \right]^{3\cdot5}$$

$$= 0\cdot834 \text{ bar}$$

At outlet from the compressor,

$$p_{02} = \left(\frac{p_{02}}{p_{01}}\right) p_{01} = 8 \cdot 0 \times 0 \cdot 834 = 6 \cdot 67 \text{ bar}$$

$$T_{02} - T_{01} = \frac{T_{01}}{\eta_c}\left[\left(\frac{p_{02}}{p_{01}}\right)^{(\gamma-1)/\gamma} - 1\right] = \frac{292}{0 \cdot 87}[8 \cdot 0^{1/3 \cdot 5} - 1] = 272 \cdot 5 \text{ K}$$

$$T_{02} = 292 + 272 \cdot 5 = 564 \cdot 5 \text{ K}$$

$W_t = W_c/\eta_m$ and hence

$$T_{03} - T_{04} = \frac{c_{pa}(T_{02} - T_{01})}{c_{pg}\eta_m} = \frac{1 \cdot 005 \times 272 \cdot 5}{1 \cdot 148 \times 0 \cdot 99} = 241 \text{ K}$$

$$T_{04} = 1200 - 241 = 959 \text{ K}$$

$$p_{03} \fallingdotseq p_{02}\left(1 - \frac{\Delta p_b}{p_{02}}\right) = 6 \cdot 67\,(1 - 0 \cdot 04) = 6 \cdot 40 \text{ bar}$$

$$T'_{04} = T_{03} - \frac{1}{\eta_t}(T_{03} - T_{04}) = 1200 - \frac{241}{0 \cdot 90} = 932 \cdot 2 \text{ K}$$

$$p_{04} = p_{03}\left(\frac{T'_{04}}{T_{03}}\right)^{\gamma/(\gamma-1)} = 6 \cdot 40\left(\frac{932 \cdot 2}{1200}\right)^4 = 2 \cdot 327 \text{ bar}$$

The nozzle pressure ratio is therefore

$$\frac{p_{04}}{p_a} = \frac{2 \cdot 327}{0 \cdot 5405} = 4 \cdot 31$$

The critical pressure ratio, from equation (3.14), is

$$\frac{p_{04}}{p_c} = \frac{1}{\left[1 - \frac{1}{\eta_j}\left(\frac{\gamma-1}{\gamma+1}\right)\right]^{\gamma/(\gamma-1)}} = \frac{1}{\left[1 - \frac{1}{0 \cdot 95}\left(\frac{0 \cdot 333}{2 \cdot 333}\right)\right]^4} = 1 \cdot 914$$

Since $p_{04}/p_a > p_{04}/p_c$ the nozzle is choking.†

$$T_5 = T_c = \left(\frac{2}{\gamma+1}\right) T_{04} = \frac{2 \times 959}{2 \cdot 333} = 822 \text{ K}$$

$$p_5 = p_c = p_{04}\left(\frac{1}{p_{04}/p_c}\right) = \frac{2 \cdot 327}{1 \cdot 914} = 1 \cdot 215 \text{ bar}$$

$$\rho_5 = \frac{p_c}{RT_c} = \frac{100 \times 1 \cdot 215}{0 \cdot 287 \times 822} = 0 \cdot 515 \text{ kg/m}^3$$

$$C_5 = (\gamma R T_c)^{\frac{1}{2}} = (1 \cdot 333 \times 0 \cdot 287 \times 822 \times 1000)^{\frac{1}{2}} = 560 \cdot 8 \text{ m/s}$$

$$\frac{A_5}{m} = \frac{1}{\rho_5 C_5} = \frac{1}{0 \cdot 515 \times 560 \cdot 8} = 0 \cdot 00\,3462 \text{ m}^2\text{s/kg}$$

† The next example in this chapter shows how the calculation is performed when the nozzle is unchoked.

The specific thrust is

$$F_s = (C_5 - C_a) + \frac{A_5}{m}(p_c - p_a)$$

$$= (560 \cdot 8 - 270) + 0 \cdot 00\,3462\,(1 \cdot 215 - 0 \cdot 5405)10^5$$
$$= 290 \cdot 8 + 234 = 525 \text{ N s/kg}$$

From Fig. 2.15, with $T_{02} = 564 \cdot 5$ K and $T_{03} - T_{02} = 1200 - 564 \cdot 5 = 635 \cdot 5$ K, we find that the theoretical fuel/air ratio required is 0·0174. Thus the actual fuel/air ratio is

$$f = \frac{0 \cdot 0174}{0 \cdot 98} = 0 \cdot 0178$$

The specific fuel consumption is therefore

$$\text{s.f.c.} = \frac{f}{F_s} = \frac{0 \cdot 0178 \times 3600}{525} = 0 \cdot 122 \text{ kg/h N}$$

## Optimization of the turbojet cycle

When considering the design of a turbojet the basic thermodynamic parameters at the disposal of the designer are the turbine inlet temperature and the compressor pressure ratio. It is common practice to carry out a series of design point calculations covering a suitable range of these two variables, using fixed polytropic efficiencies for the compressor and turbine, and to plot s.f.c. versus specific thrust with turbine inlet temperature $T_{03}$ and compressor pressure ratio $r_c$ as parameters. Such calculations may be made for several appropriate flight conditions of forward speed and altitude. Typical results applying to a subsonic cruise condition are shown in Fig. 3.11. The effects of turbine inlet temperature and compressor pressure ratio will be considered in turn.

It can be seen that specific thrust is strongly dependent on the value of $T_{03}$, and utilization of the highest possible temperature is desirable in order to keep the engine as small as possible for a given thrust. At a constant pressure ratio, however, an increase in $T_{03}$ will cause some increase in s.f.c. This is in contrast to the effect of $T_{03}$ on shaft power cycle performance, where increasing $T_{03}$ improves both specific power and s.f.c. as discussed in section 2.4. Nevertheless the gain in specific thrust with increasing temperature is invariably more important than the penalty in increased s.f.c., particularly at high flight speeds where small engine size is essential to reduce both weight and drag.

The effect of increasing the pressure ratio $r_c$ is clearly to reduce the s.f.c. At a fixed value of $T_{03}$, increasing the pressure ratio initially results in an increase in specific thrust but eventually leads to a decrease; and the optimum pressure ratio for maximum specific thrust increases as the value of $T_{03}$ is increased. Evidently the effects of pressure ratio follow the pattern already observed to hold for shaft power cycles and need no further comment.

Figure 3.11 applies to a particular subsonic cruise condition. When such calculations are repeated for a higher cruising speed at the same altitude it is found that in general, for any given values of $r_c$ and $T_{03}$, the s.f.c. is increased and the specific thrust is reduced. These effects are due to the combination of an increase in inlet momentum drag and an increase in compressor work consequent upon the rise in inlet temperature. Corresponding curves for different altitudes show an increase in specific thrust and a decrease in s.f.c.

**FIG. 3.11  Typical turbojet cycle performance**

with increasing altitude, due to the fall in temperature and the resulting reduction in compressor work. Perhaps the most notable effect of an increase in the design cruising speed is that the optimum compressor pressure ratio for maximum specific thrust is reduced. This is because of the larger ram compression in the intake. The higher temperature at the compressor inlet and the need for a higher jet velocity make the use of a high turbine inlet temperature desirable—and indeed essential for economic operation of supersonic aircraft.

The thermodynamic optimization of the turbojet cycle cannot be isolated from mechanical design considerations, and the choice of cycle parameters depends very much on the type of aircraft. While high turbine temperatures are thermodynamically desirable they mean the use of expensive alloys and cooled turbine blades leading to an increase in complexity and cost, or to the acceptance of a decrease in engine life. The thermodynamic gains of increased pressure ratio must be considered in the light of increased weight, complexity and cost due to the need for more compressor and turbine stages and perhaps the need for a multispool configuration. Figure 3.12 illustrates the relation between performance and design considerations. In the case of a small business jet, for example, the need is for a simple, reliable engine of low initial cost; s.f.c. is not critical because of the relatively small amount of flying done, and a low pressure ratio turbojet of modest turbine inlet temperature would be satisfactory. To take another example, specialized lifting engines have been developed where the prime requirement is for maximum thrust per unit weight and volume, with s.f.c. less critical because of the very low running times; these conditions can be met by a low pressure ratio unit with a very high turbine inlet temperature (permissible because of the short life required). The compressor pressure ratio would probably be governed

Specific thrust

**FIG. 3.12  Performance and design considerations**

by the maximum that could be handled by a single-stage turbine. Lastly, high pressure ratio turbojets were used in early jet transports because of the need for long range and hence low s.f.c.; in this case the increased engine weight was acceptable because of the large reduction in (engine + fuel) weight for a long range. Turbojets have now been superseded by turbofans for commercial subsonic aircraft, but are still suitable and competitive for supersonic transports. It is worth noting that the subsonic fuel consumption of a supersonic transport is of great importance because a considerable portion of any journey is flown at subsonic speeds: it follows that the optimization procedure cannot be carried out around one single cruising flight condition

*Variation of thrust and s.f.c. with flight conditions for a given engine*
The reader is reminded that we have been discussing the results of design

point cycle calculations. Curves such as those of Fig. 3.11 do not represent what happens to the performance of a particular engine when the turbine inlet temperature, forward speed or altitude differ from the design values. The method of arriving at such data is described in Chapter 8: here we will merely note some of the more important aspects of the behaviour of a turbojet.

At different flight conditions both the thrust and s.f.c. will vary, due to the change in air mass flow with density and the variation of momentum drag with forward speed. Furthermore, even if the engine were run at a fixed rotational speed, the compressor pressure ratio and turbine inlet temperature will change with intake conditions. Typical variations of thrust and s.f.c. with change in altitude and Mach number, for a simple turbojet operating at its maximum rotational speed, are shown in Fig. 3.13. It can be seen that

FIG. 3.13  **Variation of thrust and s.f.c. with Mach number and altitude for typical turbojet engine**

thrust decreases significantly with increasing altitude, due to the decrease in ambient pressure and density, even though the *specific* thrust increases with altitude due to the favourable effect of the lower intake temperature. Specific fuel consumption, however, shows some improvement with increasing altitude. It will be shown in Chapter 8 that s.f.c. is dependent upon ambient temperature, but not pressure, and hence its change with altitude is not so marked as that of thrust. It is obvious from the variation in thrust and s.f.c. that the fuel consumption will be greatly reduced at high altitudes. Reference to Fig. 3.13 shows that with increase of Mach number at a fixed altitude the thrust initially decreases, due to increasing momentum drag, and then starts to increase due to the beneficial effects of the ram pressure ratio; at supersonic Mach numbers this increase in thrust is substantial.

## 3.4   The turbofan engine

The turbofan engine was originally conceived as a method of improving the propulsive efficiency of the jet engine by reducing the mean jet velocity, particularly for operation at high subsonic speeds. It was soon realized that reducing the jet velocity had a considerable effect on jet noise, a matter that became critical when large numbers of jet propelled aircraft entered commercial service. In this type of engine a portion of the total flow *by-passes* part of the compressor, combustion chamber, turbine and nozzle before being ejected through a separate nozzle as shown in Fig. 3.14. Thus the thrust

**FIG. 3.14 Twin-spool turbofan engine**

is made up of two components, the cold stream (or *fan*) thrust and the hot stream thrust. Figure 3.14 shows an engine with separate exhausts, but it is sometimes desirable to mix the two streams and eject them as a single jet of reduced velocity.

Turbofan engines are usually described in terms of *by-pass ratio*, defined as the ratio of the flow through the by-pass duct (cold stream) to the flow at entry to the high pressure compressor (hot stream).† With the notation of Fig. 3.14, the by-pass ratio $B$ is given by

$$B = \frac{m_c}{m_h}$$

It immediately follows that

$$m_c = \frac{mB}{B+1}, \; m_h = \frac{m}{B+1} \text{ and } m = m_c + m_h$$

For the particular case where both streams are expanded to atmospheric pressure in the propelling nozzles, the net thrust is given by

$$F = (m_c C_{jc} + m_h C_{jh}) - mC_a$$

The design point calculations for the turbofan are similar to those for the turbojet; in view of this, only the differences in calculation will be outlined.

---

† The terms turbofan and by-pass engine may both be encountered, often referring to the same engine. Early engines with a small portion of the flow by-passing the combustion chamber (low value of by-pass ratio) were initially referred to as by-pass engines. As the by-pass ratio is increased, the optimum pressure ratio for the by-pass stream is reduced and can eventually be provided by a single compressor stage. The term turbofan was originally used for engines of high by-pass ratio but is increasingly employed for all by-pass engines.

(a)  Overall pressure ratio and turbine inlet temperature are specified as before, but it is also necessary to specify the by-pass ratio (B) and the fan pressure ratio (FPR).

(b)  From the inlet conditions and FPR, the pressure and temperature of the flow leaving the fan and entering the by-pass duct can be calculated. The mass flow down the by-pass duct can be established from the total flow and the by-pass ratio. The cold stream thrust can then be calculated as for the jet engine, noting that air is the working fluid. It is necessary to check whether the fan nozzle is choked or unchoked; if choked the pressure thrust must be calculated.

(c)  In the two-spool configuration shown in Fig. 3.14 the fan is driven by the LP turbine. Calculations for the HP compressor and turbine are quite standard, and conditions at entry to the LP turbine can then be found. Considering the work requirement of the low pressure rotor,

$$mc_{pa}\Delta T_{012} = \eta_m m_h c_{pg}\Delta T_{056}$$

and hence

$$\Delta T_{056} = \frac{m}{m_h} \times \frac{c_{pa}}{\eta_m c_{pg}} \times \Delta T_{012} = (B+1) \times \frac{c_{pa}}{\eta_m c_{pg}} \times \Delta T_{012}$$

The value of B may range from 0·3 to 8 or more, and its value has a major effect on the temperature drop and pressure ratio required from the LP turbine. Knowing $T_{05}$, $\eta_t$ and $\Delta T_{056}$, the LP turbine pressure ratio can be found and conditions at entry to the hot stream nozzle can be established; the calculation of the hot stream thrust is then quite straightforward.

(d)  If the two streams are mixed it is necessary to find the conditions after mixing by means of an enthalpy and momentum balance; details of this calculation are given in Ref. (6). Mixing shows gains at low by-pass ratios, and is essential for a reheated turbofan when maximum thrust boosting is required to avoid the need for two reheat combustion systems.

## EXAMPLE

The following data apply to a twin-spool turbofan engine, with the fan driven by the LP turbine and the compressor by the HP turbine. Separate cold and hot nozzles are used.

| | |
|---|---|
| Overall pressure ratio | 19·0 |
| Fan pressure ratio | 1·65 |
| By-pass ratio $m_c/m_h$ | 3·0 |
| Turbine inlet temperature | 1300 K |
| Fan, compressor and turbine *polytropic* efficiency | 0·90 |
| Isentropic efficiency of each propelling nozzle | 0·95 |
| Mechanical efficiency of each spool | 0·99 |

Combustion pressure loss                                          1·25 bar

Total air mass flow                                               115 kg/s

It is required to find the thrust under sea level static conditions where the ambient pressure and temperature are 1·0 bar and 288 K.

The values of $(n-1)/n$ for the polytropic compression and expansion are:

for compression, $\dfrac{n-1}{n} = \dfrac{1}{\eta_{\infty c}}\left(\dfrac{\gamma-1}{\gamma}\right)_a = \dfrac{1}{0.9 \times 3.5} = 0.3175$

for expansion, $\dfrac{n-1}{n} = \eta_{\infty t}\left(\dfrac{\gamma-1}{\gamma}\right)_g = \dfrac{0.9}{4} = 0.225$

Under static conditions $T_{01} = T_a$ and $p_{01} = p_a$ so that, using the nomenclature of Fig. 3.14,

$\dfrac{T_{02}}{T_{01}} = \left(\dfrac{p_{02}}{p_{01}}\right)^{(n-1)/n}$  yields $T_{02} = 288 \times 1.65^{0.3175} = 337.7$ K

$T_{02} - T_{01} = 337.7 - 288 = 49.7$ K

$\dfrac{p_{03}}{p_{02}} = \dfrac{19.0}{1.65} = 11.51$

$T_{03} = T_{02}\left(\dfrac{p_{03}}{p_{02}}\right)^{(n-1)/n} = 337.7 \times 11.51^{0.3175} = 734$ K

$T_{03} - T_{02} = 734 - 337.7 = 396.3$ K

The cold nozzle pressure ratio is

$\dfrac{p_{02}}{p_a} = \text{FPR} = 1.65$

and the critical pressure ratio for this nozzle is

$\dfrac{p_{02}}{p_c} = \dfrac{1}{\left[1 - \dfrac{1}{\eta_j}\left(\dfrac{\gamma-1}{\gamma+1}\right)\right]^{\gamma/(\gamma-1)}} = \dfrac{1}{\left[1 - \dfrac{1}{0.95}\left(\dfrac{0.4}{2.4}\right)\right]^{3.5}} = 1.965$

Thus the cold nozzle is not choking, so that $p_8 = p_a$ and the cold thrust $F_c$ is given simply by

$F_c = m_c C_8$

The nozzle temperature drop, from equation (3.12), is

$T_{02} - T_8 = \eta_j T_{02}\left[1 - \left(\dfrac{1}{p_{02}/p_a}\right)^{(\gamma-1)/\gamma}\right]$

$= 0.95 \times 337.7\left[1 - \left(\dfrac{1}{1.65}\right)^{1/3.5}\right] = 42.8$ K and hence

$$C_8 = [2c_p(T_{02} - T_8)]^{\frac{1}{2}} = (2 \times 1 \cdot 005 \times 42 \cdot 8 \times 1000)^{\frac{1}{2}} = 293 \text{ m/s}$$

Since the by-pass ratio $B$ is $3 \cdot 0$,

$$m_c = \frac{mB}{B+1} = \frac{115 \times 3 \cdot 0}{4 \cdot 0} = 86 \cdot 25 \text{ kg/s}$$

$$F_c = 86 \cdot 25 \times 293 = 25\,300 \text{ N}$$

Considering the work requirement of the HP rotor,

$$T_{04} - T_{05} = \frac{c_{pa}}{\eta_m c_{pg}} (T_{03} - T_{02}) = \frac{1 \cdot 005 \times 396 \cdot 3}{0 \cdot 99 \times 1 \cdot 148} = 350 \cdot 3 \text{ K}$$

and for the LP rotor

$$T_{05} - T_{06} = (B+1) \frac{c_{pa}}{\eta_m c_{pg}} (T_{02} - T_{01}) = \frac{4 \cdot 0 \times 1 \cdot 005 \times 49 \cdot 7}{0 \cdot 99 \times 1 \cdot 148} = 176 \text{ K}$$

Hence

$$T_{05} = T_{04} - (T_{04} - T_{05}) = 1300 - 350 \cdot 3 = 949 \cdot 7$$
$$T_{06} = T_{05} - (T_{05} - T_{06}) = 949 \cdot 7 - 176 = 773 \cdot 7$$

$p_{06}$ may then be found as follows.

$$\frac{p_{04}}{p_{05}} = \left(\frac{T_{04}}{T_{05}}\right)^{n/(n-1)} = \left(\frac{1300}{949 \cdot 7}\right)^{1/0 \cdot 225} = 4 \cdot 02$$

$$\frac{p_{05}}{p_{06}} = \left(\frac{T_{05}}{T_{06}}\right)^{n/(n-1)} = \left(\frac{949 \cdot 7}{773 \cdot 7}\right)^{1/0 \cdot 225} = 2 \cdot 48$$

$$p_{04} = p_{03} - \Delta p_b = 19 \cdot 0 \times 1 \cdot 0 - 1 \cdot 25 = 17 \cdot 75 \text{ bar}$$

$$p_{06} = \frac{p_{04}}{(p_{04}/p_{05})(p_{05}/p_{06})} = \frac{17 \cdot 75}{4 \cdot 02 \times 2 \cdot 48} = 1 \cdot 78 \text{ bar}$$

Thus the hot nozzle pressure ratio is

$$\frac{p_{06}}{p_a} = 1 \cdot 78$$

while the critical pressure ratio is

$$\frac{p_{06}}{p_c} = \frac{1}{\left[1 - \dfrac{1}{0 \cdot 95}\left(\dfrac{0 \cdot 333}{2 \cdot 333}\right)\right]^4} = 1 \cdot 914$$

This nozzle is also unchoked, and hence $p_7 = p_a$.

$$T_{06} - T_7 = \eta_j T_{06}\left[1 - \left(\frac{1}{p_{06}/p_a}\right)^{(\gamma-1)/\gamma}\right]$$

$$= 0.95 \times 773.7 \left[ 1 - \left( \frac{1}{1.78} \right)^{\frac{1}{4}} \right] = 98.5 \text{ K}$$

$$C_7 = [2c_p(T_{06} - T_7)]^{\frac{1}{2}} = [2 \times 1.148 \times 98.5 \times 1000]^{\frac{1}{2}} = 476 \text{ m/s}$$

$$m_h = \frac{m}{B+1} = \frac{115}{4.0} = 28.75 \text{ kg/s}$$

$$F_h = 28.75 \times 476 = 13\,700 \text{ N}$$

Thus the total thrust is

$$F_c + F_h = 25\,300 + 13\,700 = 39\,000 \text{ N or } 39.0 \text{ kN}$$

The fuel flow can readily be calculated from the known temperatures in the combustor and the airflow through the combustor, i.e. $m_h$. The combustion temperature rise is $(1300 - 734) = 566$ K and the combustion inlet temperature is 734 K. From Fig. 2.15 the ideal fuel:air ratio is found to be 0.0161 and the actual fuel:air ratio is then $(0.0161/0.99) = 0.0163$. Hence the fuel flow is given by

$$m_f = 0.0163 \times 28.75 \times 3600 = 1683.2 \text{ kg/h}$$

and

$$\text{s.f.c.} = \frac{1683.2}{39\,000} = 0.0432 \text{ kg/h N}$$

This example illustrates the method followed when a propelling nozzle is unchoked, while the previous example showed how a choked nozzle may be dealt with.

Note that at static conditions the by-pass stream contributes approximately 65 per cent of the total thrust. At a forward speed of 60 m/s, which is approaching a normal take-off speed, the momentum drag $mC_a$ will be $115 \times 60$ or 6900 N; the ram pressure ratio and temperature rise will be negligible and thus the net thrust is reduced to 32 100 N. The drop in thrust during take-off is even more marked for engines of higher by-pass ratio and for this reason it is preferable to quote turbofan thrusts at a typical take-off speed rather than at static conditions.

## Optimization of the turbofan cycle

In the case of the turbofan cycle the designer has four thermodynamic parameters at his disposal; overall pressure ratio and turbine inlet temperature (as for the simple turbojet) and also by-pass ratio and fan pressure ratio. Optimization of the cycle is somewhat complex but the basic principles are easily understood.

Let us consider an engine with the overall pressure ratio and by-pass ratio both fixed. If we select a value of turbine inlet temperature the energy input is fixed because the combustion chamber air flow and entry temperature are determined by the chosen operating conditions. The remaining variable is the fan pressure ratio and as a first step it is necessary to consider the variation of specific thrust and specific fuel consumption with FPR. If we start with a low value of FPR, the fan thrust will be small and the work extracted from the LP turbine will also be small; thus little energy will be extracted from the hot stream and a large value of hot thrust will result. As the FPR is raised it is evident that the fan thrust will increase and the hot thrust will decrease. A typical variation of specific thrust and s.f.c. with FPR, for a range of turbine temperature, is shown in Fig. 3.15. It can be seen that for any value of turbine inlet temperature there will be an optimum value of FPR;

FIG. 3.15  **Optimization of fan pressure ratio**

optimum values of FPR for minimum s.f.c. and maximum specific thrust coincide because of the fixed energy input. Taking the values of s.f.c. and specific thrust for each of these values of FPR in turn, a curve of s.f.c. against specific thrust may be plotted as shown in Fig. 3.16(a). Note that each point on this curve is the result of a previous optimization, and is associated with a particular value of FPR and turbine inlet temperature.

The foregoing calculations may be repeated for a series of by-pass ratios, still at the same overall pressure ratio, to give a family of curves as shown in Fig. 3.16(b). This plot yields the optimum variation of s.f.c. with specific thrust for the selected overall pressure ratio as shown by the dotted envelope curve. The procedure can then be repeated for a range of overall pressure ratios. It will be clear that the optimization procedure is lengthy and that a large amount of detailed calculation is necessary. The qualitative results of such a series of calculations can be summarized as follows:

(a)  Increasing by-pass ratio improves s.f.c. at the expense of a significant reduction in specific thrust.

(b)  The optimum fan pressure ratio increases with turbine inlet temperature.

FIG. 3.16 Turbofan optimization

(c)   The optimum fan pressure ratio decreases with increase of by-pass ratio. (At a by-pass ratio of about 5 the FPR may be low enough to permit the use of a single-stage fan.)

The choice of cycle parameters is dependent on the aircraft application, and both high and low by-pass ratios have their place. In the case of a long range subsonic transport, for example, specific fuel consumption is of major importance and the requirement can best be met by use of a by-pass ratio of 4–6 and a high overall pressure ratio, combined with a very high turbine inlet temperature. Military aircraft with a supersonic dash capability and a requirement for good subsonic s.f.c. would have to use a much smaller by-pass ratio, perhaps 0·5–1, to keep the frontal area down, and reheat would probably be used. Short-haul commercial aircraft are not as critical as long-haul aircraft regarding s.f.c. and for many years by-pass ratios of around 1 were used; modern designs, however, use higher by-pass ratios similar to those used in long-haul aircraft. There are two reasons for this development: first, the cost of developing a new engine is so high that it pays to aim for excellent s.f.c. and, second, the increased by-pass ratio gives a significant reduction in engine noise.

From the note at the end of the preceding example it should be clear that the thrust of engines of high by-pass ratio is very sensitive to forward speed due to the large intake mass flow and hence large momentum drag. The use of a high by-pass ratio will also cause an increase in drag because of the increase in frontal area, and the drag of the nacelle itself may be significant. For a by-pass ratio of 5 the thrust may decrease by about 25 per cent as an aircraft accelerates from rest to take-off speed. Similarly, the thrust of a turbofan decreases with altitude much more rapidly than in the case of a turbojet because of the large mass flows involved, and for this reason if a turbofan and turbojet are designed for the same cruise thrust the turbofan will give a significantly higher value of take-off thrust.

It was mentioned earlier that, because of their reduced mean jet velocity, turbofans produce less exhaust noise than turbojets. At first sight it would appear that noise considerations would demand the highest possible by-pass ratio, resulting in a low jet velocity. Unfortunately, however, as by-pass ratio is increased the resulting high tip speed of the fan leads to a large increase in fan noise. Indeed at approach conditions, with the engine operating at a low thrust setting, the fan noise predominates; fan noise is essentially produced at discrete frequencies which can be much more irritating than the broad band jet noise. The problem can be alleviated by acoustic treatment of the intake duct, avoiding the use of inlet guide vanes, and careful choice of axial spacing between the fan rotor and stator blades.

## Turbofan configurations

The cycle parameters for a turbofan have a much greater effect on the mechanical design of the engine than in the case of the turbojet. This is because variation in by-pass ratio implies variation in component diameters and rotational speeds, and the configuration of engines of low and high by-pass ratio may be completely different.

Some early turbofans were directly developed from existing turbojets, and this led to the 'aft fan' configuration shown in Fig. 3.17. A combined turbine-

**FIG. 3.17  Aft-fan configuration**

fan was mounted downstream of the gas generator turbine. Two major problems arise with this configuration. Firstly, the blading of the turbine-fan unit must be designed to give turbine blade sections for the hot stream and compressor blade sections for the cold stream. This obviously leads to blading of high cost and, because the entire blade must be made from the turbine material, of high weight. The other problem is that of sealing between the two streams. It is unlikely that the aft-fan configuration will be widely used in the future.

For moderate by-pass and overall pressure ratios the simple two-spool arrangement of Fig. 3.14 is adequate. At very high by-pass ratios, especially when combined with high overall pressure ratios, design problems arise because the fan rotational speed must be much lower than that of the high pressure system; the important limitations on blade tip speed will be discussed in Chapter 5.

Four different configurations may be used to obtain high by-pass ratio and high overall pressure ratio, as shown in Fig. 3.18. The configuration of Fig. 3.18(a) suffers from the fact that the later stages on the LP rotor contribute very little because of their low rotational speed. Scheme (b) is more attractive, but requires a very high pressure ratio from the HP compressor which leads to the instability problems referred to in section 1.2. The three-spool arrangement of Fig. 3.18(c) is in many ways the most attractive concept, with a modest pressure ratio from each compressor. A geared-fan arrangement as in (d) is possible for smaller engines, and units of this sort have been developed from a turboprop background. The power requirements for the fan of a large turbofan may be about 30 000 kW and clearly a gearbox to handle this power would be prohibitively heavy.

(a) Two-spool                    (b) Two-spool

(c) Three-spool                  (d) Two-spool geared fan

FIG. 3.18  Configurations for high by-pass ratio turbofans

## 3.5   The turboprop engine

The turboprop engine differs from the shaft power unit in that some of the useful output appears as jet thrust. In this case, therefore, it is necessary to combine shaft power and jet thrust. This can be done in a number of ways, but in all cases a knowledge of the aircraft speed is involved.

Power must eventually be delivered to the aircraft in the form of thrust power, just as it is with a piston engine driving a propeller. The *thrust power* (TP) can be expressed in terms of shaft power (SP), propeller efficiency $\eta_{pr}$ and jet thrust $F$ by

$$\text{TP} = (\text{SP})\eta_{pr} + FC_a$$

In practice the shaft power will account for a large proportion of the enthalpy drop available at the gas generator exit, and thrust power is therefore largely dependent on the propeller efficiency which may vary significantly with flight conditions.

It is desirable to find some way of expressing the power so that it can be readily compared with that of a piston engine, and so that it is not quite so dependent on propeller efficiency. The same basic engine may be used in conjunction with a variety of propellers for different applications, and it is the performance of the engine itself which is then of most interest to us. A more suitable way of expressing the power is to quote the *equivalent* (or *effective*) *power* EP defined as

$$\text{EP} = \frac{\text{TP}}{\eta_{pr}} = \text{SP} + \frac{FC_a}{\eta_{pr}}$$

$\eta_{pr}$ now affects only the smaller term. The equivalent power is an arbitrarily defined quantity and it should not be quoted without reference to the flight speed. Note that by definition the EP and SP are equal at static conditions, although there is some beneficial jet thrust. Allowance must be made for this when comparing engines under take-off conditions. Experiments have shown that an average propeller produces a thrust of about 8·5 N per kW of power input under static conditions, so that the *take-off* EP is conventionally taken as SP + (F/8.5) with SP in kW and F in newtons.

Turboprops are usually rated on the basis of equivalent power at take-off conditions, and the specific fuel consumption and specific power are often expressed in terms of that equivalent power. It is nevertheless desirable to quote both the shaft power and jet thrust available at any condition of interest and it should be recognized that the equivalent power is merely a useful, but artificial, concept. In view of the similarity between the turboprop and the shaft power units discussed in Chapter 2 it is not necessary to elaborate on cycle requirements. The only basic difference is that the designer can choose the proportions of the available enthalpy drop used to produce shaft power and jet thrust. It can be shown that there is an optimum division for any given flight speed and altitude and details can be found in Ref. (3). The proportion can be adjusted by varying the area of the propelling nozzle.

The combined efficiency of the power turbine, propeller, and the necessary reduction gear, is well below that of an equivalent propelling nozzle. It follows that $\eta_e$ for a turboprop engine is lower than that of a turbojet or turbofan engine. The turboprop has held its position as a power plant for slow-speed aircraft because the propulsive efficiency $\eta_p$ is so much higher than that of the turbojet; the turboprop is extremely widely used in business aircraft, mostly at power levels of 500–1000 kW. The propeller efficiency, using conventional propeller design methods, decreases drastically at flight speeds above M 0·6 and for this reason the turboprop did not become widely used for longer haul aircraft, being superseded by turbofans of equivalent propulsive efficiency; an exception was long endurance patrol aircraft, which fly to the search area at M 0·6 and then loiter at much lower flight speeds. In the early eighties, however, considerable attention was focused on the design of propellers with the goal of obtaining propeller efficiencies of 0·80 at flight Mach numbers of 0·8. The successful development of such propellers would give large savings in fuel compared to existing turbofans. The design of the propeller was markedly different from conventional designs, using highly swept supersonic blading and 8–10 blades; these devices were called 'propfans' to differentiate them from conventional propellers. Studies showed that propfan-powered aircraft would require much higher powers than previously experienced with turboprops, and at power levels in excess of 8000 kW the gearbox design becomes difficult. A further major problem is the transmission of propeller noise to the passenger cabin, and it was widely recognized that a substantial increase in noise over existing turbofan levels would not be acceptable to passengers; it is probable that this can be overcome only by the use of 'pusher' configurations with the propellers mounted behind the passenger cabin.

An alternative configuration being actively pursued by General Electric is the 'unducted fan'. In this approach, two counter-rotating variable pitch fans are directly coupled to counter-rotating turbines with no stators, entirely dispensing with the need for a gear box. The schematic arrangement depicted in Fig. 3.19 shows a similarity to the aft fan configuration described earlier. This revolutionary approach could lead to a much reduced fuel

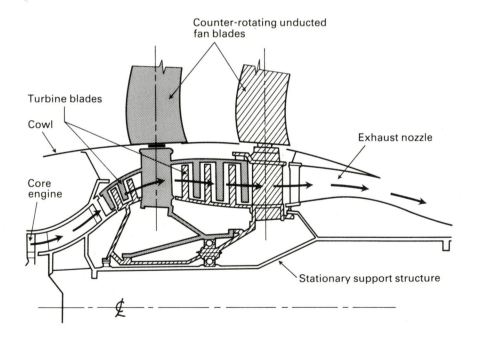

FIG. 3.19 Unducted fan engine

consumption but not all engine manufacturers are convinced that the gearbox can be eliminated.

The status of advanced turboprops, propfans and unducted fans in the mid eighties is very similar to that of turbofans in the mid fifties; at that time there was a heated controversy regarding the relative merits of turboprops and turbofans, with the latter being the clear winner. The current controversy will not be settled until the mid 1990 period and future levels of fuel prices will have a major influence on the outcome.

The conventional turboprop will certainly continue to dominate the market for smaller aircraft of 10–50 seats with flight speeds in the range of 400–600 km/h. The turboshaft engine, in which the output power drives a helicopter rotor, is also of great importance and is almost universally used because of its low weight. In the helicopter application, free turbine engines are used. The helicopter rotor speed is kept constant by changing pitch and the power is varied by changing the gas generator speed. While at first sight the design requirements of turboprops and turboshafts appear identical, the former may be optimized for cruise at an altitude of 6–10 000 m while the latter is optimized for operation at very low altitudes. A number of engines are available both as turboprops and turboshafts, but invariably one application is much more successful than the other because of the different requirements.

## 3.6   Thrust augmentation

If the thrust of an engine has to be increased above the original design value several alternatives are available. Increase of turbine inlet temperature, for example, will increase the specific thrust and hence the thrust for a given engine size. Alternatively the mass flow through the engine could be increased without altering the cycle parameters. Both of these methods imply some redesign of the engine, and either or both may be used to uprate an existing engine.

Frequently, however, there will be a requirement for a temporary increase in thrust, e.g. for take-off, for acceleration from subsonic to supersonic speed or during combat manoeuvres; the problem then becomes one of *thrust augmentation*. Numerous schemes for thrust augmentation have been proposed, but the two methods most widely used are *liquid injection* and *reheat* (or *afterburning*).

Liquid injection is primarily useful for increasing take-off thrust. Substantial quantities of liquid are required, but if the liquid is consumed during take-off and initial climb the weight penalty is not significant. Spraying water into the compressor inlet causes evaporation of the water droplets, resulting in extraction of heat from the air; the effect of this is equivalent to a drop in compressor inlet temperature. Chapter 8 will show that reducing the temperature at entry to a turbojet will increase the thrust, due to the increase in pressure ratio and mass flow resulting from the effective increase in rotational speed. In practice a mixture of water and methanol is used; the methanol lowers the freezing point of water, and in addition it will burn when it reaches the combustion chamber. Liquid is sometimes injected directly into the combustion chamber. The resulting 'blockage' forces the compressor to operate at a higher pressure ratio causing the thrust to increase. In both cases the mass of liquid injected adds to the useful mass flow, but this is a secondary effect.

In the case of reheat, fuel will be introduced into the jet pipe as shown in Fig. 3.7. In the absence of highly stressed rotating blades the temperature allowable after reheat is much higher than the turbine inlet temperature. Stoichiometric combustion is desirable for maximum thrust augmentation and final temperatures of around 2000 K are possible. Figure 3.20 shows the $T$–$s$ diagram for the simple turbojet analysed in section 3.3 with the addition of reheat to 2000 K. The large increase in fuel flow required is evident from the relative temperature rises in the combustion chamber and reheat system, and the penalty in increased s.f.c. is heavy. Assuming that a choked convergent nozzle is used, the jet velocity will correspond to the sonic velocity at the appropriate temperature in the plane of the nozzle, i.e. $T_7$ or $T_5$ depending on whether the engine is operated with or without reheat. Thus the jet velocity can be found from $(\gamma R T_c)^{\frac{1}{2}}$, with $T_c$ given either by $T_{06}/T_c = (\gamma+1)/2$ or $T_{04}/T_c = (\gamma+1)/2$. It follows that the jet velocity is proportional to $\sqrt{T_0}$ at inlet to the propelling nozzle, and that the gross momentum thrust, relative to that for the simple turbojet, will be increased in the ratio $\sqrt{(T_{06}/T_{04})}$. For the temperatures shown in Fig. 3.20 this amounts to $\sqrt{(2000/959)}$ or 1·44. As an approximation, the increase in fuel would be in the ratio $(2000-959)+(1200\text{-}565)$ with reheat, to $(1200\text{-}565)$ without reheat, i.e. 2·64. Thus a 44 per cent increase in thrust is obtained at the expense of a 164 per cent increase in fuel flow, and clearly reheat should be used only for short periods. This might be the thrust augmentation under take-off conditions where the

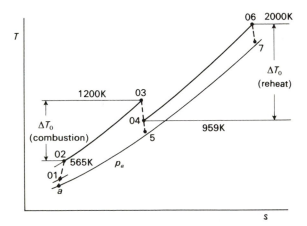

**FIG. 3.20  Cycle of reheated turbojet**

gross thrust is equal to the net thrust. At high forward speeds, however, the gain is much greater and is often well over 100 per cent. This is because for a fixed momentum drag an increase in *gross* thrust represents a considerably greater increase in *net* thrust. Reheat offers even greater gains for turbofans because of the relatively low temperature after mixing of the hot and cold streams and the larger quantity of excess air available for combustion.

It is essential for reheated turbojets to incorporate a variable area nozzle, because of the large change in density of the flow approaching the nozzle due to the large change in temperature. Reheat will normally be brought into operation when the engine is running at its maximum rotational speed, corresponding to its maximum unaugmented thrust. The reheat system should be designed so that the engine will continue to operate at the same speed when reheat is applied, and hence the nozzle must pass the same mass flow at a much reduced density. This can only be achieved if a variable nozzle is fitted permitting a significant increase in nozzle area. Note that the pressure thrust will also be increased due to the enlarged nozzle area.

The reheat pressure loss can be significant. Combustion pressure losses are discussed in Chapter 6, where it is shown that the pressure loss is due both to fluid friction and momentum changes resulting from heat addition. In combustion chambers the former predominates, but in reheat systems the losses due to momentum changes are much more important. The temperature rise in the reheat system is determined by the turbine outlet temperature and the stoichiometric combustion, and the pressure loss due to momentum changes can be determined using the Rayleigh functions and the method outlined in Appendix A.4. This pressure loss is found to be a function of both the temperature ratio across the reheat system and the Mach number at inlet to the reheat duct. If the inlet Mach number is too high, heat release can result in the downstream Mach number reaching 1·0 and this places an upper limit

FIG. 3.21 **Momentum pressure loss**

on the allowable heat release: the phenomenon is referred to as *thermal choking*. Figure 3.21 shows values of the pressure loss due to momentum changes and emphasises the need for a low Mach number. Typically, the exit Mach number from the turbine of a jet engine will be about 0·5 and it is necessary to introduce a diffuser between the turbine and reheat system to reduce the Mach number to about 0·25–0·30 before introducing the reheat fuel.

Even when not in use, a reheat system incurs some penalty in pressure loss due to the presence of the burners and flame stabilizing devices. Another disadvantage of this method of thrust augmentation is that the very high jet velocities resulting from a large degree of reheat result in a noisy exhaust. Fortunately, reheat systems for commercial supersonic transports are only required to produce about 10 per cent increase in thrust and the increase in noise level is not as serious as might be expected from experience with military aircraft.

## NOMENCLATURE

$A$     cross-sectional area
$B$     by-pass ratio $(m_c/m_h)$
$F$     net thrust
$F_s$    specific thrust
$K_F$   specific thrust coefficient
$M$    Mach number
$\eta_e$   efficiency of energy conversion
$\eta_i$   intake efficiency
$\eta_j$   nozzle efficiency
$\eta_m$  mechanical efficiency
$\eta_o$   overall efficiency
$\eta_p$   propulsive (Froude) efficiency
$\eta_{pr}$  propeller efficiency
$\eta_r$   ram efficiency
$\eta_\infty$  polytropic efficiency

*Suffixes*

$c$     critical condition, cold stream
$h$     hot stream
$j$     jet

# 4

# *Centrifugal compressors*

Very rapid progress in the development of gas turbines was made during the Second World War, where attention was focused on the simple turbojet unit. German efforts were based on the axial flow compressor, but British developments used the centrifugal compressor—Refs (1) and (2). It was recognized in Britain that development time was critical and much experience had already been gained on the design of small high-speed centrifugal compressors for supercharging reciprocating engines. Centrifugal compressors were used in both early British and American fighter aircraft and also in the original Comet airliners which were the first gas turbine powered civil aircraft in regular service. As power requirements grew, however, it became clear that the axial flow compressor was more suitable for large engines. The result was that a very high proportion of development funding was diverted to the axial type leading to the availability of axial compressors with an appreciably higher isentropic efficiency than could be achieved by their centrifugal counterparts.

By the late fifties, however, it became clear that smaller gas turbines would have to use centrifugal compressors, and serious research and development work started again. Small turboprops, turboshafts and auxiliary power units (APUs) have been made in very large numbers and have nearly all used centrifugal compressors; notable examples include the Pratt and Whitney Canada PT-6, the Garrett-Ai Research 331 and the large stable of APUs built by the latter organization. They are also used for the high-pressure spools in small turbofans, see Fig. 1.12(a). Centrifugals were used primarily for their suitability for handling small volume flows, but other advantages include a shorter length than an equivalent axial compressor, better resistance to Foreign Object Damage (FOD), less susceptibility to loss of performance by build-up of deposits on the blade surfaces and the ability to operate over a wider range of mass flow at a particular rotational speed. The importance of the latter feature, in alleviating problems of matching operating conditions with those of the associated turbine, will be made clear in Chapter 8.

A pressure ratio of around 4:1 can readily be obtained from a single-stage compressor made of aluminium alloys, and in section 2.4 it was shown that this is adequate for a heat-exchange cycle when the turbine inlet temperature is in the region of 1000–1200 K. Many proposals for vehicular gas turbines were based on this arrangement and manufacturers such as Leyland, Ford, General Motors and Chrysler built development engines which never went into production. The advent of titanium alloys, permitting much higher tip speeds, combined with advances in aerodynamics now permit pressure ratios of greater than 8:1 to be achieved in a single stage. When higher pressure ratios are required, the centrifugal compressor may be used in conjunction with an axial flow compressor (Fig. 1.11), or as a two-stage centrifugal (Fig. 1.10). Even though the latter arrangement involves rather complex ducting between stages, it is still regarded as a practical proposition. Reference (2) of Chapter 1 describes the design process leading to the choice of a twin-spool all-centrifugal compressor for the Pratt and Whitney Canada PW100 turboprop which entered service in 1984.

## 4.1   Principle of operation

The centrifugal compressor consists essentially of a stationary casing containing a rotating *impeller* which imparts a high velocity to the air, and a number of fixed diverging passages in which the air is decelerated with a consequent rise in static pressure. The latter process is one of diffusion, and consequently the part of the compressor containing the diverging passages is known as the *diffuser*. Figure 4.1(a) is a diagrammatic sketch of a centrifugal compressor. The impeller may be single- or double-sided as in 1(b) or 1(c), but the fundamental theory is the same for both. The double-sided impeller was required in early aero-engines because of the relatively small flow capacity of the centrifugal compressor for a given overall diameter.

Air is sucked into the impeller eye and whirled round at high speed by the vanes on the impeller disc. At any point in the flow of air through the impeller, the centripetal acceleration is obtained by a pressure head, so that the static pressure of the air increases from the eye to the tip of the impeller. The remainder of the static pressure rise is obtained in the diffuser, where the very high velocity of the air leaving the impeller tip is reduced to somewhere in the region of the velocity with which the air enters the impeller eye; it should be appreciated that friction in the diffuser will cause some loss in *stagnation* pressure. The normal practice is to design the compressor so that about half the pressure rise occurs in the impeller and half in the diffuser.

It will be appreciated that owing to the action of the vanes in carrying the air around with the impeller, there will be a slightly higher static pressure on the forward face of a vane than on the trailing face. The air will thus tend to flow round the edges of the vanes in the clearance space between the impeller and the casing. This naturally results in a loss of efficiency, and the clearance must be kept as small as possible. A shroud attached to the vanes, Fig. 4.1(d), would eliminate such a'loss, but the manufacturing difficulties

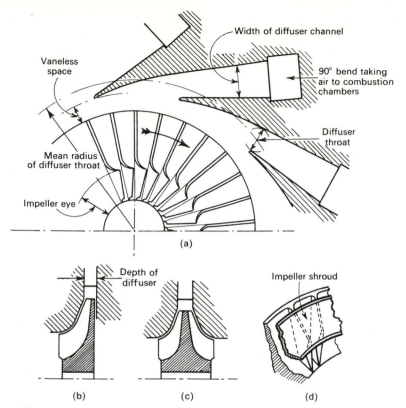

**FIG. 4.1 Diagrammatic sketches of centrifugal compressors**

are vastly increased and there would be a disc friction or 'windage' loss associated with the shroud. Although shrouds have been used on superchargers, they are not used on impellers for gas turbines.

The impellers of modern centrifugal compressors operate with very high tip speeds resulting in very high stress levels. It will be shown in the next section that backswept curved vanes are desirable for compressors of high pressure ratio, but for many years designers were forced to use radial vanes because of the tendency for curved vanes to straighten out under the action of the considerable centrifugal force involved, setting up undesirable bending stresses in the vanes. Modern methods of stress analysis combined with stronger materials, however, now permit backswept vanes to be used in high-performance compressors.

## 4.2   Work done and pressure rise

Since no work is done on the air in the diffuser, the energy absorbed by the compressor will be determined by the conditions of the air at the inlet and outlet of the impeller. Figure 4.2 shows the nomenclature employed.

**FIG. 4.2 Nomenclature**

In the first instance it will be assumed that the air enters the impeller eye in the axial direction, so that the initial angular momentum of the air is zero. The axial portion of the vanes must be curved so that the air can pass smoothly into the eye. The angle which the leading edge of a vane makes with the tangential direction $\alpha$ will be given by the direction of the relative velocity of the air at inlet, $V_1$, as shown in Fig. 4.2.

If the air leaves the impeller tip with an absolute velocity $C_2$, it will have a tangential or whirl component $C_{w2}$, and a comparatively small radial component $C_{r2}$. Under ideal conditions $C_2$ would be such that the whirl component is equal to the impeller tip speed $U$, as shown by the velocity triangle at the top of Fig. 4.2. Due to its inertia, the air trapped between the impeller vanes is reluctant to move round with the impeller, and we have already noted that this results in a higher static pressure on the leading face of a vane than on the trailing face. It also prevents the air from acquiring a whirl velocity equal to the impeller speed. This effect is known as *slip*. How far the whirl velocity at the impeller tip falls short of the tip speed depends largely upon the number of vanes on the impeller. The greater the number of vanes, the smaller the slip, i.e. the more nearly $C_{w2}$ approaches $U$. It is necessary in design to assume a value for the *slip factor* $\sigma$, where $\sigma$ is defined as the ratio $C_{w2}/U$. Various approximate analyses of the flow in an impeller channel have led to formulae for $\sigma$: the one appropriate to radial-vaned impellers which seems to agree best with experiment is that due to Stanitz, Ref. (4):

$$\sigma = 1 - \frac{0 \cdot 63\pi}{n}$$

where $n$ is the number of vanes.

As explained in any elementary text on applied thermodynamics, the theoretical torque which must be applied to the impeller will be equal to the rate of change of angular momentum experienced by the air. Considering unit mass flow of air, this torque is given by

$$\text{theoretical torque} = C_{w2}r_2 \tag{4.1}$$

If $\omega$ is the angular velocity, the work done on the air will be

$$\text{theoretical work done} = C_{w2}r_2\omega = C_{w2}U$$

Or, introducing the slip factor,

$$\text{theoretical work done} = \sigma U^2 \tag{4.2}$$

For convenience, in both the chapters on compressors we shall treat the work done on the air as a positive quantity.

Due to friction between the casing and the air carried round by the vanes, and other losses which have a braking effect such as disc friction or 'windage', the applied torque and therefore the actual work input is greater than this theoretical value. A *power input factor* $\psi$ can be introduced to take account of this, so that the actual work done on the air becomes

$$\text{work done} = \psi\sigma U^2 \tag{4.3}$$

If $(T_{03} - T_{01})$ is the stagnation temperature rise across the whole compressor then, since no energy is added in the diffuser, this must be equal to the stagnation temperature rise $(T_{02} - T_{01})$ across the impeller alone. It will therefore be equal to the temperature equivalent of the work done on the air given by equation (4.3), namely

$$T_{03} - T_{01} = \frac{\psi\sigma U^2}{c_p} \tag{4.4}$$

where $c_p$ is the mean specific heat over this temperature range. Typical values for the power input factor lie in the region of 1·035–1·04.

So far we have merely considered the work which must be put into the compressor. If a value for the *overall* isentropic efficiency $\eta_c$ be assumed, then it is known how much of the work is usefully employed in raising the pressure of the air. The *overall* stagnation pressure ratio follows as

$$\frac{p_{03}}{p_{01}} = \left(\frac{T'_{03}}{T_{01}}\right)^{\gamma/(\gamma-1)} = \left[1 + \frac{\eta_c(T_{03} - T_{01})}{T_{01}}\right]^{\gamma/(\gamma-1)}$$

$$= \left[1 + \frac{\eta_c\psi\sigma U^2}{c_p T_{01}}\right]^{\gamma/(\gamma-1)} \tag{4.5}$$

The distinction between the power input factor and the slip factor should be clearly understood: they are neither independent of one another nor of $\eta_c$. The power input factor represents an increase in the work input, the whole of which is absorbed in overcoming frictional loss and therefore degraded into thermal energy. The fact that the outlet temperature is raised by this loss, and incidentally by other frictional losses as well, enables the maximum cycle temperature to be reached without burning so much fuel, so that as far as the efficiency of the whole gas turbine unit is concerned these losses are not entirely wasteful. Nevertheless, this effect is outweighed by the fact that more turbine work is used in driving the compressor and isentropic (i.e. frictionless adiabatic) compression is the ideal at which to aim. It follows that the power input factor should be as low as possible, a low value of $\psi$ implying simultaneously a high value of $\eta_c$. It should be appreciated that $\eta_c$ depends also upon the friction loss in the diffuser which does not affect the argument up to equation (4.4). For this reason it is not helpful to consider $\psi$ implicitly as part of $\eta_c$.

The slip factor, on the other hand, is a factor limiting the work capacity of the compressor even under isentropic conditions, and this quantity should be as great as possible. Clearly the more nearly $C_{w2}$ approaches $U$, the greater becomes the rate at which work can usefully be put into a compressor of given size. Unfortunately an increase in the number of vanes, which would increase $\sigma$, entails an increase in the solidity of the impeller eye, i.e. a decrease in the effective flow area. Additional friction losses arise because, for the same mass flow or 'throughput', the inlet velocity must be increased. Thus the additional work input that can be employed by increasing the number of vanes may not result in an increase in that portion which is usefully employed in raising the pressure of the air; it may only increase the thermal energy produced by friction resulting in an increase in $\psi$ and reduction in $\eta_c$. A suitable compromise must be found, and present-day practice is to use the number of vanes which give a slip factor of about 0·9, i.e. about 19 or 21 vanes (see also under heading 'Mach number in the diffuser', p. 107).

From equation (4.5) it will be seen that the remaining factors influencing the pressure ratio for a given working fluid are the impeller tip speed $U$, and the inlet temperature $T_{01}$. Any lowering of the inlet temperature $T_{01}$ will clearly increase the pressure ratio of the compressor for a given work input, but it is not a variable under the control of the designer. Reference to texts on strength of materials will show that the centrifugal stresses in a rotating disc are proportional to the square of the rim speed. For single-sided impellers of light alloy, $U$ is limited to about 460 m/s by the maximum allowable centrifugal stresses in the impeller: such a speed yields a pressure ratio of about 4:1. Higher speeds can be used with more expensive materials such as titanium and pressure ratios of over 8:1 are now possible. Because of the additional disc loading, lower speeds must be used for double-sided impellers.

## EXAMPLE

The following data are suggested as a basis for the design of a single-sided centrifugal compressor:

| | |
|---|---|
| power input factor $\psi$ | 1·04 |
| slip factor $\sigma$ | 0·9 |

rotational speed $N$      290 rev/s
overall diameter of impeller      0·5 m
eye tip diameter      0·3 m
eye root diameter      0·15 m
air mass flow $m$      9 kg/s
inlet stagnation temperature $T_{01}$      295 K
inlet stagnation pressure $p_{01}$      1·1 bar
isentropic efficiency $\eta_c$      0·78

Requirements are (a) to determine the pressure ratio of the compressor and the power required to drive it assuming that the velocity of the air at inlet is axial; (b) to calculate the inlet angle of the impeller vanes at the root and tip radii of the eye, assuming that the axial inlet velocity is constant across the eye annulus; and (c) to estimate the axial depth of the impeller channels at the periphery of the impeller.

(a)  Impeller tip speed $U = \pi \times 0\cdot5 \times 290 = 455\cdot5$ m/s
Temperature equivalent of the work done on unit mass flow of air is,

$$T_{03} - T_{01} = \frac{\psi \sigma U^2}{c_p} = \frac{1\cdot04 \times 0\cdot9 \times 455\cdot5^2}{1\cdot005 \times 10^3} = 193 \text{ K}$$

$$\frac{p_{03}}{p_{01}} = \left[ 1 + \frac{\eta_c(T_{03} - T_{01})}{T_{01}} \right]^{\gamma/(\gamma-1)} = \left( 1 + \frac{0\cdot78 \times 193}{295} \right)^{3\cdot5} = 4\cdot23$$

Power required $= m c_p(T_{03} - T_{01}) = 9 \times 1\cdot005 \times 193 = 1746 \text{ kW}$

(b)  To find the inlet angle of the vanes it is necessary to determine the inlet velocity which in this case is axial, i.e. $C_{a1} = C_1$. $C_{a1}$ must satisfy the continuity equation $m = \rho_1 A_1 C_{a1}$, where $A_1$ is the flow area at inlet. Since the density $\rho_1$ depends upon $C_1$, and both are unknown, a trial and error process is required.

The iterative procedure is not critically dependent on the initial value assumed for the axial velocity, but clearly it is desirable to have some rational basis for obtaining an estimated value for starting the iteration. The simplest way of obtaining a reasonable estimate of the axial velocity is to calculate the density on the basis of the known *stagnation* temperature and pressure; in practice this will give a density that is too high and a velocity that is too low. Having obtained an initial estimate of the axial velocity, the density can be recalculated and thence the actual velocity from the continuity equation; if the assumed and calculated velocities do not agree it is necessary to iterate until agreement is reached (only the final trial is shown below). Note that it is normal to design for an axial velocity of about 150 m/s, this providing a suitable compromise between high flow per unit frontal area and low frictional losses in the intake.

$$\text{Annulus area of impeller eye, } A_1 = \frac{\pi(0\cdot3^2 - 0\cdot15^2)}{4} = 0\cdot053 \text{ m}^2$$

Based on stagnation conditions:

$$\rho_1 \simeq \frac{p_{01}}{RT_{01}} = \frac{1 \cdot 1 \times 100}{0 \cdot 287 \times 295} = 1 \cdot 30 \text{ kg/m}^3$$

$$C_{a1} = \frac{m}{\rho_1 A_1} = \frac{9}{1 \cdot 30 \times 0 \cdot 053} = 131 \text{ m/s}$$

Since $C_1 = C_{a1}$, the equivalent dynamic temperature is

$$\frac{C_1^2}{2c_p} = \frac{131^2}{2 \times 1 \cdot 005 \times 10^3} = \frac{1 \cdot 31^2}{0 \cdot 201} = 8 \cdot 5 \text{ K}$$

$$T_1 = T_{01} - \frac{C_1^2}{2c_p} = 295 - 8 \cdot 5 = 286 \cdot 5 \text{ K}$$

$$p_1 = \frac{p_{01}}{(T_{01}/T_1)^{\gamma/(\gamma-1)}} = \frac{1 \cdot 1}{(295/286 \cdot 5)^{3 \cdot 5}} = 0 \cdot 992 \text{ bar}$$

$$\rho_1 = \frac{p_1}{RT_1} = \frac{0 \cdot 992 \times 100}{0 \cdot 287 \times 286 \cdot 5} = 1 \cdot 21 \text{ kg/m}^3$$

Check on $C_{a1}$:

$$C_{a1} = \frac{m}{\rho_1 A_1} = \frac{9}{1 \cdot 21 \times 0 \cdot 053} = 140 \text{ m/s}$$

Final trial:

Try $C_{a1} = C_1 = 145 \text{ m/s}$

Equivalent dynamic temperature is

$$\frac{C_1^2}{2c_p} = \frac{145^2}{2 \times 1 \cdot 005 \times 10^3} = \frac{1 \cdot 45^2}{0 \cdot 201} = 10 \cdot 5 \text{ K}$$

$$T_1 = T_{01} - \frac{C_1^2}{2c_p} = 295 - 10 \cdot 5 = 284 \cdot 5 \text{ K}$$

$$p_1 = \frac{p_{01}}{(T_{01}/T_1)^{\gamma/(\gamma-1)}} = \frac{1 \cdot 1}{(295/284 \cdot 5)^{3 \cdot 5}} = 0 \cdot 968 \text{ bar}$$

$$\rho_1 = \frac{p_1}{RT_1} = \frac{0 \cdot 968 \times 100}{0 \cdot 287 \times 284 \cdot 5} = 1 \cdot 185 \text{ kg/m}^3$$

Check on $C_{a1}$:

$$C_{a1} = \frac{m}{\rho_1 A_1} = \frac{9}{1 \cdot 185 \times 0 \cdot 053} = 143 \text{ m/s}$$

This is a good agreement and a further trial using $C_{a1} = 143$ m/s is unnecessary because a small change in $C$ has little effect upon $\rho$. For this reason it is more accurate to use the final value 143 m/s, rather than the mean of 145 m/s (the trial value) and 143 m/s. The vane angles can now be calculated as follows:

> Peripheral speed at the impeller eye tip radius
> $= \pi \times 0.3 \times 290 = 273$ m/s
> and at eye root radius $= 136.5$ m/s
> $\alpha$ at root $= \tan^{-1} 143/136.5 = 46°20'$
> $\alpha$ at tip $= \tan^{-1} 143/273 = 27°39'$

(c)  The shape of the impeller channel between eye and tip is very much a matter of trial and error. The aim is to obtain as uniform a change of flow velocity up the channel as possible, avoiding local decelerations up the trailing face of the vane which might lead to flow separation. Only tests on the machine can show whether this has been achieved: the flow analyses already referred to [Ref. (4)] are for inviscid flow and are not sufficiently realistic to be of direct use in design. To calculate the required depth of the impeller channel at the periphery we must make some assumptions regarding both the radial component of velocity at the tip, and the division of losses between the impeller and the diffuser so that the density can be evaluated. The radial component of velocity will be relatively small and can be chosen by the designer; a suitable value is obtained by making it approximately equal to the axial velocity at inlet to the eye.

To estimate the density at the impeller tip, the static pressure and temperature are found by calculating the absolute velocity at this point and using it in conjunction with the stagnation pressure which is calculated from the assumed loss up to this point. Figure 4.3 may help the reader to follow the calculation.

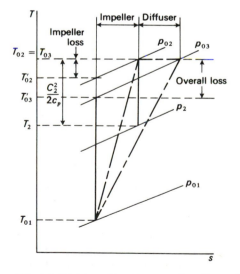

FIG. 4.3  Division of loss between impeller and diffuser

Making the choice $C_{r2} = C_{a1}$, we have $C_{r2} = 143$ m/s

$C_{w2} = \sigma U = 0.9 \times 455.5 = 410$ m/s

$$\frac{C_2^2}{2c_p} = \frac{C_{r2}^2 + C_{w2}^2}{2c_p} = \frac{1.43^2 + 4.10^2}{0.201} = 93.8 \text{ K}$$

Assuming that half the total loss, i.e. $0.5(1 - \eta_c) = 0.11$, occurs in the impeller, the effective efficiency of compression from $p_{01}$ to $p_{02}$ will be 0.89 so that

$$\frac{p_{02}}{p_{01}} = \left(1 + \frac{0.89 \times 193}{295}\right)^{3.5} = 1.582^{3.5}$$

Now $(p_2/p_{02}) = (T_2/T_{02})^{3.5}$, and $T_{02} = T_{03} = 193 + 295 = 488$ K, so that

$$T_2 = T_{02} - \frac{C_2^2}{2c_p} = 488 - 93.8 = 394.2 \text{ K}$$

$$\frac{p_2}{p_{02}} = \left(\frac{394.2}{488}\right)^{3.5}$$

Hence, since $(p_2/p_{01}) = (p_2/p_{02})(p_{02}/p_{01})$,

$$\frac{p_2}{p_{01}} = \left(1.582 \times \frac{394.2}{488}\right)^{3.5} = 2.35$$

$$p_2 = 2.35 \times 1.1 = 2.58 \text{ bar}$$

$$\rho_2 = \frac{p_2}{RT_2} = \frac{2.58 \times 100}{0.287 \times 394.2} = 2.28 \text{ kg/m}^3$$

The required area of cross-section of flow in the radial direction at the impeller tip is

$$A = \frac{m}{\rho_2 C_{r2}} = \frac{9}{2.28 \times 143} = 0.0276 \text{ m}^2$$

Hence the depth of impeller channel

$$= \frac{0.0276}{\pi \times 0.5} = 0.0176 \text{ m or } 1.76 \text{ cm}$$

This result will be used when discussing the design of the diffuser in the next section.

Before leaving the subject of the impeller, it is worth noting the effect of using backswept curved vanes, which we said at the end of section 4.1 are increasingly being used for high-performance compressors. The velocity triangle at the tip section for a backswept impeller is shown in Fig. 4.4, drawn for the ideal case of zero slip for ease of understanding. The corresponding triangle for the radial-vaned impeller is shown by dotted lines. Assuming the radial component of velocity to be the same, implying the same mass flow, it can be seen that the velocity relative to the tip, $V_2$, is increased while the

**FIG. 4.4 Impeller with backswept vanes**

absolute velocity of the fluid, $C_2$, is reduced. These changes imply less stringent diffusion requirements in both the impeller and diffuser, tending to increase the efficiency of both components. The backsweep angle $\beta$ may be in the region of 30–40 degrees. The work-absorbing capacity of the rotor is reduced, however, because $C_{w2}$ is lower and the temperature rise will be less than would be obtained with the radial-vaned impeller. This effect is countered by the increased efficiency and it must be remembered that the ultimate goal is high pressure ratio and efficiency rather than high temperature rise. The use of backswept vanes also gives the compressor a wider operating range of air-flow at a given rotational speed, which is important for matching the compressor to its driving turbine; this will be discussed in Chapter 8.

## 4.3   The diffuser

In Chapter 6 it will be seen that the problem of designing an efficient combustion system is eased if the velocity of the air entering the combustion chamber is as low as possible. It is necessary, therefore, to design the diffuser so that only a small part of the stagnation temperature at the compressor outlet corresponds to kinetic energy. Usually the velocity of the air at the compressor outlet is in the region of 90 m/s.

As will be emphasized throughout this book, it is much more difficult to arrange for an efficient deceleration of flow than it is to obtain an efficient acceleration. There is a natural tendency in a diffusing process for the air to break away from the walls of the diverging passage, reverse its direction, and flow back in the direction of the pressure gradient—see Fig. 4.5(a). If the divergence is too rapid, this may result in the formation of eddies with consequent transfer of some kinetic energy into internal energy and a reduction in useful pressure rise. A small angle of divergence, however, implies a long diffuser and a high value of skin friction loss. Experiments have shown that the optimum included angle of divergence is about 7 degrees, although angles of up to twice this value can be used with diffusers of low length/width (or radius) ratio without incurring a serious increase in stagnation pressure loss. Empirical design curves for diffusers of both circular and rectangular cross-section can be found in Ref. (8). During acceleration in a converging passage, on the other hand, the gas naturally tends to fill the passage and follow the boundary walls closely, as in Fig. 4.5(b), however rapid the rate of convergence. Only the normal frictional losses will be incurred in this case.

**FIG. 4.5 Diffusing and accelerating flow**

In order to control the flow of air effectively and carry out the diffusion process in as short a length as possible, the air leaving the impeller is divided into a number of separate streams by fixed diffuser vanes. Usually the passages formed by the vanes are of constant depth, the width diverging in accordance with the shape of the vanes, as shown in Fig. 4.1. The angle of the diffuser vanes at the leading edge must be designed to suit the direction of the absolute velocity of the air at the radius of the leading edges, so that the air will flow smoothly over the vanes. As there is always a radial gap between the impeller tip and the leading edges of the vanes, this direction will not be that with which the air leaves the impeller tip. The reason for the vaneless space after the impeller will be explained in section 4.4, when the effects of compressibility in this region are discussed.

To find the correct inlet angle for the diffuser vanes, the flow in the vaneless space must be considered. No further energy is supplied to the air after it leaves the impeller so that, neglecting the effect of friction, the angular momentum $C_w r$ must be constant. Hence $C_w$ decreases from impeller tip to diffuser vane ideally in inverse proportion to the radius. For a channel of constant depth, the area of flow in the radial direction is directly proportional to the radius. The radial velocity $C_r$ will therefore also decrease from impeller tip to diffuser vane, in accordance with the equation of continuity. If both $C_r$ and $C_w$ decrease, then the resultant velocity $C$ will decrease from the impeller tip, and some diffusion evidently takes place in the vaneless space. The consequent increase in density means that $C_r$ will not decrease in inverse proportion to the radius as does $C_w$, and the way in which $C_r$ varies must be found from the continuity equation. An example at the end of this section will show how this may be done. When $C_r$ and $C_w$ have been calculated at the radius of the leading edges of the diffuser vanes, then the direction of the resultant velocity can be found and hence the inlet angle of the vanes.

It will be apparent that the direction of the air flow in the vaneless space will vary with mass flow and pressure ratio, so that when the compressor is operating under conditions other than those of the design point the air may not flow smoothly into the diffuser passages in which event some loss of efficiency will result. In gas turbines where weight and complexity are not so important as high part-load efficiency, it is possible to incorporate adjustable diffuser vanes so that the inlet angle is correct over a wide range of operating conditions.

For a given pressure and temperature at the leading edge of the diffuser vanes, the mass flow passed will depend upon the total throat area of the diffuser passages (see Fig. 4.1). Once the number of vanes and the depth of passage have been decided upon, the throat width can be calculated to suit the mass flow required under given conditions of temperature and pressure. For reasons given later in section 4.6, the number of diffuser vanes is appreciably less than the number of impeller vanes. The length of the diffuser passages will of course be determined by the maximum angle of divergence permissible, and the amount of diffusion required, and use would be made of the data in Ref. (8) to arrive at an optimum design. Although up to the throat the vanes must be curved to suit the changing direction of flow, after the throat the air flow is fully controlled and the walls of the passage may be straight. Note that diffusion can be carried out in a much shorter flow path once the air is controlled than it can be in a vaneless space where the air follows an approximately logarithmic spiral path (for an incompressible fluid $\tan^{-1}(C_r/C_w) = $ constant).

After leaving the diffuser vanes, the air may be passed into a volute (or scroll) and thence to a single combustion chamber (via a heat-exchanger if one is employed). This would be done only in an industrial gas turbine unit: indeed, in some small industrial units the diffuser vanes are omitted and the volute alone is used. For aircraft gas turbines, where volume and frontal area are important, the individual streams of air may be retained, each diffuser passage being connected to a separate combustion chamber. Alternatively the streams can be fed into an annular combustion chamber surrounding the shaft connecting the compressor and turbine.

## EXAMPLE

Consider the design of a diffuser for the compressor dealt with in the previous example. The following additional data will be assumed:

| | |
|---|---|
| radial width of vaneless space | 5 cm |
| approximate mean radius of diffuser throat | 0·33 m |
| depth of diffuser passages | 1·76 cm |
| number of diffuser vanes | 12 |

Required are (a) the inlet angle of the diffuser vanes, and (b) the throat width of the diffuser passages which are assumed to be of constant depth. For simplicity it will be assumed that the additional friction loss in the short distance between impeller tip and diffuser throat is small and therefore that 50 per cent of the overall loss can be considered to have occurred up to the diffuser throat. *For convenience suffix 2 will be used to denote any plane in the flow after the impeller tip, the context making it clear which plane is under consideration.*

(a)   Consider conditions at the radius of the diffuser vane leading edges, i.e. at $r_2 = 0.25 + 0.05 = 0.3$ m. Since in the vaneless space $C_w r = $ constant for constant angular momentum,

$$C_{w2} = 410 \times \frac{0.25}{0.30} = 342 \text{ m/s}$$

The radial component of velocity can be found by trial and error. The iteration may be started by assuming that the temperature equivalent of the resultant velocity is that corresponding to the whirl velocity, but only the final trial is given here.

Try $C_{r2} = 97$ m/s

$$\frac{C_2^2}{2c_p} = \frac{3.42^2 + 0.97^2}{0.201} = 62.9 \text{ K}$$

Ignoring any additional loss between the impeller tip and the diffuser vane leading edges at 0·3 m radius, the stagnation pressure will be that calculated for the impeller tip, namely it will be that given by

$$\frac{p_{02}}{p_{01}} = 1.582^{3.5}$$

Proceeding as before we have:

$$T_2 = 488 - 62.9 = 425.1 \text{ K}$$

$$\frac{p_2}{p_{02}} = \left(\frac{425.1}{488}\right)^{3.5}$$

$$\frac{p_2}{p_{01}} = \left(1.582 \times \frac{425.1}{488}\right)^{3.5} = 3.07$$

$$p_2 = 3 \cdot 07 \times 1 \cdot 1 = 3 \cdot 38 \text{ bar}$$

$$\rho_2 = \frac{3 \cdot 38 \times 100}{0 \cdot 287 \times 425 \cdot 1} = 2 \cdot 77 \text{ kg/m}^3$$

Area of cross-section of flow in radial direction

$$= 2\pi \times 0 \cdot 3 \times 0 \cdot 0176 = 0 \cdot 0332 \text{ m}^2$$

Check on $C_{r2}$:

$$C_{r2} = \frac{9}{2 \cdot 77 \times 0 \cdot 0332} = 97 \cdot 9 \text{ m/s}$$

Taking $C_r$ as 97·9 m/s, the angle of the diffuser vane leading edge for zero incidence should be

$$\tan^{-1}(C_{r2}/C_{w2}) = \tan^{-1}\frac{97 \cdot 9}{342} = 16°$$

(b) The throat width of the diffuser channels may be found by a similar calculation for the flow at the assumed throat radius of 0·33 m.

$$C_{w2} = 410 \times \frac{0 \cdot 25}{0 \cdot 33} = 311 \text{ m/s}$$

Try $C_{r2} = 83$ m/s

$$\frac{C_2^2}{2c_p} = \frac{3 \cdot 11^2 + 0 \cdot 83^2}{0 \cdot 201} = 51 \cdot 5 \text{ K}$$

$$T_2 = 488 - 51 \cdot 5 = 436 \cdot 5 \text{ K}$$

$$\frac{p_2}{p_{01}} = \left(1 \cdot 582 \times \frac{436 \cdot 5}{488}\right)^{3 \cdot 5} = 3 \cdot 37$$

$$p_2 = 3 \cdot 37 \times 1 \cdot 1 = 3 \cdot 71 \text{ bar}$$

$$\rho_2 = \frac{3 \cdot 71 \times 100}{0 \cdot 287 \times 436 \cdot 5} = 2 \cdot 96 \text{ kg/m}^3$$

As a first approximation we may neglect the thickness of the diffuser vanes, so that the area of flow in the radial direction

$$= 2\pi \times 0 \cdot 33 \times 0 \cdot 0176 = 0 \cdot 0365 \text{ m}^2$$

Check on $C_{r2}$:

$$C_{r2} = \frac{9}{2 \cdot 96 \times 0 \cdot 0365} = 83 \cdot 3 \text{ m/s}$$

$$\text{Direction of flow} = \tan^{-1}\frac{83 \cdot 3}{311} = 15°$$

Now $m = \rho_2 A_{r2} C_{r2} = \rho_2 A_2 C_2$, or $A_2 = A_{r2} C_{r2}/C_2$, and hence area of flow in the direction of resultant velocity, i.e. total throat area of the diffuser passages, is

$$0.0365 \sin 15° = 0.00945 \text{ m}^2$$

With 12 diffuser vanes, the width of the throat in each passage of depth 0.0176 m is therefore

$$\frac{0.00945}{12 \times 0.0176} = 0.0440 \text{ m or } 4.40 \text{ cm}$$

For any required outlet velocity, knowing the total loss for the whole compressor, it is a simple matter to calculate the final density and hence the flow area required at outlet: the length of the passage after the throat then depends on the chosen angle of divergence.

On the basis of such calculations as these it is possible to produce a preliminary layout of the diffuser. Some idea of the thickness of the vanes at various radii can then be obtained enabling more accurate estimates of flow areas and velocities to be made. When arriving at a final value of the throat area, an empirical contraction coefficient will also be introduced to allow for the blockage effect of the boundary layer. The design of the diffuser is essentially a process of successive approximation: sufficient information has been given to indicate the method of approach.

Recently it has been shown that a diffuser comprising channels of circular cross-section is more efficient than the conventional type with rectangular channels. The passages are formed by drilling holes tangentially through a ring surrounding the impeller, and after the throats the passages are conical. Overall isentropic efficiencies of over 85 per cent are claimed for compressors using this new 'pipe' diffuser—see Ref. (5).

Earlier it was stated that the air leaving the diffuser passages might be collected in a volute. The simplest method of volute design assumes that the angular momentum of the flow remains constant, i.e. friction in the volute is neglected. Using the nomenclature of Fig. 4.6, the cross-sectional

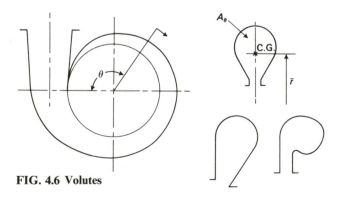

FIG. 4.6 Volutes

area $A_\theta$ at any angle $\theta$ can then be shown to be given by

$$\frac{A_\theta}{\bar{r}} = V\frac{\theta}{2\pi K}$$

where $K$ is the constant angular momentum $C_w r$, $V$ is the volume flow, and $\bar{r}$ is the radius of the centroid of the cross-section of the volute. The shape of the cross-section has been shown to have little effect upon the friction loss: the various shapes shown in Fig. 4.6 all yield volutes of similar efficiency. For further information and useful references the reader may turn to Ref. (6).

## 4.4  Compressibility effects

It is now well known (see Appendix A) that breakdown of flow and excessive pressure loss can be incurred if the velocity of a compressible fluid, relative to the surface over which it is moving, reaches the speed of sound in the fluid. This is a most important phenomenon in a diffusing process where there is always a tendency for the flow to break away from the boundary even at low speeds. When an effort is made to obtain the maximum possible mass flow from the smallest possible compressor, as is done most particularly in the design of aircraft gas turbines, the air speeds are very high. It is of the utmost importance that the Mach numbers at certain points in the flow do not exceed the value beyond which the losses increase rapidly due to the formation of shock waves. The critical Mach number is usually less than unity when calculated on the basis of the mean velocity of the fluid relative to the boundary, because the actual relative velocity near the surface of a curved boundary may be in excess of the mean velocity. As a general rule, unless actual tests indicate otherwise, the Mach numbers are restricted to about 0·8.

Consider now the Mach numbers at vital points in the compressor, beginning with the intake.

### Impeller intake Mach number

At the intake, the air is deflected through a certain angle before it passes into the radial channels on the impeller. There is always a tendency for the air to break away from the convex face of the curved part of the impeller vane. Here then is a point at which the Mach number will be extremely important; a shock wave might occur as shown in Fig. 4.7(a).

The inlet velocity triangle for the impeller eye is shown in Fig. 4.7(b). The full lines represent the case of axial inlet velocity which we have considered up to now. It has also been assumed that the axial velocity is uniform from the root to the tip of the eye. In this case, the velocity of the air relative to the vane, $V_1$, will reach a maximum at the eye tip where the vane speed is

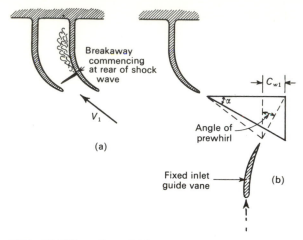

**FIG. 4.7 Effect of prewhirl**

greatest. The inlet Mach number will be given by

$$M = \frac{V_1}{\sqrt{(\gamma R T_1)}}$$

where $T_1$ is the static temperature at the inlet.

Even though this Mach number be satisfactory under ground level atmospheric conditions, if the compressor is part of an aircraft gas turbine the Mach number may well be too high at altitude. It is possible to reduce the relative velocity, and hence the Mach number, by introducing *prewhirl* at the intake. This is achieved by allowing the air to be drawn into the impeller eye over curved *inlet guide vanes* attached to the compressor casing. For the same axial velocity, and hence approximately the same mass flow, the relative velocity is reduced as shown by the dotted triangle in Fig. 4.7(b). An additional advantage of using prewhirl is that the curvature of the impeller vanes at the inlet is reduced, i.e. the inlet angle $\alpha$ is increased.

This method of reducing the Mach number unfortunately reduces the work capacity of the compressor. The air now has an initial whirl component $C_{w1}$, so that the rate of change of angular momentum per unit mass flow of air is

$$C_{w2}r_2 - C_{w1}r_1$$

If $C_{w1}$ is constant over the eye of the impeller, then the initial angular momentum will increase from root to tip of the eye. The amount of work done on each kilogramme of air will therefore depend upon the radius at which it enters the eye of the impeller. A mean value of the work done per kg can be found by using the initial angular momentum of the air at the mean radius of the eye. There is, however, no point in reducing the work capacity of the compressor unnecessarily, and the Mach number is only high at the tip. It is clearly preferable to vary the prewhirl, gradually reducing it from a maximum at the tip to zero at the root of the eye. This may be done if the inlet guide vanes are suitably twisted.


## EXAMPLE

Consider the inlet Mach number for the compressor which has been dealt with in the previous examples.

| | |
|---|---|
| Inlet velocity | $= 143$ m/s and is axial |
| Eye tip speed | $= 273$ m/s |
| Relative velocity at tip | $= \sqrt{(143^2 + 273^2)} = 308$ m/s |
| Velocity of sound | $= \sqrt{(1 \cdot 4 \times 0 \cdot 287 \times 284 \cdot 5 \times 10^3)}$ |
| | $= 338$ m/s |
| Maximum Mach number at inlet | $= 308/338 = 0 \cdot 91$ |

This would not be considered satisfactory even if it were actually the maximum value likely to be reached. But if the compressor is part of an aircraft engine required to operate at an altitude of 11 000 m, where the atmospheric temperature is only about 217 K, we must calculate the Mach number under these conditions. Since there will be a temperature rise due to the ram effect in the intake when the aircraft has a forward speed, the effect of drop in atmospheric temperature will not be quite so great as might be expected. We will take 90 m/s as being the minimum speed likely to be reached at high altitude.

| | |
|---|---|
| Temperature equivalent of forward speed | $= 4$ K |
| Inlet stagnation temperature | $= 217 + 4 = 221$ K |
| Temperature equivalent of axial inlet velocity from the first example† | $= 10 \cdot 5$ K |
| Inlet static temperature at altitude | $= 210 \cdot 5$ K |
| Inlet Mach number at altitude | $= 0 \cdot 91 \left( \dfrac{284 \cdot 5}{210 \cdot 5} \right)^{\frac{1}{2}} = 1 \cdot 06$ |

This is clearly too high and we will find the Mach number when 30 degrees of prewhirl are introduced. In this case the absolute inlet velocity will be slightly higher than before, so that the inlet static temperature will be slightly lower. A new value for the axial velocity must be found by the usual trial and error process. Reverting to the original sea-level static case:

$$\text{Try } C_{a1} = 150 \text{ m/s}$$

$$C_1 = \frac{150}{\cos 30} = 173 \cdot 2 \text{ m/s}$$

† This assumes that the mass flow through the compressor at altitude will be such as to give correct flow direction into the eye. Then if the rotational speed is unaltered the axial velocity must be the same.

Temperature equivalent of $C_1 = 14.9$ K

$$T_1 = 295 - 14.9 = 280.1 \text{ K}$$
$$p_1 = 0.918 \text{ bar and } \rho_1 = 1.14 \text{ kg/m}^3$$

Check on $C_{a1} = \dfrac{9}{1.14 \times 0.053} = 149$ m/s

Whirl velocity at inlet, $C_{w1}$ = 149 tan 30
$$= 86 \text{ m/s}$$
Maximum relative velocity $= \sqrt{[149^2 + (273 - 86)^2]}$
$$= 239 \text{ m/s}$$
Hence maximum inlet Mach number when $T_{01} = 295$ K is

$$\frac{239}{\sqrt{(1.4 \times 0.287 \times 280.1 \times 10^3)}} = 0.71$$

Under altitude conditions this would rise to little more than 0·8 and 30 degrees of prewhirl can be regarded as adequate.

To show the effect of the 30 degrees of prewhirl on the pressure ratio, we will take the worst case and assume that the prewhirl is constant over the eye of the impeller.

Speed of impeller eye at mean radius, $U_e = \dfrac{273 + 136.5}{2}$

$$= 204.8 \text{ m/s}$$

Actual temperature rise $= \dfrac{\psi}{c_p}(\sigma U^2 - C_{w1} U_e)$

$$= \frac{1.04\,(0.9 \times 455.5^2 - 86 \times 204.8)}{1.005 \times 10^3} = 175 \text{ K}$$

$$\frac{p_{03}}{p_{01}} = \left[1 + \frac{0.78 \times 175}{295}\right]^{3.5} = 3.79$$

This pressure ratio may be compared with the figure of 4·23 obtained with no prewhirl. It is sometimes advantageous to use adjustable inlet guide vanes to improve the performance under off-design conditions.

## Mach number in the diffuser

Before discussing the effects of high air velocities in the diffuser, it will be helpful to have an idea of the magnitude of the Mach numbers in this part of the compressor. Obviously the maximum value will occur at the entry to the diffuser, that is, at the impeller tip.

Once again the values calculated in the previous examples of this chapter will be used. In section 4.2 the temperature equivalent of the resultant velocity of the air leaving the impeller was found to be

$$\frac{C_2^2}{2c_p} = 93 \cdot 8 \text{ K}$$

and hence

$C_2 = 434 \text{ m/s}$
$T_2$ was found to be $394 \cdot 2$ K, and thus the Mach number at the impeller tip equals

$$\frac{434}{\sqrt{(1 \cdot 4 \times 0 \cdot 287 \times 394 \cdot 2 \times 10^3)}} = 1 \cdot 09$$

Now consider the leading edge of the diffuser vanes. In section 4.3 the whirl velocity was found to be 342 m/s and the radial component 97·6 m/s. The resultant velocity at this radius is therefore 355 m/s. The static temperature was 425·1 K at this radius so that the Mach number is

$$\frac{355}{\sqrt{(1 \cdot 4 \times 0 \cdot 287 \times 425 \cdot 1 \times 10^3)}} = 0 \cdot 86$$

In the particular design under consideration, the Mach number is 1·09 at the impeller tip and 0·86 at the leading edges of the diffuser vanes. It has been found that as long as the radial velocity component is subsonic, Mach numbers greater than unity can be used at the impeller tip without loss of efficiency: it appears that supersonic diffusion can occur without the formation of shock waves if it is carried out at constant angular momentum with vortex motion in the vaneless space, Ref. (7). But the Mach number at the leading edge of the diffuser vanes is rather high and it would probably be advisable to increase the radial width of the vaneless space or the depth of the diffuser to reduce the velocity at this radius.

High Mach numbers at the leading edges of the diffuser vanes are undesirable, not only because of the danger of shock losses, but because they imply high air speeds and comparatively large pressures at the stagnation points where the air is brought to rest locally at the leading edges of the vanes. This causes a circumferential variation in static pressure, which is transmitted upstream in a radial direction through the vaneless space to the impeller tip. Although the variation will have been considerably reduced by the time it reaches the impeller, it may well be large enough to excite the impeller vanes and cause mechanical failure due to vibrational fatigue cracks in the vanes. This will occur when the exciting frequency, which depends on the rotational speed and relative number of impeller and diffuser vanes, is of the same order as one of the natural frequencies of the impeller vanes. To reduce the likelihood of this, care is taken to see that the number of vanes in the impeller is not an exact multiple of the number in the diffuser; it is common practice to use a prime number for the impeller vanes and an even number for the diffuser vanes.

The reason for the vaneless space will now be apparent: both the dangers of shock losses and excessive circumferential variation in static pressure would be considerably increased if the leading edges of the diffuser vanes were too near the impeller tip where the Mach numbers are very high.

## 4.5   Non-dimensional quantities for plotting compressor characteristics

The performance of a compressor may be specified by curves of delivery pressure and temperature plotted against mass flow for various fixed values of rotational speed. These characteristics, however, are dependent on other variables such as the conditions of pressure and temperature at entry to the compressor and the physical properties of the working fluid. Any attempt to allow for full variations of all these quantities over the working range would involve an excessive number of experiments and make a concise presentation of the results impossible. Much of this complication may be eliminated by using the technique of dimensional analysis, by which the variables involved may be combined to form a smaller and more manageable number of dimensionless groups. As will be shown, the complete characteristics of any particular compressor may then be specified by only two sets of curves.

Before embarking on the dimensional analysis of the behaviour of a compressor, the following special points should be noted.

(a)   When considering the dimensions of temperature, it is convenient always to associate with it the gas constant $R$ so that the combined variable $RT$, being equal to $p/\rho$, has the dimensions $\mathbf{ML}^{-1}\mathbf{T}^{-2}/\mathbf{ML}^{-3} = \mathbf{L}^2\mathbf{T}^{-2}$. Thus they are the same as those of (velocity)$^2$. When the same gas, e.g. air, is being employed during both the testing and subsequent use of the compressor, $R$ can finally be eliminated, but if for any reason there is a change from one gas to another it must be retained in the final expressions.

(b)   A physical property of the gas which undoubtedly influences the behaviour of the compressor is its density $\rho$ but, if pressure $p$ and the $RT$ product are also cited, its inclusion is superfluous because $\rho = p/RT$.

(c)   A further physical property of the gas which in theory would also influence the problem would be its viscosity. The presence of this variable would ultimately result in the emergence of a dimensionless group having the character of a Reynolds number. In the highly turbulent conditions which generally prevail in machines of this type, it is found from experience that the influence of this group is negligibly small over the normal operating range† and it is customary to exclude it from turbomachinery analysis.

---

† To reduce the power requirements when testing large compressors, it is often necessary to throttle the flow at inlet, giving a drop in inlet pressure and hence mass flow. The resultant drop in density lowers the Reynolds number, and it is known that the performance of compressors drops off with a substantial reduction in Reynolds number.

Bearing the above points in mind, it is now possible to consider the various quantities which will both influence the behaviour of the compressor and depend upon it. The solution of the problem may then be stated in the form of an equation in which a function of all these variables is equated to zero. Thus in the present instance, it would be reasonable to state that

$$\text{Function } (D, N, m, p_{01}, p_{02}, RT_{01}, RT_{02}) = 0 \tag{4.6}$$

where $D$ is a characteristic linear dimension of the machine (usually taken as the impeller diameter) and $N$ is the rotational speed. Note that now we are simply considering the performance of the compressor as a whole, *suffix 2 will be used to refer to conditions at the compressor outlet for the remainder of this chapter.*

By the principle of dimensional analysis, often referred to as the Pi theorem, it is known that the function of seven variables expressed by equation (4.6) is reducible to a different function of $7-3 = 4$ non-dimensional groups formed from these variables. The reduction by 3 is due to the presence of the three fundamental units, $\mathbf{M, L, T}$, in the dimensions of the original variables. Various techniques exist for the formation of these non-dimensional groups and it is possible in theory to obtain an infinite variety of self-consistent sets of these groups. Generally there is little difficulty in deriving them by inspection, having decided on the most suitable quantities to 'non-dimensionalize' by using the others. In this case, it is most useful to have the non-dimensional forms of $p_{02}$, $T_{02}$, $m$ and $N$ and the groups emerge as

$$\frac{p_{02}}{p_{01}}, \frac{T_{02}}{T_{01}}, \frac{m\sqrt{(RT_{01})}}{D^2 p_{01}}, \frac{ND}{\sqrt{(RT_{01})}}$$

This now means that the performance of the machine in regard to the variations of delivery pressure and temperature with mass flow, rotational speed and inlet conditions can be expressed in the form of a function of these groups. When we are concerned with the performance of a machine of fixed size compressing a specified gas, $R$ and $D$ may be omitted from the groups so that

$$\text{Function} \left( \frac{p_{02}}{p_{01}}, \frac{T_{02}}{T_{01}}, \frac{m\sqrt{T_{01}}}{p_{01}}, \frac{N}{\sqrt{T_{01}}} \right) = 0 \tag{4.7}$$

The quantities $m\sqrt{T_{01}}/p_{01}$ and $N/\sqrt{T_{01}}$ are usually termed the 'non-dimensional' mass flow and rotational speed respectively, although they are not truly dimensionless.

A function of this form can be expressed graphically by plotting one group against another for various fixed values of a third. In this particular case, experience has shown that the most useful plots are those of the pressure and temperature ratios $p_{02}/p_{01}$ and $T_{02}/T_{01}$ against the non-dimensional mass flow $m\sqrt{T_{01}}/p_{01}$ using the non-dimensional rotational speed $N/\sqrt{T_{01}}$ as a parameter. As will be shown in the next section, it is also possible to replace the temperature ratio by a derivative of it, namely the isentropic efficiency.

Finally, it is worth noting one particular physical interpretation of the non-dimensional mass flow and rotational speed parameters. The former can be written as

$$\frac{m\sqrt{(RT)}}{D^2 p} = \frac{\rho A C\sqrt{(RT)}}{D^2 p} = \frac{\rho A C\sqrt{(RT)}}{RTD^2 p} \propto \frac{C}{\sqrt{(RT)}} \propto M_F$$

and the latter as

$$\frac{ND}{\sqrt{(RT)}} = \frac{U}{\sqrt{(RT)}} \propto M_R$$

Thus the parameters can be regarded as a flow Mach number $M_F$ and a rotational speed Mach number $M_R$. All operating conditions covered by a pair of values of $m\sqrt{T}/p$ and $N/\sqrt{T}$ should give rise to similar velocity triangles, so that the vane angles and air flow directions will match and the compressor will yield the same performance in terms of pressure ratio, temperature ratio and isentropic efficiency. This is what the non-dimensional method of plotting the characteristics implies.

## 4.6   Compressor characteristics

The value of the dimensional analysis is now evident. It is only necessary to plot two sets of curves in order to describe the performance of a compressor completely. The stagnation pressure and temperature ratios are plotted separately against 'non-dimensional' mass flow in the form of a family of curves, each curve being drawn for a fixed value of the 'non-dimensional' rotational speed. From these two sets of curves it is possible to construct constant speed curves of isentropic efficiency plotted against 'non-dimensional' mass flow because this efficiency is given by

$$\eta_c = \frac{T'_{02} - T_{01}}{T_{02} - T_{01}} = \frac{(p_{02}/p_{01})^{(\gamma - 1)/\gamma} - 1}{(T_{02}/T_{01}) - 1}$$

Before describing a typical set of characteristics, it will be as well to consider what might be expected to occur when a valve, placed in the delivery line of a compressor running at constant speed, is slowly opened. The variation in pressure ratio is shown in Fig. 4.8. When the valve is shut and the mass flow is zero, the pressure ratio will have some value $A$, corresponding to the centrifugal pressure head produced by the action of the impeller on the air trapped between the vanes. As the valve is opened and flow commences, the diffuser begins to contribute its quota of pressure rise, and the pressure ratio increases. At some point $B$, where the efficiency approaches its maximum value, the pressure ratio will reach a maximum, and any further increase in mass flow will result in a fall of pressure ratio. For mass flows greatly in excess of that corresponding to the design mass flow, the air angles will be

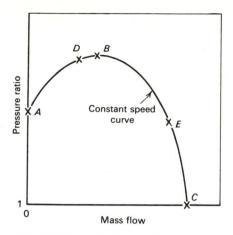

**FIG. 4.8 Theoretical characteristic**

widely different from the vane angles, breakaway of the air will occur, and the efficiency will fall off rapidly. In this hypothetical case the pressure ratio drops to unity at $C$, when the valve is fully opened and all the power is absorbed in overcoming internal frictional resistance.

In actual practice, though point $A$ could be obtained if desired, most of the curve between $A$ and $B$ could not be obtained owing to the phenomenon of *surging*. Surging is associated with a sudden drop in delivery pressure, and with violent aerodynamic pulsation which is transmitted throughout the whole machine. It may be explained as follows. If we suppose that the compressor is operating at some point $D$ on the part of the characteristic having positive slope, then a decrease in mass flow will be accompanied by a fall of delivery pressure. If the pressure of the air downstream of the compressor does not fall quickly enough, the air will tend to reverse its direction and flow back in the direction of the resulting pressure gradient. When this occurs, the pressure ratio drops rapidly. Meanwhile, the pressure downstream of the compressor has fallen also, so that the compressor will now be able to pick up again to repeat the cycle of events which occurs at high frequency.

This surging of the air may not happen immediately the operating point moves to the left of $B$ in Fig. 4.8, because the pressure downstream of the compressor may first of all fall at a greater rate than the delivery pressure. Sooner or later, as the mass flow is reduced, the reverse will apply and the conditions are inherently unstable between $A$ and $B$. As long as the operating point is on the part of the characteristic having a negative slope, however, decrease of mass flow is accompanied by a rise of delivery pressure and stability of operation is assured. In a gas turbine, the actual point at which surging occurs depends upon the swallowing capacity of the components downstream of the compressor, e.g. the turbine, and the way in which this swallowing capacity varies over the range of operating conditions.

Surging probably starts to occur in the diffuser passages, where the flow is retarded by frictional forces near the vanes: certainly the tendency to surge appears to increase with the number of diffuser vanes. This is due to the fact that it is very difficult to split the flow of air so that the mass flow is the same in each passage. When there are several diffuser channels to every impeller channel, and these deliver into a common outlet pipe, there is a tendency for the air to flow up one channel and down another when the conditions are conducive to surging. If this occurs in only one pair of channels the delivery pressure will fall, and thus increase the likelihood of surging. For this reason, the number of diffuser vanes is usually less than the number of impeller vanes. The conditions of flow are then approximately the same in each diffuser passage, because if each is supplied with air from several impeller channels the variations in pressure and velocity between the channels will be evened out by the time the air reaches the diffuser. Surging is therefore not likely to occur until the instability has reached a point at which reversal of flow will occur in most of the diffuser passages simultaneously.

There is one other important cause of instability and poor performance, which may contribute to surge but can exist in the nominally stable operating range: this is the *rotating stall*. When there is any non-uniformity in the flow or geometry of the channels between vanes or blades, breakdown in the flow in one channel, say *B* in Fig. 4.9, causes the air to be deflected in such a way that channel *C* receives fluid at a reduced angle of incidence and channel *A* at an increased incidence. Channel *A* then stalls, resulting in a reduction of

FIG. 4.9 Rotating staff

incidence to channel *B* enabling the flow in that channel to recover. Thus the stall passes from channel to channel: at the impeller eye it would rotate in a direction opposite to the direction of rotation of the impeller. Rotating stall may lead to aerodynamically induced vibrations resulting in fatigue failures in other parts of the gas turbine.

Returning now to consider the hypothetical constant speed curve *ABC* in Fig. 4.8, there is an additional limitation to the operating range, between *B* and *C*. As the mass flow increases and the pressure decreases, the density is reduced and the radial component of velocity must increase. At constant rotational speed this must mean an increase in resultant velocity and hence in angle of incidence at the diffuser vane leading edge. Sooner or later, at some point $E$ say, the position is reached where no further increase in mass flow can be obtained and choking is said to have occurred. This point represents the maximum delivery obtainable at the particular rotational speed for which the curve is drawn. Other curves may be obtained for different speeds, so that the actual variation of pressure ratio over the com-

FIG. 4.10 Centrifugal compressor characteristics

plete range of mass flow and rotational speed will be shown by curves such as those in Fig. 4.10(a). The left-hand extremities of the constant speed curves may be joined up to form what is known as the *surge line*, while the right-hand extremities represent the points where choking occurs.

The temperature ratio is a simple function of the pressure ratio and isentropic efficiency, so that the form of the curves for temperature ratio plotted on the same basis will be similar to Fig. 4.10(a); there is no need to give a separate diagram here. From these two sets of curves the isentropic efficiency may be plotted as in Fig. 4.10(b) or, alternatively, contour lines for various values of the efficiency may be superimposed upon Fig. 4.10(a). The efficiency varies with mass flow at a given speed in a similar manner to the pressure ratio, but the maximum value is approximately the same at all speeds. A curve representing the locus of operating points for maximum efficiency can be obtained as shown by the dotted curve in Fig. 4.10(a). Ideally, the gas turbine should be so designed that the compressor will always be operating on this curve. Chapter 8 will describe a method for estimating the position of the operating line on the compressor characteristics.

In conclusion, mention must be made of two other parameters sometimes used in preference to $m\sqrt{T_{01}}/p_{01}$ and $N/\sqrt{T_{01}}$ when plotting compressor characteristics. These are the *equivalent flow* $m\sqrt{\theta}/\delta$ and *equivalent speed* $N/\sqrt{\theta}$, where $\theta = T_{01}/T_{\text{ref}}$ and $\delta = p_{01}/p_{\text{ref}}$. The reference ambient state is normally that corresponding to the I.S.A. at sea level, namely 288 K and 1·013 bar. When the compressor is operating with the reference intake condition, $m\sqrt{\theta}/\delta$ and $N/\sqrt{\theta}$ are equal to the actual mass flow and rotational speed respectively. With this method of plotting the characteristics, the numbers on the axes are recognizable quantities.

## 4.7  Computerized design procedures

Although the approach to the design of centrifugal compressors given here is adequate as an introduction to the subject, the reader should be aware that more sophisticated methods are available. For example, Ref. (9) outlines a computerized design procedure developed at the National Gas Turbine Establishment (now Royal Aircraft Establishment). This makes use of the Marsh 'matrix throughflow' aerodynamic analysis for determining the shape of the impeller channels. It includes a program for checking the stresses in the impeller, and concludes with one which predicts the performance characteristics of the compressor. The procedure leads to a numerical output of co-ordinates for defining the shape of the impeller and diffuser vanes, in such a form that it can be fed directly to numerically controlled machine tools used for manufacturing these components. Tests have suggested that compressors designed in this way have an improved performance.

Reference (10) describes a method of performance prediction which has been compared with test results from seven different compressors, including one designed on the basis of Ref. (9). The form of the pressure ratio v. mass flow characteristics, and the choking flow, were predicted satisfactorily. The efficiency was within $\pm 1$–$2$ per cent of the experimental value at the design speed, with somewhat larger discrepancies at low speeds. No general means have yet been devised for predicting the surge line.

## NOMENCLATURE

$n$     number of vanes
$N$     rotational speed
$r$     radius
$U$     impeller speed at tip
$U_e$   impeller speed at mean radius of eye
$V$     relative velocity, volume flow
$\alpha$   vane angle
$\sigma$   slip factor
$\psi$   power input factor
$\omega$   angular velocity

*Suffixes*

$a$     axial component, ambient
$r$     radial component
$w$     whirl component

# 5

*Axial flow compressors*

The importance of a high overall pressure ratio in reducing specific fuel consumption was referred to in sections 2.4 and 3.3, and the difficulties of obtaining a high pressure ratio with the centrifugal compressor were pointed out in Chapter 4. From an early stage in the history of the gas turbine, it was recognized that the axial flow compressor had the potential for both higher pressure ratio and higher efficiency than the centrifugal compressor. Another major advantage, especially for jet engines, was the much larger flow rate possible for a given frontal area. These potential gains have now been fully realized as the result of intensive research into the aerodynamics of axial compressors: the axial flow machine dominates the field for large powers and the centrifugal compressor is restricted to the lower end of the power spectrum where the flow is too small to be handled efficiently by axial blading.

Early axial flow units had pressure ratios of around 5:1 and required about 10 stages. Over the years the overall pressure ratios available have risen dramatically, and some turbofan engines have pressure ratios of around 30:1. Continued aerodynamic development has resulted in a steady increase in stage pressure ratio, with the result that the number of stages for a given overall pressure ratio has been greatly reduced. There has been in consequence a reduction in engine weight for a specified level of performance, which is particularly important for aircraft engines. It should be realized, however, that high-stage pressure ratios imply high Mach numbers and large gas deflections in the blading which would not generally be justifiable in an industrial gas turbine where weight is not critical; industrial units, built on a much more restricted budget than an aircraft engine, will inevitably use more conservative design techniques resulting in more stages.

In accordance with the introductory nature of this book, attention will be focused on the design of subsonic compressors. True supersonic compressors, i.e. those for which the velocity at entry is everywhere supersonic, have not proceeded beyond the experimental stage. *Transonic* compressors, however, in which the velocity relative to a moving row of blades is supersonic over *part* of the blade height, are now successfully used in both aircraft and industrial gas turbines. The reader must turn to more advanced texts for a full discussion of these topics.

## 5.1  Basic operation

The axial flow compressor consists of a series of stages, each stage comprising a row of rotor blades followed by a row of stator blades: the individual stages can be seen in Fig. 5.1. The working fluid is initially accelerated by the rotor blades, and then decelerated in the stator blade passages wherein the kinetic energy transferred in the rotor is converted to static pressure. The process is repeated in as many stages as are necessary to yield the required overall pressure ratio.

FIG. 5.1 16-stage high-pressure ratio compressor [by courtesy of General Electric]

The flow is always subject to an adverse pressure gradient, and the higher the pressure ratio the more difficult becomes the design of the compressor. The process consists of a series of diffusions, both in the rotor and stator blade passages: although the *absolute* velocity of the fluid is increased in the rotor, it will be shown that the fluid velocity *relative* to the rotor is decreased, i.e. there is diffusion within the rotor passages. The need for moderate rates of change of cross-sectional area in a diffusing flow has already been emphasized in the previous chapter. This limit on the diffusion in each stage means that a single compressor stage can provide only a relatively small pressure ratio, and very much less than can be used by a turbine with its advantageous pressure gradient, converging blade passages, and accelerating flow (Fig. 5.2). This is why a single turbine stage can drive a large number of compressor stages.

Careful design of compressor blading based on both aerodynamic theory and experiment is necessary, not only to prevent wasteful losses, but also to ensure a minimum of stalling troubles which are all too prevalent in axial compressors, especially if the pressure ratio is high. Stalling, as in the case of isolated aerofoils, arises when the difference between the flow direction and the blade angle (i.e. the angle of incidence) becomes excessive. The fact that the pressure gradient is acting against the flow direction is always a danger to the stability of the flow, and flow reversals may easily occur at conditions of mass flow and rotational speed which are different from those for which the blades were designed.

The compressor shown in Fig. 5.1 makes use of inlet guide vanes (IGVs), which guide the flow into the first stage. Many industrial units have variable

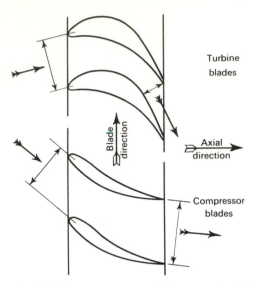

**FIG. 5.2 Comparison of typical forms of turbine and compressor rotor blades**

IGVs, permitting the flow angle entering the first stage to vary with rotational speed to improve the off-design performance. Most aircraft engines have now dispensed with IGVs, however, mainly to obtain the maximum possible flow per unit area and minimum engine weight. Other benefits include an easing of noise and icing problems.

Figure 5.1 shows the marked change in blade size from front to rear in a high pressure ratio compressor. For reasons which will appear later, it is desirable to keep the axial velocity approximately constant throughout the compressor. With the density increasing as the flow progresses through the machine, it is therefore necessary to reduce the flow area and hence the blade height. When the machine is running at a lower speed than design, the density in the rear stages will be far from the design value, resulting in incorrect axial velocities which will cause blade stalling and compressor surge. Several methods may be used to overcome this problem, all of which entail increased mechanical complexity. The approach of Rolls-Royce and Pratt and Whitney has been to use multi-spool configurations, whereas General Electric has favoured the use of variable stator blades; the IGVs and first six stator rows of a GE compressor can be seen to be pivoted in Fig. 5.1. For turbofan engines the large difference in diameter between the fan and the rest of the compressor requires the use of multi-spool units; Pratt and Whitney and General Electric have used two spools, but Rolls-Royce have used three spools on the RB211. Another possibility is the use of blow-off valves, and on advanced engines it is sometimes necessary to include all these schemes. It is important to realize that the designer must consider at the outset the performance of the compressor at conditions far from design, although detailed discussion of these matters is left to the end of the chapter.

In early axial compressors the flow was entirely subsonic, and it was found necessary to use aerofoil section blading to obtain a high efficiency. The need to pass higher flow rates at high pressure ratios increased the Mach numbers, which became especially critical at the tip of the first row of rotor blades. Eventually it became necessary to design blading for transonic compressors, where the flow over part of the blade is supersonic. It was found that the most effective blading for transonic stages consisted of sections based on circular arcs, often referred to as biconvex blading. As Mach numbers were further increased it was found that blade sections based on parabolas became more effective, and most high-performance stages no longer use aerofoil sections.

Before looking at the basic theory of the axial flow compressor it should be emphasized that successful compressor design is very much an art, and all the major engine manufacturers have developed a body of knowledge which is kept proprietary for competitive reasons. Bearing in mind the introductory nature of this text, the aim of this chapter will be to present only the fundamentals of compressor design.

## 5.2   Elementary theory

The working fluid in an axial flow compressor is normally air, but for closed-cycle gas turbines other gases such as helium or carbon dioxide might be used. The analysis which follows is applicable to any gas, but unless otherwise noted it will be assumed that the working fluid is air.

A sketch of a typical stage is shown in Fig. 5.3. Applying the steady flow energy equation to the rotor, and recognizing that the process can be assumed to be adiabatic, it can readily be seen that the power input is given by

$$W = mc_p(T_{02} - T_{01}) \tag{5.1}$$

Repeating with the stator, where the process can again be assumed adiabatic and there is zero work input, it follows that $T_{02} = T_{03}$. All the power is absorbed in the rotor, and the stator merely transforms kinetic energy to an increase in static pressure with the stagnation temperature remaining constant. The increase in stagnation pressure is accomplished wholly within the rotor and, in practice, there will be some decrease in stagnation pressure in the stator due to fluid friction. Losses will also occur in the rotor and the stagnation pressure rise will be less than would be obtained with an isentropic compression and the same power input. A $T$–$s$ diagram for the stage, showing the effect of losses in both rotor and stator, is also shown in Fig. 5.3.

Obtaining the power input to the stage from simple thermodynamics is no help in designing the blading. For this purpose we need to relate the power input to the stage velocity triangles. Initially attention will be focused on a simple analysis of the flow at the mean height of a blade where the peripheral speed is $U$, assuming the flow to occur in a tangential plane at the mean radius. This two-dimensional approach means that in general the flow velocity will

Entropy *s*

**FIG. 5.3 Compressor stage and *T–s* diagram**

have two components, one axial (denoted by subscript *a*) and one tangential (denoted by subscript *w*, implying a whirl velocity). This simplified analysis is reasonable for the later stages of an axial flow compressor where the blade height is small and the blade speeds at root and tip are similar. At the front end of the compressor, however, the blades are much longer, there are marked variations in blade speed from root to tip, and it becomes essential to consider three-dimensional effects in analysing the flow: these will be treated in a later section.

The velocity vectors and associated velocity diagram for a typical stage are shown in Fig. 5.4. The air approaches the rotor with a velocity $C_1$ at an angle $\alpha_1$ from the axial direction; combining $C_1$ vectorially with the blade speed $U$ gives the velocity relative to the blade, $V_1$, at an angle $\beta_1$ from the axial direction. After passing through the rotor, which increases the absolute velocity of the air, the fluid leaves the rotor with a relative velocity $V_2$ at an angle $\beta_2$ determined by the rotor blade outlet angle. Assuming that the design is such that the axial velocity $C_a$ is kept constant, the value of $V_2$ can be obtained and the outlet velocity triangle constructed by combining $V_2$ and $U$ vectorially to give $C_2$ at angle $\alpha_2$. The air leaving the rotor at $\alpha_2$ then passes to the stator where it is diffused to a velocity $C_3$ at angle $\alpha_3$; typically the design is such that $C_3 \approx C_1$ and $\alpha_3 \approx \alpha_1$ so that the air is prepared for entry to another similar stage.

Assuming that $C_a = C_{a1} = C_{a2}$, two basic equations follow immediately

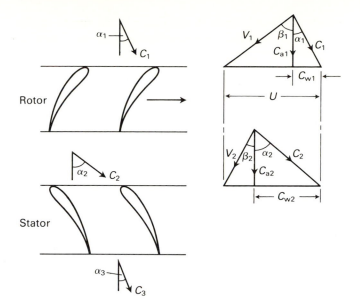

**FIG. 5.4 Velocity triangles for one stage**

from the geometry of the velocity triangles. These are

$$\frac{U}{C_a} = \tan \alpha_1 + \tan \beta_1 \tag{5.2}$$

$$\frac{U}{C_a} = \tan \alpha_2 + \tan \beta_2 \tag{5.3}$$

By considering the change in angular momentum of the air in passing through the rotor, the following expression for the power input to the stage can be deduced:

$$W = mU(C_{w2} - C_{w1})$$

where $C_{w1}$ and $C_{w2}$ are the tangential components of fluid velocity before and after the rotor. This expression can be put in terms of the axial velocity and air angles to give

$$W = mUC_a(\tan \alpha_2 - \tan \alpha_1)$$

It is more useful, however, to express the power in terms of the rotor blade air angles, $\beta_1$ and $\beta_2$. It can readily be seen from equations (5.2) and (5.3) that $(\tan \alpha_2 - \tan \alpha_1) = (\tan \beta_1 - \tan \beta_2)$. Thus the power input is given by

$$W = mUC_a(\tan \beta_1 - \tan \beta_2) \tag{5.4}$$

This input energy will be absorbed usefully in raising the pressure of the air and wastefully in overcoming various frictional losses. But regardless of the losses, or in other words of the efficiency of compression, the whole of this input will reveal itself as a rise in stagnation temperature of the air. Equating (5.1) and (5.4), the stagnation temperature rise in the stage, $\Delta T_{os}$, is given by

$$\Delta T_{os} = T_{03} - T_{01} = T_{02} - T_{01} = \frac{UC_a}{c_p} (\tan \beta_1 - \tan \beta_2) \qquad (5.5)$$

The pressure rise obtained will be strongly dependent on the efficiency of the compression process, as should be clear from Fig. 5.3. Denoting the isentropic efficiency of the stage by $\eta_s$, where $\eta_s = (T'_{03} - T_{01})/(T_{03} - T_{01})$, the stage pressure ratio is then given by

$$R_s = \frac{p_{03}}{p_{01}} = \left[ 1 + \frac{\eta_s \Delta T_{os}}{T_{01}} \right]^{\gamma/(\gamma - 1)} \qquad (5.6)$$

It can now be seen that to obtain a high temperature rise in a stage, which is desirable to minimize the number of stages for a given overall pressure ratio, the designer must combine

(i)   high blade speed
(ii)  high axial velocity
(iii) high fluid deflection $(\beta_1 - \beta_2)$ in the rotor blades.

Blade stresses will obviously limit the blade speed, and it will be seen in the next section that aerodynamic considerations and the previously mentioned adverse pressure gradient combine to limit (ii) and (iii).

## 5.3   Factors affecting stage pressure ratio

### Tip speed

The centrifugal stress in the rotor blades depends on the rotational speed, the blade material and the length of the blade. The maximum centrifugal tensile stress, which occurs at the blade root, can be seen to be given by

$$(\sigma_{ct})_{max} = \frac{\rho_b \omega^2}{a_r} \int_r^t ar \, dr$$

where $\rho_b$ is the density of the blade material, $\omega$ is the angular velocity, $a$ is the cross-sectional area of the blade at any radius, and suffixes $r$ and $t$ refer to root and tip of the blade. If, for simplicity, the blade cross-section is assumed to be constant from root to tip

$$(\sigma_{ct})_{max} = \frac{\rho_b}{2} (2\pi N)^2 (r_t^2 - r_r^2) = 2\pi N^2 \rho_b A$$

where $N$ is the rotational speed and $A$ the annulus area. The tip speed, $U_t$, is given by $2\pi N r_t$ so the equation for centrifugal tensile stress can be written also as

$$(\sigma_{ct})_{max} = \frac{\rho_b}{2} U_t^2 \left[ 1 - \left( \frac{r_r}{r_t} \right)^2 \right]$$

The ratio $r_r/r_t$ is normally referred to as the *hub–tip ratio*. It is immediately apparent that the centrifugal stress is proportional to the square of the tip speed , and that a reduction of hub–tip ratio increases the blade stress. It will be seen later that the hub–tip ratio is also very important with regard to aerodynamic considerations.

In practice, the blade sectional area will be decreased with radius to relieve the blade root stress and the loading on the disc carrying the blades, so that the integral would be evaluated numerically or graphically. A simple analytical expression can be deduced, however, for a linear variation of cross-sectional area from root to tip. If the hub–tip ratio is $b$ and the ratio of the cross-sectional area at the tip to that at the root is $d$, the stress in a tapered blade is given by

$$(\sigma_{ct})_{max} = \frac{\rho_b}{2} U_t^2 (1 - b^2) K$$

where $K = 1 - [(1-d)(2-b-b^2)/3(1-b^2)]$

Typical values of $K$ would range from 0.55 to 0.65 for tapered blades.

Direct centrifugal tensile stresses are not often of major concern in compressor blades. The first-stage blades, being the longest, are the most highly stressed, but the later stages often appear to be very moderately stressed. The designer should not be lulled into a false sense of security by this, because the blades are also subject to fluctuating gas bending stresses which may cause fatigue failure. Although the problem is significantly more difficult for compressors than turbines, because of the high probability of aerodynamic vibration resulting from flow instability when some stages are operating in a stalled condition, a discussion of methods of predicting gas bending stresses is left to Chapter 7.

For tip speeds of around 350 m/s, stress problems are not usually critical in the sizing of the annulus. Tip speeds of 450 m/s, however, are common in the fans of high by-pass ratio turbofans, and with their low hub–tip ratios the design of the disc to retain the long heavy blades becomes critical. The inner radius of the annulus may then be dictated by disc stressing constraints.

## Axial velocity

The expression for stage temperature rise, which in conjunction with an isentropic efficiency determines the stage pressure ratio, showed the desirability of using a high value of axial velocity. A high axial velocity is also required to provide a high flow rate per unit frontal area, which is important for turbojet and turbofan engines.

The axial velocity at inlet, however, must be limited for aerodynamic reasons. Considering a first stage with no IGVs, the entry velocity will be purely axial in direction and the velocity diagram at entry to the rotor will be as shown by the right-angled triangle in Fig. 5.5. The velocity relative to the rotor is given by $V_1^2 = C_1^2 + U^2$ and, assuming the axial velocity to be constant

**FIG. 5.5 Relative Mach number at rotor entry**

over the blade height, the maximum relative velocity will occur at the tip, i.e. $V_{1t}$. The static temperature at rotor entry, $T_1$, is given by $T_{01} - (C_1^2/2c_p)$ and the local acoustic velocity by $a = \sqrt{(\gamma R T_1)}$. The Mach number relative to the rotor tip is $V_{1t}/a$, and for a given speed $U_t$ it is therefore determined by the axial velocity at entry to the stage as indicated by the curves in Fig. 5.5. Axial velocities for an industrial gas turbine will usually be of the order of 150 m/s whereas for advanced aero engines they could be up to 200 m/s.

On early compressors the design had to be such that the Mach number at the rotor tip was subsonic, but in the early fifties it became possible to use transonic Mach numbers up to about 1·1 without introducing excessive losses. With fans of large by-pass ratio the Mach number at the rotor tip may be of the order of 1·5. The dotted velocity triangle in Fig. 5.5 shows how the Mach number at entry may be slightly reduced by means of IGVs, and IGV's were aerodynamically necessary until the ability to operate at transonic Mach number at entry may be slightly reduced by means of IGVs, and IGVs were spool compressor goes some way towards alleviating the tip Mach number problem, because the LP compressor runs at a lower speed than the HP compressor, thereby reducing the tip speed at entry. The acoustic velocity increases in successive stages because of the progressive increase in static temperature, so the Mach number problem diminishes for the later stages. We have already seen that the later stages are also not so critical from the mechanical point of view, short blades implying low stresses.

For a given frontal diameter, the flow area can be increased by decreasing the hub–tip ratio. As this ratio is reduced to very small values, however, the incremental gain in flow area becomes less and less, and as mentioned previously the mechanical design of the first-stage disc becomes difficult; it will be seen later that the Mach number at the tip of the first-stage stator blades also becomes important. For these reasons, hub–tip ratios much below 0·4 are not used for aero engines and significantly higher values are normal for industrial gas turbines.

When air is the working fluid it is generally found that compressibility effects become critical before stress considerations, i.e. $U_t$ and $C_a$ are limited by the need to keep the relative gas velocity to an acceptable level. Closed-cycle gas turbines, with external combustion, are not restricted to using air as the working fluid and both helium and carbon dioxide are possible working fluids. With a light gas such as helium, the gas constant is much higher than for air and the acoustic velocity is correspondingly greater; in this case, it is found that the Mach numbers are low and stress limitations predominate. For a heavier gas such as carbon dioxide, the opposite is true.

## High fluid deflections in the rotor blades

Recalling equation (5.5) for convenience, the stage temperature rise was given by $\Delta T_{0s} = UC_a(\tan \beta_1 - \tan \beta_2)/c_p$. For most compressor stages it can be assumed with little error that the value of $U$ is the same at inlet and outlet of a rotor blade for any particular streamline, and the velocity triangles can conveniently be drawn on a common base as in Fig. 5.6. The amount of deflection required in the rotor is shown by the directions of the relative velocity vectors $V_1$ and $V_2$, and the change in whirl velocity is $\Delta C_w$. Considering a fixed value of $\beta_1$, it is obvious that increasing the deflection by reducing $\beta_2$ entails a reduction in $V_2$. In other words, high fluid deflection implies a high rate of diffusion. The designer must have some method for assessing the allowable diffusion, and one of the earliest criteria used was the *de Haller number*, defined as $V_2/V_1$; a limit of $V_2/V_1 \nless 0\cdot72$ was set, lower values leading to excessive losses. Because of its extreme simplicity, the de Haller number is still used in preliminary design work, but for final design calculations a criterion called the *diffusion factor* is preferred. The latter concept was developed by NACA (the precursor of NASA), and it is widely used on both sides of the Atlantic.

To explain the diffusion factor it is necessary to make a small foray into cascade testing, which will be discussed more fully in a later section. Figure 5.7

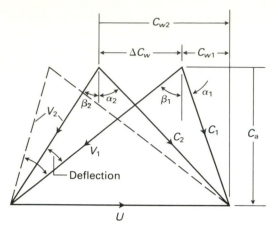

**FIG. 5.6 Effect of increasing fluid deflection**

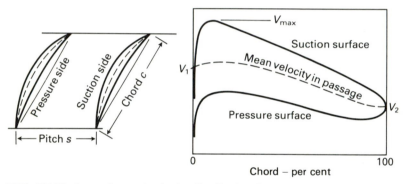

**FIG. 5.7 Blade spacing and velocity distribution through passage**

shows a pair of typical blades which have a pitch $s$ and a chord $c$. The air passing over an aerofoil will accelerate to a higher velocity on the convex surface, and in a stationary row this will give rise to a drop in static pressure; for this reason the convex surface is known as the suction side of the blade. On the concave surface, the pressure side, the fluid will be decelerated. The velocity distribution through the blade passage will be of the form shown: the maximum velocity on the suction surface will occur at around 10–15 per cent of the chord from the leading edge and will then fall steadily until the outlet velocity is reached. The losses in a blade row arise primarily from the growth of boundary layers on the suction and pressure sides of the blade. These surface boundary layers come together at the blade trailing edge to form a wake, giving rise to a local drop in stagnation pressure. Relatively thick surface boundary layers, resulting in high losses, have been found to occur in regions where rapid changes of velocity are occurring, i.e. in regions of high velocity gradient. Figure 5.7 would suggest that these would be most likely to occur on the suction surface. The derivation of the NACA diffusion factor is based on the establishment of the velocity gradient on the suction surface in terms of $V_1$, $V_2$ and $V_{max}$ in conjunction with results from cascade tests. From the cascade tests it was deduced that the maximum velocity $V_{max} \approx V_1 + 0.5(\Delta C_w s/c)$. In simplified form,† the diffusion factor, $D$, can be expressed as

$$D \approx \frac{V_{max} - V_2}{V_1} \approx \frac{V_1 + \dfrac{\Delta C_w}{2}\cdot\dfrac{s}{c} - V_2}{V_1}$$

$$\approx 1 - \frac{V_2}{V_1} + \frac{\Delta C_w}{2V_1}\cdot\frac{s}{c} \qquad\qquad (5.7)$$

The variation of friction loss with $D$ obtained from a large number of NACA tests [Ref. (6)] is shown in Fig. 5.8. These tests were carried out over a wide range of cascade geometries for a particular aerofoil section, and were found

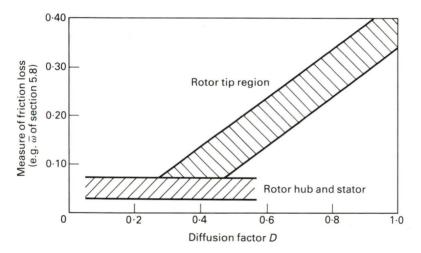

FIG. 5.8 **Variation of friction loss with diffusion factor**

to be generally applicable as long as the maximum local Mach numbers were subsonic or only slightly supersonic. It can be seen that for the rotor hub region and stators the losses are unaffected by variation in $D$ up to 0.6; in the rotor tip region, however, the losses increase rapidly at values of $D$ above 0.4. The great merit of the diffusion factor, as a criterion for placing a limit on the permissible gas deflection, is its relative simplicity and the fact that it can be established as soon as the velocity diagram has been constructed and a value

† The full derivation of the diffusion factor is given in Ref. (4).

of the pitch/chord ratio has been chosen. In American practice the term *solidity*, the inverse of pitch/chord ratio, is used.

## 5.4   Blockage in the compressor annulus

Because of the adverse pressure gradient in compressors, the boundary layers along the annulus walls thicken as the flow progresses. The main effect is to reduce the area available for flow below the geometric area of the annulus. This will have a considerable effect on the axial velocity through the compressor and must be allowed for in the design process. The flow is extremely complex, with successive accelerations and decelerations combined with changes in tangential flow direction; the effects of tip clearance are also significant, making the calculation of boundary layer growth extremely difficult. For this reason, compressor designers normally make use of empirical correction factors based on experimental data from compressor tests.

Early British experiments revealed that the stage temperature rise was always less than would be given by equation (5.5). The explanation of this is based on the fact that the radial distribution of axial velocity is not constant across the annulus, but becomes increasingly 'peaky' as the flow proceeds, settling down to a fixed profile at about the fourth stage. This is illustrated in Fig. 5.9 which shows typical axial velocity profiles in the first and fourth stages. To show how the change in axial velocity affects the work-absorbing capacity of the stage, equation (5.4) can be recast using equation (5.2) as

$$W = mU[(U - C_a \tan \alpha_1) - C_a \tan \beta_2]$$
$$= mU[U - C_a(\tan \alpha_1 + \tan \beta_2)] \tag{5.8}$$

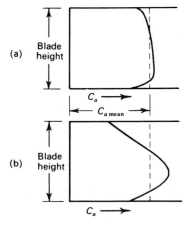

FIG. 5.9 Axial velocity distributions: (a) at first stage, (b) at fourth stage

The outlet angles of the stator and rotor blades determine the values of $\alpha_1$ and $\beta_2$ which can therefore be regarded as fixed (unlike $\alpha_2$ which varies with $C_a$). Equation (5.8) shows that an increase in $C_a$ will result in a decrease of $W$. If it is assumed that the compressor has been designed for a constant radial distribution of axial velocity as shown by the dotted line in Fig. 5.9, the effect of an increase in $C_a$ in the central region of the annulus will be to reduce the work capacity of the blading in that area. The reduction in $C_a$ at the root and tip might be expected to compensate for this effect by increasing the work capacity of the blading close to the annulus walls. Unfortunately the influence of both the boundary layers on the annulus walls and the blade tip clearance has an adverse affect on this compensation and the net result is a decrease in total work capacity. This effect becomes more pronounced as the number of stages is increased.

The reduction in work capacity can be accounted for by use of the *work-done factor* $\lambda$, which is a number less than unity. The actual stage temperature rise is then given by

$$\Delta T_{os} = \frac{\lambda}{c_p} UC_a(\tan \beta_1 - \tan \beta_2) \tag{5.9}$$

Because of the variation of axial velocity profile through the compressor, the mean work-done factor will vary as shown in Fig. 5.10.

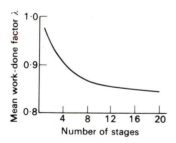

**FIG. 5.10 Variation of mean work-done factor with number of stages**

Care should be taken to avoid confusion of the work-done factor with the idea of an efficiency. If $W$ is the value of the work input calculated from equation (5.4), then $\lambda W$ is the measure of the actual work which can be supplied to the stage. Having established the actual temperature rise, application of the isentropic efficiency of the stage yields the pressure ratio in accordance with equation (5.6).

It was mentioned earlier that a high axial velocity was required for a high stage temperature rise, and at first sight the explanation of work-done factor, on the basis of a lower temperature rise in regions of high axial velocity, appears to be contradictory. This is not so for the following reason. At the design stage, where the values of $\beta_1$ and $\beta_2$ can be selected to give a satisfactory

$\Delta T_{0s}$ within the limits set by the de Haller number or diffusion factor, a high axial velocity is required. Once the design has been specified, however, and the angles $\alpha_1$, $\beta_1$ and $\beta_2$ are fixed, then equation (5.8) correctly shows that an increase in axial velocity will reduce the stage temperature rise.

An alternative approach is to assign 'blockage factors' to reduce the effective annulus area to allow for the growth in boundary layer thickness, and this is the American design practice. Both the 'work-done factor' and 'blockage factor' represent empirical corrections based on a particular organization's experience of compressor development.

## 5.5   Degree of reaction

It was shown in an earlier section that in an axial compressor stage, diffusion takes place in both rotor and stator, and there will be an increase in static pressure through both rows. The *degree of reaction* $\Lambda$ provides a measure of the extent to which the rotor contributes to the overall static pressure rise in the stage. It is normally defined in terms of enthalpy rise as follows,

$$\Lambda = \frac{\text{static enthalpy rise in the rotor}}{\text{static enthalpy rise in the stage}}$$

but, because the variation of $c_p$ over the relevant temperature ranges is negligible, the degree of reaction can be expressed more conveniently in terms of temperature rises.

The degree of reaction is a useful concept in compressor design, and it is possible to obtain a formula for it in terms of the various velocities and air angles associated with the stage. This will be done for the most common case in which it is assumed that (a) $C_a$ is constant through the stage, and (b) the air leaves the stage with the same absolute velocity with which it enters, i.e. $C_3 = C_1$ and hence $\Delta T_S = \Delta T_{0s}$. If $\Delta T_A$ and $\Delta T_B$ denote the static temperature rises in the rotor and stator respectively, then equation (5.5) gives, per unit mass flow,

$$W = c_p(\Delta T_A + \Delta T_B) = c_p \Delta T_S = UC_a(\tan \beta_1 - \tan \beta_2)$$

$$= UC_a(\tan \alpha_2 - \tan \alpha_1) \tag{5.10}$$

Since all the work input to the stage takes place in the rotor, the steady flow energy equation yields

$$W = c_p \Delta T_A + \tfrac{1}{2}(C_2^2 - C_1^2)$$

so that with equation (5.10)

$$c_p \Delta T_A = UC_a(\tan \alpha_2 - \tan \alpha_1) - \tfrac{1}{2}(C_2^2 - C_1^2)$$

But $C_2 = C_a \sec \alpha_2$ and $C_1 = C_a \sec \alpha_1$, and hence

$$c_p \Delta T_A = U C_a (\tan \alpha_2 - \tan \alpha_1) - \tfrac{1}{2} C_a^2 (\sec^2 \alpha_2 - \sec^2 \alpha_1)$$
$$= U C_a (\tan \alpha_2 - \tan \alpha_1) - \tfrac{1}{2} C_a^2 (\tan^2 \alpha_2 - \tan^2 \alpha_1)$$

From the definition of $\Lambda$,

$$\Lambda = \frac{\Delta T_A}{\Delta T_A + \Delta T_B}$$

$$= \frac{U C_a (\tan \alpha_2 - \tan \alpha_1) - \tfrac{1}{2} C_a (\tan^2 \alpha_2 - \tan^2 \alpha_1)}{U C_a (\tan \alpha_2 - \tan \alpha_1)}$$

$$= 1 - \frac{C_a}{2U} (\tan \alpha_2 + \tan \alpha_1)$$

By the addition of equations (5.2) and (5.3),

$$\frac{2U}{C_a} = \tan \alpha_1 + \tan \beta_1 + \tan \alpha_2 + \tan \beta_2$$

Hence

$$\Lambda = \frac{C_a}{2U} \left[ \frac{2U}{C_a} - \frac{2U}{C_a} + \tan \beta_1 + \tan \beta_2 \right]$$

$$= \frac{C_a}{2U} (\tan \beta_1 + \tan \beta_2) \tag{5.11}$$

Because the case of 50 per cent reaction is important in design, it is of interest to see the consequences of putting $\Lambda = 0.5$. In this case, from equation (5.11)

$$\tan \beta_1 + \tan \beta_2 = \frac{U}{C_a}$$

and it immediately follows from equations (5.2) and (5.3) that

$$\tan \alpha_1 = \tan \beta_2, \qquad \text{i.e. } \alpha_1 = \beta_2$$
$$\tan \beta_1 = \tan \alpha_2, \qquad \text{i.e. } \alpha_2 = \beta_1$$

Furthermore, since $C_a$ is constant through the stage,

$$C_a = C_1 \cos \alpha_1 = C_3 \cos \alpha_3$$

It was initially assumed that $C_1 = C_3$ and hence $\alpha_1 = \alpha_3$. Because of this equality of angles, namely $\alpha_1 = \beta_2 = \alpha_3$ and $\beta_1 = \alpha_2$, the velocity diagram of Fig. 5.6 becomes symmetrical and blading designed on this basis is sometimes referred to as *symmetrical blading*. From the symmetry of the velocity diagram, it follows that $C_1 = V_2$ and $V_1 = C_2$.

It must be pointed out that in deriving equation (5.11) for $\Lambda$, a work-done factor $\lambda$ of unity has implicitly been assumed in making use of equation (5.10). A stage designed with symmetrical blading will always be referred to as a 50 per cent reaction stage, although the value of $\Lambda$ actually achieved will differ slightly from 0·5 because of the influence of $\lambda$.

It is of interest to consider the significance of the limiting values of $\Lambda$, viz. 0 and 1·0. The degree of reaction was defined in terms of enthalpy rises but could also have been expressed in terms of changes in static pressure. Making use of the thermodynamic relation $T\,\mathrm{d}s = \mathrm{d}h - v\,\mathrm{d}p$ and assuming incompressible isentropic flow

$$0 = \mathrm{d}h - \mathrm{d}p/\rho$$

which on integration from state 1 to state 2 becomes

$$h_2 - h_1 = (p_2 - p_1)/\rho$$

showing that enthalpy and static pressure changes are related. By putting $\Lambda = 0$ in equation (5.11) it can readily be seen that $\beta_1 = -\beta_2$; the rotor blades are then of impulse type (i.e. passage area the same at inlet and outlet) and all the static pressure rise occurs in the stator. Conversely, for $\Lambda = 1\cdot0$ the stators are of impulse type. In the interests of obtaining the most efficient overall diffusion, it is desirable to share the diffusion between both components of the stage, and for this reason the use of 50 per cent reaction is attractive. We are not suggesting, however, that the efficiency is very sensitive to the precise degree of reaction chosen. In practice, as will appear later, the degree of reaction may show considerable variation across the annulus (i.e. along the blade span), especially for stages of low hub–tip ratio, and the designer will primarily be concerned to satisfy more stringent conditions set by a limiting diffusion factor and Mach number.

At this point the reader may feel he would like to see an example of a preliminary mean diameter design to consolidate the previous material. If so, he can turn to section 5.7. To enable the full design procedure to be illustrated, however, we need to consider three-dimensional effects, and these will be introduced in the next section.

## 5.6   Three-dimensional flow

The elementary theory presented in section 5.2 assumed that the flow in the compressor annulus is two-dimensional, meaning that any effect due to radial movement of the fluid is ignored. The assumption of two-dimensional flow is quite reasonable for stages in which the blade height is small relative to the mean diameter of the annulus, i.e. those of hub–tip ratio greater than about 0·8, which would be typical of the later stages of a compressor. The front stages, however, have lower values of hub–tip ratio, and values as low as 0·4 are used for the first stage of aero-engine compressors so that a high mass flow can be passed through a machine of low frontal area. When the compressor has a low hub–tip ratio in the first stage and a high hub–tip ratio in the later stages, the annulus will have a substantial taper [see Fig. 5.1] and the streamlines will not lie on a surface of revolution parallel to the axis of the rotor as previously assumed. Under these conditions, the flow must have a radial component of velocity, although it will generally be small compared with the axial and whirl components.

A second cause of radial movement occurs as follows. Because the flow has a whirl component, the pressure must increase with radius, i.e. up the blade height, to provide the force associated with the centripetal acceleration of the fluid. In the course of adjusting itself to provide a balance between the pressure forces and the inertia forces, the flow will undergo some movement in the radial direction.

With a low hub–tip ratio, the variation in blade speed from root to tip is large and this will have a major effect on the shape of the velocity triangles and the resulting air angles. Furthermore, the aforementioned change of pressure, and hence density, with radius will cause the fluid velocity vectors to change in magnitude and these too affect the shape of the velocity triangles. It follows that the air angles at the mean diameter will be far from representative of those at the root and tip of a blade row. For high efficiency it is essential that the blade angles match the air angles closely at all radii, and the blade must therefore be twisted from root to tip to suit the changing air angles.

The basic equation expressing the balance between pressure forces and inertia forces can be derived by considering the forces acting on the fluid element shown in Fig. 5.11. The whirl component of velocity is shown in Fig. 5.11(a), and the axial component and much smaller radial component resulting from the curvature of the streamline in Fig. 5.11(b). The inertia forces in the radial direction arise from

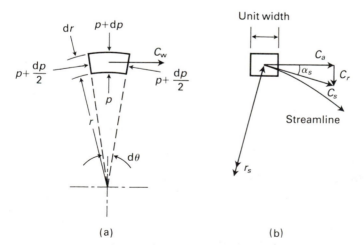

**FIG. 5.11 Radial equilibrium of fluid element**

(i)  the centripetal force associated with circumferential flow;
(ii)  the radial component of the centripetal force associated with the flow
      along the curved streamline;
(iii) the radial component of the force required to produce the linear
      acceleration along the streamline.

The total inertia force, $F_I$, must be produced by the pressure forces acting on the
element in the radial direction. (The acceleration in the radial direction may
amount to several thousand times the acceleration due to gravity so that
gravitational forces can be neglected.)

Considering a fluid element of unit width, having a density $\rho$, the three
terms of the inertia force can be expressed as follows. The centripetal force
associated with the circumferential flow is

$$F_{(i)} = \frac{mC_w^2}{r} = (\rho r \, dr \, d\theta)\frac{C_w^2}{r}$$

For the flow along the curved streamline, the radial force is given by

$$F_{(ii)} = \frac{mC_S^2}{r_S}\cos \alpha_S = (\rho r \, dr \, d\theta)\frac{C_S^2}{r_S}\cos \alpha_S$$

where the suffix $S$ refers to the component along the streamline and $r_S$ is the
radius of curvature of the streamline. For the acceleration along the
streamline, the radial component of force is

$$F_{(iii)} = m\frac{dC_S}{dt}\sin \alpha_S = (\rho r \, dr \, d\theta)\frac{dC_S}{dt}\sin \alpha_S$$

With the curvature shown in Fig. 5.11(b) the forces $F_{(ii)}$ and $F_{(iii)}$ are in the
same direction as $F_{(i)}$. Thus the total inertia force, $F_I$, is given by

$$F_I = \rho r \, dr \, d\theta\left[\frac{C_w^2}{r} + \frac{C_S^2}{r_S}\cos \alpha_S + \frac{dC_S}{dt}\sin \alpha_S\right]$$

The pressure force, $F_P$, producing this inertia force is obtained by resolving
in the radial direction to give

$$F_P = (p+dp)(r+dr)\, d\theta - pr\, d\theta - 2\left(p+\frac{dp}{2}\right)dr\,\frac{d\theta}{2}$$

The third term in the equation results from the resolution of the pressure
forces on the two sides of the element in the radial–axial plane, on which it is
assumed that the pressure is the average of the two extremes, namely
$p+(dp/2)$. Equating the forces $F_I$ and $F_P$, and neglecting second-order terms,
we are left with

$$\frac{1}{\rho}\frac{dp}{dr} = \frac{C_w^2}{r} + \frac{C_S^2}{r_S}\sin \alpha_S + \frac{dC_S}{dt}\sin \alpha_S \tag{5.12}$$

This is the complete radial equilibrium equation which includes all contributory factors.

For most design purposes it can be assumed that $r_s$ is so large, and $\alpha_s$ so small, that the last two terms of equation (5.12) can be ignored. Thus, finally, we have

$$\frac{1}{\rho}\frac{dp}{dr} = \frac{C_w^2}{r} \tag{5.13}$$

and this is what will be referred to as the *radial equilibrium equation*. In effect, the radial component of velocity is being neglected. Certainly in the spaces between the blade rows $C_r$ is very much smaller than either $C_a$ or $C_w$ and can safely be assumed to be negligible.

Equation (5.13) enables us to deduce an energy equation which expresses the variation of enthalpy with radius. The stagnation enthalpy $h_0$ at any radius $r$ where the absolute velocity is $C$ is given by

$$h_0 = h + \frac{C^2}{2} = h + \tfrac{1}{2}(C_a^2 + C_w^2)$$

and the variation of enthalpy with radius is therefore

$$\frac{dh_0}{dr} = \frac{dh}{dr} + C_a\frac{dC_a}{dr} + C_w\frac{dC_w}{dr} \tag{5.14}$$

From the thermodynamic relation $T\,ds = dh - dp/\rho$,

$$\frac{dh}{dr} = T\frac{ds}{dr} + ds\frac{dT}{dr} + \frac{1}{\rho}\frac{dp}{dr} - \frac{1}{\rho^2}\frac{d\rho}{dr}dp$$

Dropping second-order terms

$$\frac{dh}{dr} = T\frac{ds}{dr} + \frac{1}{\rho}\frac{dp}{dr}$$

Substituting for $dh/dr$ in equation (5.14)

$$\frac{dh_0}{dr} = T\frac{ds}{dr} + \frac{1}{\rho}\frac{dp}{dr} + C_a\frac{dC_a}{dr} + C_w\frac{dC_w}{dr}$$

Using the radial equilibrium equation (5.13), the second term of the right-hand side of the above equation can be replaced by $C_w^2/r$, leaving the basic equation for the analysis of flow in the compressor annulus as

$$\frac{dh_0}{dr} = T\frac{ds}{dr} + C_a\frac{dC_a}{dr} + C_w\frac{dC_w}{dr} + \frac{C_w^2}{r}$$

The term $T\,ds/dr$ represents the radial variation of loss across the annulus, and may be significant in detailed design calculations; this is especially true if Mach numbers relative to the blade are supersonic and shock losses occur. For our purposes, however, it will be assumed that the entropy gradient term can be ignored and the final form of the equation is given by

$$\frac{dh_0}{dr} = C_a \frac{dC_a}{dr} + C_w \frac{dC_w}{dr} + \frac{C_w^2}{r} \qquad (5.15)$$

This will be referred to as the *vortex energy equation.*

Apart from regions near the walls of the annulus, the stagnation enthalpy (and temperature) will be uniform across the annulus at entry to the compressor. If the frequently used design condition of *constant specific work at all radii* is applied, then although $h_0$ will increase progressively through the compressor in the axial direction, its radial distribution will remain uniform. Thus $dh_0/dr = 0$ in any plane between a pair of blade rows. Equation (5.15) then reduces to

$$C_a \frac{dC_a}{dr} + C_w \frac{dC_w}{dr} + \frac{C_w^2}{r} = 0 \qquad (5.16)$$

A special case may now be considered in which $C_a$ is maintained constant across the annulus, so that $dC_a/dr = 0$. Equation (5.16) then reduces to

$$\frac{dC_w}{dr} = -\frac{C_w}{r}, \quad \text{or} \quad \frac{dC_w}{C_w} = -\frac{dr}{r}$$

which on integration gives

$$C_w r = \text{constant} \qquad (5.17)$$

Thus the whirl velocity varies inversely with radius, this being known as the *free vortex condition.*

It can therefore be seen that the three conditions of (a) constant specific work, (b) constant axial velocity, and (c) free vortex variation of whirl velocity, naturally satisfy the radial equilibrium equation (5.13) and are therefore conductive to the design flow conditions being achieved. It would at first appear that they constitute an ideal basis for design. Unfortunately, there are certain disadvantages associated with the resultant 'free vortex blading', described in later sections, which influence the designer in considering other combinations of basic conditions. Free vortex designs, however, are widely used in axial flow compressors and will be discussed further in Chapter 7.

One disadvantage of free vortex blading will be dealt with here: the marked variation of degree of reaction with radius. In section 5.5 it was shown that for the case where $C_3 = C_1$ and $C_{a1} = C_{a2} = C_a$, the degree of reaction can be expressed by

$$\Lambda = 1 - \frac{C_a}{2U} (\tan \alpha_2 + \tan \alpha_1)$$

which can readily be written in terms of whirl velocities as

$$\Lambda = 1 - \frac{C_{w2} + C_{w1}}{2U} \qquad (5.18)$$

Remembering that $U = U_m r/r_m$, where $U_m$ is the blade speed at the mean radius of the annulus $r_m$, equation (5.18) can be written

$$\Lambda = 1 - \frac{C_{w2}r + C_{w1}r}{2U_m r^2/r_m}$$

For a free vortex design $C_w r = \text{constant}$, so that

$$\Lambda = 1 - \frac{\text{constant}}{r^2} \tag{5.19}$$

Evidently the degree of reaction increases markedly from root to tip of the blade. Even if the stage has the desirable value of 50 per cent at the mean radius, it may well be too low at the root and too high at the tip for good efficiency. Because of the lower blade speed at the root section, more fluid deflection is required for a given work input, i.e. a greater rate of diffusion is required at the root section. It is, therefore, particularly undesirable to have a low degree of reaction in this region, and the problem is aggravated as the hub–tip ratio is reduced.

When considering other possible sets of design conditions, it is usually desirable to retain the constant specific work-input condition to provide a constant stage pressure ratio up the blade height. It would be possible, however, to choose a variation of one of the other variables, say $C_w$, and determine the variation of $C_a$ from equation (5.16). The radial equilibrium requirement would still be satisfied. [It should be clear that in general a design can be based on arbitrarily chosen radial variations of any two variables and the appropriate variation of the third can be determined by inserting them into equation (5.15).]

As an illustration we shall use the normal design condition (a) constant specific work input at all radii, together with (b) an arbitrary whirl velocity distribution which is compatible with (a). To obtain constant work input, $U(C_{w2} - C_{w1})$ must remain constant across the annulus. Let us consider distributions of whirl velocity at inlet to and outlet from the rotor blade given by

$$C_{w1} = aR^n - \frac{b}{R} \quad \text{and} \quad C_{w2} = aR^n + \frac{b}{R} \tag{5.20}$$

where $a$, $b$ and $n$ are constants and $R$ is the radius ratio $r/r_m$. At any radius $r$ the blade speed is given by $U = U_m R$. It is immediately seen that $(C_{w2} - C_{w1}) = 2b/R$ and hence that

$$U(C_{w2} - C_{w1}) = 2bU_m$$

which is independent of radius. The two design conditions (a) and (b) are therefore compatible. We shall consider three special cases: where $n = -1, 1$ and 0 respectively. (It will be shown later that $a$ and $b$ are not arbitrary constants but depend upon the chosen values of degree of reaction and stage temperature rise.)

## When $n = -1$

The whirl distributions become

$$C_{w1} = \frac{a}{R} - \frac{b}{R} \quad \text{and} \quad C_{w2} = \frac{a}{R} + \frac{b}{R}$$

which are of free vortex form, $C_w r = $ constant. From what has gone before it should be clear that $C_a = $ constant must be the third design condition required to ensure radial equilibrium, i.e. to satisfy equation (5.16). It also follows that the variation of $\Lambda$ is given by equation (5.19), which since $(C_{w2}r + C_{w1}r) = 2ar_m$, can be written

$$\Lambda = 1 - \frac{2ar_m}{2U_m Rr} = 1 - \frac{a}{U_m R^2} \tag{5.21}$$

## When $n = 1$

$$C_{w1} = aR - \frac{b}{R} \quad \text{and} \quad C_{w2} = aR + \frac{b}{R}$$

Rewriting equation (5.16) in terms of the dimensionless $R$ we have

$$C_a\, dC_a + C_w\, dC_w + \frac{C_w^2}{R}\, dR = 0$$

Integrating from the mean radius $(R = 1)$ to any other radius,

$$-\tfrac{1}{2}[C_a^2]_1^R = \tfrac{1}{2}[C_w^2]_1^R + \int_1^R \frac{C_w^2}{R}\, dR$$

At exit from the rotor, where $C_w = aR + (b/R)$, the right-hand side of the equation becomes

$$-\frac{1}{2}\left[a^2 R^2 + 2ab + \frac{b^2}{R}\right]_1^R + \int_1^R \left[a^2 R + \frac{2ab}{R} + \frac{b^2}{R^3}\right] dR$$

and, finally, the radial distribution of $C_a$ is given by

$$C_{a2}^2 - (C_{a2}^2)_m = -2[a^2 R^2 + 2ab \ln R - a^2] \tag{5.22}$$

Similarly, at inlet to the rotor,

$$C_{a1}^2 - (C_{a1}^2)_m = -2[a^2 R^2 - 2ab \ln R - a^2] \tag{5.23}$$

Note that $C_{a2}$ cannot equal $C_{a1}$ except at the mean radius $(R = 1)$. It is not possible, therefore, to use the simple equation (5.18) for the degree of reaction when finding the variation of $\Lambda$ with radius. To do this it is necessary to revert to the original definition of $\Lambda$ $[= \Delta T_A/(\Delta T_A + \Delta T_B)]$ and work from first principles as follows. We shall retain the assumption that the stage is designed to give $C_3 = C_1$. Then $\Delta T_S = \Delta T_{OS}$ so that

$$c_p(\Delta T_A + \Delta T_B) = W = U(C_{w2} - C_{w1})$$

$$c_p \Delta T_A = \tfrac{1}{2}(C_1^2 - C_2^2) + U(C_{w2} - C_{w1})$$

$$= \tfrac{1}{2}[(C_{a1}^2 + C_{w1}^2) - (C_{a2}^2 + C_{w2}^2)] + U(C_{w2} - C_{w1})$$

$$\Lambda = 1 + \frac{C_{a1}^2 - C_{a2}^2}{2U(C_{w2} - C_{w1})} - \frac{C_{w2} + C_{w1}}{2U}$$

From equations (5.22) and (5.23), assuming we choose to design with $(C_{a2})_m = (C_{a1})_m$,

$$C_{a1}^2 - C_{a2}^2 = 8ab \ln R$$

From the whirl distributions,

$$C_{w2} - C_{w1} = \frac{2b}{R} \quad \text{and} \quad C_{w2} + C_{w1} = 2aR$$

Substituting in the equation for $\Lambda$, and writing $U = U_m R$, we get finally

$$\Lambda = 1 + \frac{2a \ln R}{U_m} - \frac{a}{U_m} \tag{5.24}$$

Designs with the exponent $n = 1$ have been referred to as *first power* designs.

At this point it can easily be made clear that the constants $a$ and $b$ in the whirl distributions are not arbitrary. Equation (5.24) shows that $a$ is fixed once the degree of reaction at the mean radius is chosen, i.e. at $R = 1$

$$a = U_m(1 - \Lambda_m)$$

And $b$ is fixed when the stage temperature rise is chosen because

$$c_p \Delta T_{OS} = U(C_{w2} - C_{w1}) = 2bU_m$$

and hence $b = c_p \Delta T_{OS}/2U_m$.


## When $n = 0$

Proceeding as in the previous case, but with

$$C_{w1} = a - \frac{b}{R} \quad \text{and} \quad C_{w2} = a + \frac{b}{R}$$

we find that

$$(C_{a2}^2) - (C_{a2}^2)_m = -2\left[ a^2 \ln R - \frac{ab}{R} + ab \right] \tag{5.25}$$

$$(C_{a1}^2) - (C_{a1}^2)_m = -2\left[ a^2 \ln R + \frac{ab}{R} - ab \right] \tag{5.26}$$

Blading designed on this basis is usually referred to as *exponential* blading. Again proceeding as before, and assuming $(C_{a2})_m = (C_{a1})_m$, it can be shown that

$$\Lambda = 1 + \frac{a}{U_m} - \frac{2a}{U_m R} \qquad\qquad (5.27)$$

It should be noted that for all three cases, at the mean radius ($R = 1$), the expression $a = U_m(1 - \Lambda_m)$ holds. If a value is specified for the reaction at the mean radius, $\Lambda_m$, the variation of reaction with radius can readily be obtained by substituting for $a$ in equations (5.21), (5.24) and (5.27). The results can be summarized conveniently in the following table.

| $n$ | $\Lambda$ | Blading |
|---|---|---|
| $-1$ | $1 - \dfrac{1}{R^2}(1 - \Lambda_m)$ | Free vortex |
| $0$ | $1 + \left[1 - \dfrac{2}{R}\right](1 - \Lambda_m)$ | Exponential |
| $1$ | $1 + (2\ln R - 1)(1 - \Lambda_m)$ | First power |

It is instructive to evaluate the variation of $\Lambda$ for all three cases assuming that $\Lambda_m = 0.5$, and the results are given in Fig. 5.12. It can be seen that the free

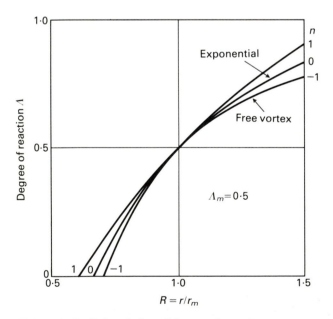

FIG. 5.12 Radial variation of degree of reaction

vortex design gives the greatest reduction in $\Lambda$ for low values of $R$ while the first-power design gives the least reduction. Furthermore, for all designs there is a lower limit to $R$ below which the degree of reaction becomes negative. A negative value implies a *reduction* in static pressure in the rotor. For the free vortex design, this limiting value of $R$ is given by $0 = 1 - (1/R^2)(1 - \Lambda_m)$, from which $R = 0.707$ when $\Lambda_m = 0.50$. As mentioned earlier, however, hub–tip ratios as low as 0.4 are common at entry to the compressor of a jet engine. Noting that

$$r/r_t = (r/r_m)(r_m/r_t) = R(r_m/r_t) = R[1 + (r_r/r_t)]/2$$

the variation in reaction can readily be evaluated in terms of $r/r_t$ for a specified value of hub–tip ratio $r_r/r_t$. Figure 5.13 shows the radial distribution of $\Lambda$ for a free vortex design of hub–tip ratio 0.4, for a series of values of $\Lambda_m$. It is evident that a stage of low hub–tip ratio must have a high value of $\Lambda_m$ to provide satisfactory conditions at the root section.

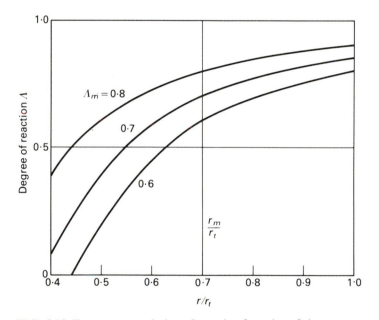

**FIG. 5.13 Free vortex variation of reaction for $r_r/r_t = 0.4$**

So far the discussion has been based upon a choice of $C_a$, $C_w$ and $h_0$ (i.e. $W$) as the primary variables. This choice is not essential, however, and another approach could be to specify a variation of degree of reaction with radius rather than $C_w$. One such design method was based on (a) axial velocity, (b) work input and (c) degree of reaction, all being independent of radius; this was referred to as a *constant reaction design*. It led to less twisted blading than the free vortex design, but consideration of the analysis leading to equation (5.17) and thence to equation (5.19) shows that radial equilibrium cannot be satisfied: conditions (a) and (b) involves a varying $\Lambda$ when equation (5.16) is satisfied. The fact is that the flow will adjust itself to satisfy radial equilibrium between the blade rows. If the design does not allow for this correctly, the actual air angles will not agree with the design values on which the blade angles are based and the efficiency is likely to be reduced.

While it is normally desirable to retain the constant specific work input condition, a fan for a turbofan engine of high by-pass ratio might provide an exception to this rule. In this case, the inner stream, feeding the high-pressure core, may be designed for a lower pressure ratio than the outer stream which is supplying air to the by-pass duct and where the blade speed is high. The work then varies with radius, $dh_0/dr$ is not zero, and equation (5.15) must be used instead of equation (5.16).

Further discussion of three-dimensional design procedures will follow in section 5.7 after an example is given to show how the material presented in the preceding sections can be used in the design of an axial flow compressor.

## 5.7 Design process

The theory presented in the previous sections will now be applied to the design of an axial flow compressor, and it will be seen that the process requires continuous judgement by the designer. The design of a compressor suitable for a simple low-cost turbojet will be considered.

A typical gas turbine design procedure was outlined in Fig. 1.20. Assuming that market research has shown there is a need for a low-cost turbojet with a take-off thrust of about 12 000 N, preliminary studies will show that a single-spool all-axial flow arrangement is satisfactory, using a low pressure ratio and a modest turbine inlet temperature to keep the cost down, as discussed in section 3.3. From cycle calculations, a suitable design point under sea-level static conditions (with $p_a = 1.01$ bar and $T_a = 288$ K) may emerge as follows:

> Compressor pressure ratio   4·15
>
> Air-mass flow                        20 kg/s
>
> Turbine inlet temperature    1100 K

With these data specified, it is now necessary to investigate the aerodynamic design of the compressor, turbine and other components of the engine. It will be assumed that the compressor has no inlet guide vanes, to keep both weight and noise down. The design of the turbine for this engine will be considered in Chapter 7.

The complete design process for the compressor will encompass the following steps:

(i)      choice of rotational speed and annulus dimensions;

(ii)    determination of number of stages, using an assumed efficiency;
(iii)   calculation of the air angles for each stage at the mean radius;
(iv)    determination of the variation of the air angles from root to tip;
(v)     investigation of compressibility effects;
(vi)    selection of compressor blading, using experimentally obtained cascade data;
(vii)   check on efficiency previously assumed, using the cascade data;
(viii)  estimation of off-design performance;
(ix)    rig testing.

Steps (i) to (v) will be outlined in this section and the remaining steps will be covered in later sections. In practice, the process will be one of continued refinement, coupled with feedback from other groups such as the designers of the combustion system and turbine, metallurgists and stress analysts, and those concerned with mechanical problems associated with whirling speeds, bearings, stiffness of structural members and so on.

## Determination of rotational speed and annulus dimensions

Reviewing the theory presented earlier, it is seen that there is no equation which enables the designer to select a suitable value of rotational speed. This can be found, however, by assuming values for the blade tip speed, and the axial velocity and hub–tip ratio at inlet to the first stage. The required annulus area at entry is obtained from the specified mass flow, assumed axial velocity, and ambient conditions, using the continuity equation.

Previous experience will suggest that a tip speed, $U_t$, of around 350 m/s will lead to acceptable stresses and that the axial velocity, $C_a$, could range from 150 to 200 m/s. Without IGVs there will be no whirl component of velocity at inlet, and this will increase the Mach number relative to the blade (see Fig. 5.5) so it may be advisable to use a modest value of 150 m/s for $C_a$. The hub–tip ratio at entry may vary between 0·4 and 0·6, and for a specified annulus area the tip radius will be a function of the hub–tip ratio. For a fixed blade speed, then, the rotational speed will also be a function of hub–tip ratio. Thus the designer will, in very short order, be presented with a wide range of solutions and must use his judgement to select the most promising. At the same time the turbine designer will be examining a suitable turbine and the compressor and turbine designers must keep in close contact while establishing preliminary designs.

To satisfy continuity

$$m = \rho_1 A C_{a1} = \rho_1 \pi r_t^2 \left[ 1 - \left[ \frac{r_r}{r_t} \right]^2 \right] C_a$$

$$r_t^2 = \frac{m}{\pi \rho_1 C_{a1} [1 - (r_r/r_t)^2]}$$

At sea-level static conditions, $T_{01} = T_a = 288$ K. Assuming no loss in the intake, $p_{01} = p_a = 1.01$ bar. With

$$C_1 = C_{a1} = 150 \text{ m/s } (C_{w1} = 0)$$

$$T_1 = 288 - \frac{150^2}{2 \times 1.005 \times 10^3} = 276.8 \text{ K}$$

$$p_1 = p_{01} \left[ \frac{T_1}{T_{01}} \right]^{\gamma/(\gamma-1)} = 1.01 \left[ \frac{276.8}{288} \right]^{3.5} = 0.879 \text{ bar}$$

$$\rho_1 = \frac{100 \times 0.879}{0.287 \times 276.8} = 1.106 \text{ kg/m}^3$$

$$r_t^2 = \frac{20}{\pi \times 1.106 \times 150[1 - (r_r/r_t)^2]} = \frac{0.03837}{[1 - (r_r/r_t)^2]}$$

The tip speed, $U_t$ is related to $r_t$ by $U_t = 2\pi r_t N$, and hence if $U_t$ is chosen to be 350 m/s,

$$N = \frac{350}{2\pi r_t}$$

Evaluating $r_t$ and $N$ over a range of hub–tip ratios the following table results:

| $r_r/r_t$ | $r_t$ (m) | $N$ (rev/s) |
|---|---|---|
| 0.40 | 0.2137 | 260.6 |
| 0.45 | 0.2194 | 253.9 |
| 0.50 | 0.2262 | 246.3 |
| 0.55 | 0.2346 | 237.5 |
| 0.60 | 0.2449 | 227.5 |

At this point it would be pertinent to consider the turbine design. The example in Chapter 7 shows that a speed of 250 rev/s results in quite an adequate single-stage unit, and the outer radius at the turbine inlet is found to be 0.239 m. Referring to the table above, a hub–tip ratio of 0.50 would give a compatible compressor tip radius of 0.2262 m although the rotational speed is 246.3 rev/s. There was nothing sacrosanct about the choice of 350 m/s for the tip speed, and the design could be adjusted for a rotational speed of 250 rev/s. With the speed slightly altered then,

$$U_t = 2\pi \times 0.2262 \times 250 = 355.3 \text{ m/s}$$

For a simple engine of the type under consideration, there would be no merit in using a low hub–tip ratio; this would merely increase the mismatch between the compressor and turbine diameters, and also complicate both the mechanical and aerodynamic design of the first stage. On the other hand, using a high hub–tip ratio would unnecessarily increase the compressor diameter and weight. But it should be realized that the choice of 0·50 for hub–tip ratio is arbitrary, and merely provides a sensible starting point; later considerations following detailed analysis could cause an adjustment, and a considerable amount of design optimization is called for.

At this stage it is appropriate to check the Mach number relative to the rotor tip at inlet to the compressor. Assuming the axial velocity to be constant across the annulus, which will be the case where there are no inlet guide vanes,

$$V_{1t}^2 = U_{1t}^2 + C_{a1}^2 = 355 \cdot 3^2 + 150^2, \quad \text{and} \quad V_{1t} = 385 \cdot 7 \text{ m/s}$$

$$a = \sqrt{(\gamma R T_1)} = \sqrt{1 \cdot 4 \times 0 \cdot 287 \times 1000 \times 276 \cdot 8} = 331 \cdot 0 \text{ m/s}$$

$$M_{1t} = \frac{V_{1t}}{a} = \frac{385 \cdot 7}{331 \cdot 0} = 1 \cdot 165$$

Thus the Mach number relative to the rotor tip is 1·165 and the first stage is transonic; this level of Mach number should not present any problem, and methods of dealing with shock losses will be covered in a later section.

With the geometry selected, i.e. a hub–tip ratio of 0.50 and a tip radius of 0·2262 m, it follows that the root radius is 0·1131 m and the mean radius is 0·1697 m. It is instructive now to estimate the annulus dimensions at exit from the compressor, and for these preliminary calculations it will be assumed that the mean radius is kept constant for all stages. The compressor delivery pressure, $p_{02} = 4 \cdot 15 \times 1 \cdot 01 = 4 \cdot 19$ bar. To estimate the compressor delivery temperature it will be assumed that the polytropic efficiency of the compressor is 0·90. Thus

$$T_{02} = T_{01} \left[ \frac{p_{02}}{p_{01}} \right]^{(n-1)/n}, \quad \text{where} \quad \frac{(n-1)}{n} = \frac{1}{0 \cdot 90} \times \frac{0 \cdot 4}{1 \cdot 4} = 0 \cdot 3175$$

so that

$$T_{02} = 288 \cdot 0 (4 \cdot 15)^{0 \cdot 3175} = 452 \cdot 5 \text{ K}$$

Assuming that the air leaving the stator of the last stage has an axial velocity of 150 m/s and no swirl, the static temperature, pressure and density at exit can readily be calculated as follows:

$$T_2 = 452 \cdot 5 - \frac{150^2}{2 \times 1 \cdot 005 \times 10^3} = 441 \cdot 3 \text{ K}$$

$$p_2 = p_{02} \left[ \frac{T_2}{T_{02}} \right]^{\gamma/(\gamma-1)} = 4 \cdot 19 \left[ \frac{441 \cdot 3}{452 \cdot 5} \right]^{3 \cdot 5} = 3 \cdot 838 \text{ bar}$$

$$\rho_2 = \frac{100 \times 3 \cdot 838}{0 \cdot 287 \times 441 \cdot 3} = 3 \cdot 03 \text{ kg/m}^3$$

The exit annulus area is thus given by

$$A_2 = \frac{20}{3 \cdot 031 \times 150} = 0 \cdot 0440 \text{ m}^2$$

With $r_m = 0 \cdot 1697$ m, the blade height at exit, $h$, is then given by

$$h = \frac{0 \cdot 044}{2 \pi r_m} = \frac{0 \cdot 044}{2 \pi \times 0 \cdot 1697} = 0 \cdot 0413 \text{ m}$$

The radii at exit from the last stator are then

$$r_t = 0 \cdot 1697 + (0 \cdot 0413/2) = 0 \cdot 1903 \text{ m}$$

$$r_r = 0 \cdot 1697 - (0 \cdot 0413/2) = 0 \cdot 1491 \text{ m}$$

At this point we have established the rotational speed and the annulus dimensions at inlet and outlet, on the basis of a constant mean diameter. To summarize:

$$N = 250 \text{ rev/s} \qquad \left. \begin{matrix} r_t = 0 \cdot 2262 \text{ m} \\ r_r = 0 \cdot 1131 \text{ m} \end{matrix} \right\} \text{inlet}$$
$$U_t = 355 \cdot 3 \text{ m/s}$$

$$C_a = 150 \text{ m/s} \qquad \left. \begin{matrix} r_t = 0 \cdot 1903 \text{ m} \\ r_r = 0 \cdot 1491 \text{ m} \end{matrix} \right\} \text{outlet}$$
$$r_m = 0 \cdot 1697 \text{ m (constant)}$$

## Estimation of number of stages

With the assumed polytropic efficiency of 0·90, the overall stagnation temperature rise through the compressor is $452 \cdot 5 - 288 = 164 \cdot 5$ K. The stage temperature rise $\Delta T_{os}$ can vary widely in different compressor designs, depending on the application and the importance or otherwise of low weight: values may vary from 10 to 30 K for subsonic stages and may be 45 K or higher in high-performance transonic stages. Rather than choosing a value at random, it is instructive to estimate a suitable $\Delta T_{os}$ based on the mean blade speed

$$U = 2 \times \pi \times 0 \cdot 1697 \times 250 = 266 \cdot 6 \text{ m/s}$$

We will adopt the simple design condition $C_{a_1} = C_{a_2} = C_a$ throughout the compressor, so the temperature rise from equation (5.8) is given by

$$\Delta T_{os} = \frac{\lambda U C_a (\tan \beta_1 - \tan \beta_2)}{c_p} = \frac{\lambda U (C_{w2} - C_{w1})}{c_p}$$

With a purely axial velocity at entry to the first stage, in the absence of IGVs,

$$\tan \beta_1 = \frac{U}{C_a} = \frac{266 \cdot 6}{150}$$

$$\beta_1 = 60 \cdot 64°$$

$$V_1 = \frac{C_a}{\cos \beta_1} = \frac{150}{\cos 60 \cdot 64} = 305 \cdot 9 \text{ m/s}$$

In order to estimate the maximum possible deflection in the rotor, we will apply the de Haller criterion $V_2/V_1 \not< 0 \cdot 72$. On this basis the minimum allowable value of $V_2 = 305 \cdot 9 \times 0 \cdot 72 = 220$ m/s, and the corresponding rotor blade outlet angle is given by

$$\cos \beta_2 = \frac{C_a}{V_2} = \frac{150}{220}, \qquad \beta_2 = 47 \cdot 01°$$

Using this deflection and neglecting the work-done factor for this crude estimate

$$\Delta T_{os} = \frac{266 \cdot 6 \times 150 (\tan 60 \cdot 64 - \tan 47 \cdot 01)}{1 \cdot 005 \times 10^3} \approx 28 \text{ K}$$

A temperature rise of 28 K per stage implies $164 \cdot 5/28 = 5 \cdot 9$ stages. It is likely, then, that the compressor will require six or seven stages and, in view of the influence of the work-done factor, seven is more likely. An attempt will therefore be made to design a seven-stage compressor.

With seven stages and an overall temperature rise of 164·5 K the average temperature rise is 23·5 K per stage. It is normal to design for a somewhat lower temperature rise in the first and last stages, for reasons which will be discussed at the end of this section. A good starting point would be to assume $\Delta T_0 \approx 20$ K for the first and last stages, leaving a requirement for $\Delta T_0 \approx 25$ K in the remaining stages.

## Stage-by-stage design

Having determined the rotational speed and annulus dimensions, and estimated the number of stages required, the next step is to evaluate the air angles for each stage *at the mean radius*. It will then be possible to check that the estimated number of stages is likely to result in an acceptable design.

From the velocity diagram, Fig. 5.6, it is seen that $C_{w_1} = C_a \tan \alpha_1$ and $C_{w_2} = C_{w_1} + \Delta C_w$. For the first stage $\alpha_1 = 0$, because these are no inlet guide vanes. The stator outlet angle for each stage, $\alpha_3$, will be the inlet angle $\alpha_1$ for the following rotor. Calculations of stage temperature rise are based on rotor considerations only, but care must be taken to ensure that the diffusion in the stator is kept to a reasonable level. The work-done factors will vary through the compressor and reasonable values for the seven stages would be 0·98 for the first stage, 0·93 for the second, 0·88 for the third and 0·83 for the remaining four stages.

## Stages 1 and 2

Recalling the equation for the stage temperature rise in terms of change in whirl velocity $\Delta C_w = C_{w_2} - C_{w_1}$, we have

$$\Delta C_w = \frac{c_p \Delta T_0}{\lambda U} = \frac{1 \cdot 005 \times 10^3 \times 20}{0 \cdot 98 \times 266 \cdot 6} = 76 \cdot 9 \text{ m/s}$$

Since $C_{w_1} = 0$, $C_{w_2} = 76.9$ m/s and hence

$$\tan \beta_1 = \frac{U}{C_a} = \frac{266 \cdot 6}{150} = 1 \cdot 7773, \qquad\qquad \beta_1 = 60 \cdot 64°$$

$$\tan \beta_2 = \frac{U - C_{w_2}}{C_a} = \frac{266 \cdot 6 - 76 \cdot 9}{150} = 1 \cdot 264, \quad \beta_2 = 51 \cdot 67°$$

$$\tan \alpha_2 = \frac{C_{w_2}}{C_a} = \frac{76 \cdot 9}{150} = 0 \cdot 513, \qquad\qquad \alpha_2 = 27 \cdot 14°$$

The velocity diagram for the first stage therefore appears as in Fig. 5.14(a).

The deflection in the rotor blades is $\beta_1 - \beta_2 = 8 \cdot 98°$, which is modest. The diffusion can readily be checked using the de Haller number as follows:

$$\frac{V_2}{V_1} = \frac{C_a/\cos \beta_2}{C_a/\cos \beta_1} = \frac{\cos \beta_1}{\cos \beta_2} = \frac{0 \cdot 490}{0 \cdot 260} = 0 \cdot 790$$

This value of de Haller number indicates a relatively light aerodynamic loading, i.e. a low rate of diffusion. It is not necessary to calculate the diffusion factor at this stage, because the de Haller number gives an adequate preliminary check. After the pitch chord ratio $(s/c)$ is determined from cascade data, as explained in a later section, the diffusion factor can be calculated readily from the known velocities.

At this point it is convenient to calculate the pressure ratio of the stage $(p_{03}/p_{01})_1$, the suffix outside the parenthesis denoting the number of the stage, and then the pressure and temperature at exit which will also be the values at inlet to the second stage. The isentropic efficiency of the stage is approximately equal to the polytropic efficiency of the compressor, which has been assumed to be 0·90, so we have

$$\left(\frac{p_{03}}{p_{01}}\right)_1 = \left(1 + \frac{0 \cdot 90 \times 20}{288}\right)^{3 \cdot 5} = 1 \cdot 236$$

$$(p_{03})_1 = 1 \cdot 01 \times 1 \cdot 236 = 1 \cdot 249 \text{ bar}$$

$$(T_{03})_1 = 288 + 20 = 308 \text{ K}$$

We have finally to choose a value for the air angle at outlet from the stator row, $\alpha_3$, which will also be the direction of flow, $\alpha_1$, into the second stage. Here it is useful to consider the degree of reaction. For this first stage, with the prescribed axial inlet velocity, $C_3$ will not equal $C_1$ (unless $\alpha_3$ is made zero) whereas our equations for $\Lambda$ were derived on the assumption of this equality of inlet and outlet velocities. Nevertheless, $C_3$ will not differ markedly from $C_1$, and we can arrive at an *approximate* value of $\Lambda$ by using equation (5.18).

$$\Lambda \approx 1 - \frac{C_{w_2} + C_{w_1}}{2U} = 1 - \frac{76 \cdot 9}{2 \times 266 \cdot 6} = 0 \cdot 856$$

The degree of reaction is high, but we have seen from Fig. 5.13 that this is necessary with low hub–tip ratios to avoid a negative value at the root radius. We shall hope to be able to use 50 per cent reaction stages from the third or fourth stage onwards, and an appropriate value of $\Lambda$ for the second stage may be about 0·70.

For the second stage $\Delta T_{os} = 25$ K and $\lambda = 0 \cdot 93$, and we can determine $\beta_1$ and $\beta_2$ using equations (5.9) and (5.11). From (5.9)

$$25 = \frac{0 \cdot 93 \times 266 \cdot 2 \times 150}{1 \cdot 005 \times 10^3} (\tan \beta_1 - \tan \beta_2)$$

$$(\tan \beta_1 - \tan \beta_2) = 0 \cdot 6756$$

And from (5.11)

$$0 \cdot 70 \approx \frac{150}{2 \times 266 \cdot 6} (\tan \beta_1 + \tan \beta_2)$$

$$(\tan \beta_1 + \tan \beta_2) \approx 2 \cdot 4883$$

Solving these simultaneous equations we get

$$\beta_1 = 57 \cdot 70^\circ \qquad \text{and} \qquad \beta_2 = 42 \cdot 19^\circ$$

Finally, using (5.2) and (5.3)

$$\alpha_1 = 11 \cdot 06^\circ \qquad \text{and} \qquad \alpha_2 = 41.05^\circ$$

The whirl velocities at inlet and outlet are readily found from the velocity diagram,

$$C_{w_1} = C_a \tan \alpha_1 = 150 \tan 11 \cdot 06 = 29 \cdot 3 \text{ m/s}$$

$$C_{w_2} = C_a \tan \alpha_2 = 150 \tan 41 \cdot 05 = 130 \cdot 6 \text{ m/s}$$

The required change in whirl velocity is 101·3 m/s, compared with 76·9 m/s for the first stage; this is due to the higher stage temperature rise and the lower work-done factor. The fluid deflection in the rotor blades has increased to 15·51 degrees. It appears that $\alpha_3$ for the first stage should be 11·06 degrees. This design gives a de Haller number for the second-stage rotor blades of $\cos 57 \cdot 70 / \cos 42 \cdot 19 = 0 \cdot 721$, which is satisfactory. With the stator outlet angle for the first-stage stator now known, the de Haller number for the first-stage stator would be

$$\frac{C_3}{C_2} = \frac{\cos \alpha_2}{\cos \alpha_3} = \frac{\cos 27 \cdot 15}{\cos 11 \cdot 06} = 0 \cdot 907$$

implying a small amount of diffusion. This is a consequence of the high degree of reaction in the first stage.

The velocity diagram for the second stage appears as in Fig. 5.14(b) and the outlet pressure and temperature become

$$\left(\frac{p_{03}}{p_{01}}\right)_2 = \left(1 + \frac{0\cdot90 \times 25}{308}\right)^{3\cdot3} = 1\cdot280$$

$$(p_{03})_2 = 1\cdot249 \times 1\cdot280 = 1\cdot599 \text{ bar}$$

$$(T_{03})_2 = 308 + 25 = 333 \text{ K}$$

At this point we do not know $\alpha_3$ for the second stage, but it will be determined from the fact that it is equal to $\alpha_1$ for the third stage, which we will now proceed to consider.

Before doing so, it is useful to point out that the degree of reaction is directly related to the shape of the velocity diagram. It was previously shown that for 50 per cent reaction the velocity diagram is symmetrical. Writing $C_{w_m} = (C_{w_1} + C_{w_2})/2$, equation (5.18) can be rewritten in the form $\Lambda = 1 - (C_{w_m}/U)$. Referring to Fig. 5.14(a) and (b) it can be seen that when $C_{w_m}/U$ is small, and the corresponding reaction is high, the velocity diagram is highly skewed; the

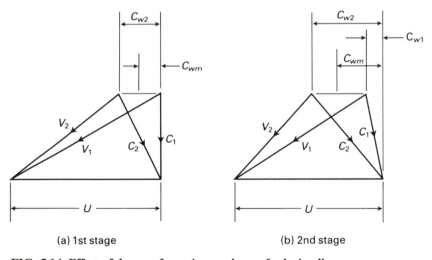

(a) 1st stage                                    (b) 2nd stage

**FIG. 5.14  Effect of degree of reaction on shape of velocity diagram**

high degree of reaction in the first stage is a direct consequence of the decision to dispense with inlet guide vanes and use a purely axial inlet velocity. The degree of reaction is reduced in the second stage, and we would eventually like to achieve 50 per cent reaction in the later stages where the hub–tip ratios are higher.

## Stage 3

Using a stage temperature rise of 25 K and a work-done factor of 0·88, an attempt will be made to use a 50 per cent reaction design for the third stage.
Proceeding as before

$$\tan \beta_1 - \tan \beta_2 = \frac{\Delta T_{0s} c_p}{\lambda U C_a} = \frac{25 \times 1 \cdot 005 \times 10^3}{0 \cdot 88 \times 266 \cdot 6 \times 150} = 0 \cdot 7140$$

$$\tan \beta_1 + \tan \beta_2 = \Lambda \frac{2U}{C_a} = \frac{0 \cdot 5 \times 2 \times 266 \cdot 6}{150} = 1 \cdot 7773$$

yielding $\beta_1 = 51.24°$ and $\beta_2 = 28 \cdot 00°$. The corresponding value of de Haller number is given by $\cos 51 \cdot 24 / \cos 28 \cdot 00 = 0 \cdot 709$. This is rather low, but could be deemed satisfactory for a preliminary design. It is instructive, however, to investigate the possibilities available to the designer for reducing the diffusion. One possibility is to consider changing the degree of reaction, but it is found that the de Haller number is not strongly influenced by the degree of reaction chosen; as $\Lambda$ had a value of $\approx 0 \cdot 70$ for the second stage it might appear that a suitable value for the third stage might be between 0·70 and 0·50. Repeating the above calculations for a range of $\Lambda$, however, shows that $\Lambda = 0 \cdot 55$ results in a further *decrease* of de Haller number to 0·706; referring again to Fig. 5.14 it can be observed that for a specified axial velocity, the required diffusion increases with reaction. A de Haller number of 0·725 can be achieved for $\Lambda = 0 \cdot 40$, but it is undesirable to use such a low degree of reaction. A more useful approach might be to accept a slightly lower temperature rise in the stage, and reducing $\Delta T_{0s}$ from 25 K to 24 K while keeping $\Lambda = 0 \cdot 50$ gives

$$\tan \beta_1 - \tan \beta_2 = 0 \cdot 6854$$

yielding $\beta_1 = 50 \cdot 92°$, $\beta_2 = 28 \cdot 63°$ and a de Haller number of 0·718, which is satisfactory for this preliminary design. Other methods of reducing the aerodynamic loading include increases in blade speed or axial velocity, which could readily be accommodated.

With a stage temperature rise of 24 K, the performance of the third stage is then given by

$$\left( \frac{p_{03}}{p_{01}} \right)_3 = \left( 1 + \frac{0 \cdot 90 \times 24}{333} \right)^{3 \cdot 5} = 1 \cdot 246$$

$$(p_{03})_3 = 1 \cdot 599 \times 1 \cdot 246 = 1 \cdot 992 \text{ bar}$$

$$(T_{03})_3 = 333 + 24 = 357 \text{ K}$$

From the symmetry of the velocity diagram $\alpha_1 = \beta_2 = 28 \cdot 63°$ and $\alpha_2 = \beta_1 = 50 \cdot 92°$. The whirl velocities are given by

$$C_{w_1} = 150 \tan 28 \cdot 63 = 81 \cdot 9 \text{ m/s}$$

$$C_{w_2} = 150 \tan 50 \cdot 92 = 184 \cdot 7 \text{ m/s}$$

## Stages 4, 5 and 6

A work-done factor of 0·83 is appropriate for all stages from the fourth onwards, and 50 per cent reaction can be used. The design can be simplified by using the same mean diameter velocity diagrams for stages four to six, although each blade will have a different length due to the continuous increase in density. The seventh, and final stage can then be designed to give the required overall pressure ratio. It is not necessary to repeat all the calculations for stages 4–6, but it should be noted that the reduction in work-done factor to 0·83, combined with the desired stage temperature rise of 25 K, results in an unacceptably low de Haller number of 0·695. Reducing the stage temperature rise to 24 K increases the de Haller number to 0·705, which will be considered to be just acceptable for the preliminary design.

Proceeding as before,

$$\tan \beta_1 - \tan \beta_2 = \frac{24 \times 1·005 \times 10^3}{0·83 \times 266·6 \times 150} = 0·7267$$

$$\tan \beta_1 + \tan \beta_2 = 0·5 \times 2 \times \frac{266·6}{150} = 1·7773$$

yielding $\beta_1 = 51·38°$ $(= \alpha_2)$ and $\beta_2 = 27·71°$ $(= \alpha_1)$. The performance of the three stages can be summarized below:

| Stage | 4 | 5 | 6 |
|---|---|---|---|
| $p_{01}$ (bar) | 1·992 | 2·447 | 2·968 |
| $T_{01}$ (K) | 357 | 381 | 405 |
| $(p_{03}/p_{01})$ | 1·228 | 1·213 | 1·199 |
| $p_{03}$ (bar) | 2·447 | 2·968 | 3·560 |
| $T_{03}$ (K) | 381 | 405 | 429 |
| $p_{03} - p_{01}$ (bar) | 0·455 | 0·521 | 0·592 |

It should be noted that although each stage is designed for the same temperature rise, the pressure ratio decreases with stage number; this is a direct consequence of the increasing inlet temperature as flow progresses through the compressor. The pressure *rise*, however, increases steadily.

## Stage 7

At entry to the final stage the pressure and temperature are 3·560 bar and 429 K. The required compressor delivery pressure is $4·15 \times 1·01 = 4·192$ bar.

The pressure ratio of the seventh stage is thus given by

$$\left(\frac{p_{03}}{p_{01}}\right)_7 = \frac{4 \cdot 192}{3 \cdot 560} = 1 \cdot 177$$

The temperature rise required to give this pressure ratio can be determined from

$$\left(1 + \frac{0 \cdot 90 \Delta T_{0s}}{429}\right)^{3 \cdot 5} = 1 \cdot 177$$

giving $\Delta T_{0s} = 22 \cdot 8$ K.

The corresponding air angles, assuming 50 per cent reaction, are then $\beta_1 = 50 \cdot 98°\ (= \alpha_2)$, $\beta_2 = 28 \cdot 52°\ (= \alpha_1)$ with a satisfactory de Haller number of $0 \cdot 717$.

With a 50 per cent reaction design used for the final stage, the fluid will leave the last stator with an angle $\alpha_3 = \alpha_1 = 28 \cdot 52°$, whereas ideally the flow should be axial at entry to the combustion chamber. The flow can be straightened by incorporating vanes after the final compressor stage and these can form part of the necessary diffuser at entry to the combustion chamber.

All the preliminary calculations have been carried out on the basis of a constant *mean* diameter. Another problem now arises: a sketch, approximately to scale, of the compressor and turbine annuli (Fig. 5.15) shows that the combustor will be an awkward shape, the required changes in

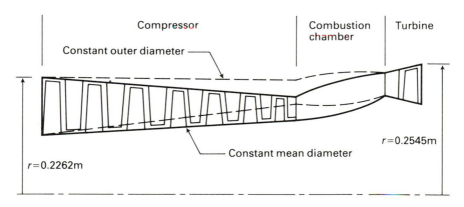

**FIG. 5.15 Annulus shape**

flow direction causing additional pressure losses. A more satisfactory solution might be to design the compressor for a constant *outer* diameter; both solutions are shown in the figure. The use of a constant outer diameter results in the mean blade speed increasing with stage number, and this in turn implies that for a given temperature rise $\Delta C_w$ is reduced. The fluid deflection is correspondingly reduced with a beneficial increase in de Haller number. Alternatively, because of the higher blade speed a higher temperature rise could be achieved in the later stages; this might permit the required pressure ratio to be obtained in six stages rather than seven. The reader should be aware that the simple equations derived on the basis of $U = $ constant are then not valid, and it would be necessary to use the appropriate values of $U_1$ and $U_2$; the stage temperature rise would then be given by $\lambda(U_2 C_{w_2} - U_1 C_{w_1})/c_p$.

Compressors which use constant inner diameter, constant mean diameter or constant outer diameter will all be found in service. The use of a constant *inner* diameter is often found in industrial units, permitting the use of rotor discs of the same diameter which lowers the cost. It would be important to minimize the number of turbine stages, again for reasons of cost, and with a subsonic compressor it is very probable that the turbine diameter would be noticeably larger than the compressor diameter. The difference in turbine and compressor diameter is not critical, however, because frontal area is unimportant and with reverse flow combustion chambers large differences in diameter can be easily accommodated. Constant *outer* diameter compressors are used where the minimum number of stages is required, and these are commonly found in aircraft engines.

The compressor annulus of the Olympus 593 engine used in Concorde employs a combination of these approaches; the LP compressor annulus has basically a constant inner diameter, while the HP compressor has a constant outer diameter. The accessories are packed around the HP compressor annulus and the engine when fully equipped is almost cylindrical in shape, with the compressor inlet and turbine exit diameters almost equal. In this application, frontal area is of critical importance because of the high supersonic speed.

## Variation of air angles from root to tip

In section 5.6 various distributions of whirl velocity with radius were considered, and it was shown that the designer had quite a wide choice. In the case of the first stage, however, the choice is restricted because of the absence of IGVs; this means that there is no whirl component at entry to the compressor and the inlet velocity will be constant across the annulus. For all other stages the whirl velocity at entry to the rotor blades will be determined by the axial velocity and the stator outlet angle from the previous stage, giving more freedom in the aerodynamic design of the stage.

The material developed in section 5.6 will be applied to the first stage, which is a special case because of the axial inlet velocity, and the third stage, which is typical of the later stages. The first stage will be investigated using a free vortex design, noting that the condition $C_w r = $ constant is satisfied for $C_w = 0$. Attention will then be turned to the design of the third stage, recalling that the mean radius design was based on $\Lambda_m = 0.50$. The third stage will be investigated for three different design approaches, viz. (i) free vortex, $\Lambda_m = 0.50$, (ii) constant reaction, $\Lambda_m = 0.50$ with radial equilibrium ignored and (iii) exponential blading, $\Lambda_m = 0.50$.

Considering the first stage, the rotor blade angle at inlet ($\beta_1$) is obtained directly from the axial velocity (150 m/s) and the blade speed. The blade speeds at root, mean and tip, corresponding to radii of 0·1131, 0·1697 and 0·2262 m, are 177·7, 266·6 and 355·3 m/s respectively. Thus

$$\tan \beta_{1_r} = \frac{177\cdot7}{150}, \qquad \beta_{1_r} = 49\cdot83°$$

$$\tan \beta_{1_m} = \frac{266\cdot6}{150}, \qquad \beta_{1_m} = 60\cdot64°$$

$$\tan \beta_{1_t} = \frac{355\cdot3}{150}, \qquad \beta_{1_t} = 67\cdot11°$$

Air angles at any radius can be calculated as above. For our purposes the calculations will be restricted to root, mean and tip radii, although for detailed design of the blading it would be necessary to calculate the angles at intermediate radii also.

To calculate the air angles $\beta_2$ and $\alpha_2$ it is necessary to determine the radial variation of $C_{w_2}$. For the free vortex condition $C_{w_2} r = \text{constant}$, and the value of $C_{w_{2m}}$ was previously determined to be 76·9 m/s. Because of the reduction of annulus area through the compressor, the blade height at exit from the rotor will be slightly less than at inlet, and it is necessary to calculate the tip and root radii at exit from the rotor blades to find the relevant variation of $C_{w_2}$. The stagnation pressure and temperature at exit from the first stage were found to be 1·249 bar and 308 K. Recalling that a stator exit angle of 11·06 degrees was established.,

$$C_3 = \frac{150}{\cos 11\cdot06} = 152\cdot8 \text{ m/s}$$

$$T_3 = 308 - (152\cdot8)^2/(2 \times 1\cdot005 \times 10^3) = 296\cdot4 \text{ K}$$

$$p_3 = 1\cdot249 \left(\frac{296\cdot4}{308}\right)^{3\cdot5} = 1\cdot092 \text{ bar}$$

$$\rho_3 = \frac{100 \times 1\cdot092}{296\cdot4 \times 0\cdot287} = 1\cdot283 \text{ kg/m}^3$$

$$A_3 = \frac{m}{\rho_3 C_{a_3}} = \frac{20}{1\cdot283 \times 150} = 0\cdot1039 \text{ m}^2$$

$$h = \frac{0\cdot1039}{2\pi \times 0\cdot1697} = 0\cdot0974 \text{ m}$$

$$r_t = 0\cdot1697 + \frac{0\cdot0974}{2} = 0\cdot2184 \text{ m}$$

$$r_r = 0\cdot1697 - \frac{0\cdot0974}{2} = 0\cdot1210 \text{ m}$$

These radii refer to conditions at the *stator* exit. With negligible error it can be assumed that the radii at exit from the rotor blades are the mean of those at rotor inlet and stator exit.† Thus at exit from the rotor,

$$r_t = \frac{0.2262 + 0.2184}{2} = 0.2223 \text{ m}, \qquad U_t = 349.2 \text{ m/s}$$

$$r_r = \frac{0.1131 + 0.1210}{2} = 0.1171 \text{ m}, \qquad U_r = 183.9 \text{ m/s}$$

From the free vortex condition,

$$C_{w2r} = 76.9 \times \frac{0.1697}{0.1171} = 111.4 \text{ m/s}$$

$$C_{w2t} = 76.9 \times \frac{0.1697}{0.2223} = 58.7 \text{ m/s}$$

The stator inlet angle is given by $\tan \alpha_2 = C_{w2}/C_a$, and the rotor exit angle by $\tan \beta_2 = (U - C_{w2})/C_a$. Hence,

$$\tan \alpha_{2r} = \frac{111.4}{150}, \qquad\qquad \alpha_{2r} = 36.60°$$

$$\tan \alpha_{2m} = \frac{76.9}{150}, \qquad\qquad \alpha_{2m} = 27.14°$$

$$\tan \alpha_{2t} = \frac{58.7}{150}, \qquad\qquad \alpha_{2t} = 21.37°$$

$$\tan \beta_{2r} = \frac{183.9 - 111.4}{150}, \qquad \beta_{2r} = 25.80°$$

$$\tan \beta_{2m} = \frac{266.6 - 76.9}{150}, \qquad \beta_{2m} = 51.67°$$

$$\tan \beta_{2t} = \frac{349.2 - 58.7}{150}, \qquad \beta_{2t} = 62.69°$$

The radial variation of air angles is shown in Fig. 5.16, which shows both the increased deflection $(\beta_1 - \beta_2)$ at the root and the requirement for considerable blade twist along the height of the blade to ensure that the blade angles are in agreement with the air angles.

The degree of reaction at the root can be approximated (approximate because $C_3 \neq C_1$) by $\Lambda_r = 1 - (C_{w2}/2U)$ giving

$$\Lambda_r \approx 1 - \frac{111.4}{2 \times 183.9} \approx 0.697$$

† For turbines, where there is considerably greater flare in the annulus, it is necessary to make an allowance for the spacing between the rotor and stator; this is done in the turbine design example presented in Chapter 7.

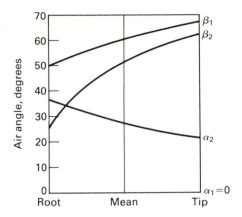

FIG. 5.16 **Radial variation of air angles, 1st stage**

As expected, with the high value of $\Lambda_m$ of 0·856 there is no problem with too low a degree of reaction at the root.

When calculating the annulus area at exit from the stage it was assumed that the density at the mean radius could be used in the continuity equation. Although the *stagnation* pressure and temperature are assumed to be constant across the height of the blade, the *static* pressure and temperature, and hence the density, will vary. This effect is small in compressor stages, with their low pressure ratio, but is more pronounced in turbine stages. The method of dealing with radial variation of density will be described in the turbine example of Chapter 7. The density variation from root to tip is about 4 per cent for the first stage of the compressor and would be even less for later stages of higher hub–tip ratio.

The velocity diagrams for the first stage are shown (to scale) in Fig. 5.17. The increase in fluid deflection and diffusion at the root section are readily visible from the vectors $V_1$ and $V_2$. Since $V_1$ is a maximum at the tip and $C_2$ a maximum at the root, it is apparent that the maximum relative Mach numbers occur at $r_t$ for the rotor blade and at $r_r$ for the stator blade. Compressibility effects will be discussed in the next sub-section.

Turning our attention to the third stage, it is useful to summarize conditions at inlet and outlet before examining different distributions of whirl velocity.

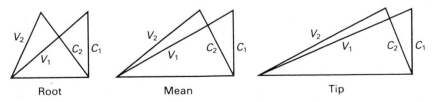

FIG. 5.17 **Velocity diagrams, 1st stage**

From the mean radius design, with $\Lambda_m = 0.50$

$$p_{0_1} = 1.599 \text{ bar}, \qquad p_{0_3}/p_{0_1} = 1.246, \qquad p_{0_3} = 1.992 \text{ bar}$$

$$T_{0_1} = 333 \text{ K}, \qquad T_{0_3}\,(= T_{0_2}) = 357 \text{ K}$$

$$\alpha_1 = 26.63^\circ \ (= \beta_2), \qquad \beta_1 = 50.92^\circ \ (= \alpha_2)$$

$$C_{w_1} = 81.9 \text{ m/s}, \qquad C_{w_2} = 184.7 \text{ m/s}, \qquad \Delta C_w = 102.8 \text{ m/s}$$

It is not necessary to repeat the calculations for rotor radii and blade speed at root and tip for this stage, and it will be sufficient to summarize the results as follows:

|        | $r_r$ (m) | $r_t$ (m) | $r_r/r_t$ | $U_r$ (m/s) | $U_t$ (m/s) |
|--------|-----------|-----------|-----------|-------------|-------------|
| Inlet  | 0.1279    | 0.2115    | 0.605     | 200.9       | 332.2       |
| Exit   | 0.1341    | 0.2053    | 0.653     | 205.8       | 322.5       |

The resulting air angles for a *free vortex design* are obtained from the calculations tabulated below, and the values already obtained at the mean radius are included for completeness.

|                                              | Root                          | Mean   | Tip                           |
|----------------------------------------------|-------------------------------|--------|-------------------------------|
| $C_{w_1} = C_{w_1} r_m/r$ (m/s)              | 108.7                         | 81.9   | 65.7                          |
| $\tan \alpha_1 = C_{w_1}/C_a$                | 108.7/150                     |        | 65.7/150                      |
| $\alpha_1$                                   | 35.93                         | 28.63  | 23.65                         |
| $\tan \beta_1 = (U - C_{w_1})/C_a$           | $\dfrac{(200.9 - 108.7)}{150}$ |        | $\dfrac{(332.2 - 65.7)}{150}$  |
| $\beta_1$                                    | 31.58                         | 50.92  | 60.63                         |
| $C_{w_2} = C_{w_2} r_m/r$ (m/s)              | 233.7                         | 184.7  | 152.7                         |
| $\tan \alpha_2 = C_{w_2}/C_a$                | 233.7/150                     |        | 152.7/150                     |
| $\alpha_2$                                   | 57.31                         | 50.92  | 45.51                         |
| $\tan \beta_2 = (U - C_{w_2})/C_a$           | $\dfrac{(205.8 - 233.7)}{150}$ |        | $\dfrac{(322.5 - 152.7)}{150}$ |
| $\beta_2$                                    | −10.54                        | 28.63  | 48.54                         |
| $\beta_1 - \beta_2$                          | 42.12                         | 22.29  | 12.09                         |
| $\Lambda = 1 - \dfrac{1}{R^2}(1 - \Lambda_m)$ | 0.161                        | 0.50   | 0.668                         |

The radial variation of air angles shown in Fig. 5.18 clearly indicates the very large fluid deflection $(\beta_1 - \beta_2)$ required at the root and the substantial blade twist from root to tip.

It is instructive to consider a *constant reaction design*, with radial equilibrium ignored. With $\Lambda_m = 0.50$, conditions at the mean radius will be the

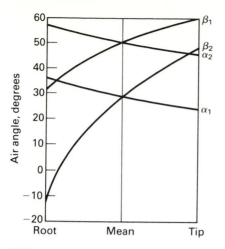

**FIG. 5.18 Radial variation of air angles: free vortex, 3rd stage**

same as those previously considered for the free vortex design. As before, it will be assumed that 50 per cent reaction gives a symmetrical velocity diagram and this will hold across the annulus. From symmetry, $U = \Delta C_w + 2C_{w_1}$, and from constant work input at all radii $U\Delta C_w = U_m \Delta C_{wm} = $ constant.

Considering the root section at inlet to the blade,

$$\Delta C_{w_r} = \Delta C_{wm}\frac{r_m}{r_r} = 102\cdot8 \times \frac{0\cdot1697}{0\cdot1279} = 136\cdot4 \text{ m/s}$$

$$C_{w_{1r}} = \frac{U_r - \Delta C_{w_r}}{2} = \frac{205\cdot8 - 136\cdot40}{2} = 34\cdot7 \text{ m/s}$$

$$\tan\alpha_{1_r} = \frac{C_{w_{1r}}}{C_a} = \frac{34\cdot70}{150}, \qquad \alpha_{1_r} = 13\cdot03^\circ = \beta_{2_r}$$

$$C_{w_{2r}} = C_{w_{1r}} + \Delta C_{w_r} = 34\cdot70 + 136\cdot4 = 171\cdot1 \text{ m/s}$$

$$\tan\alpha_{2_r} = \frac{C_{w_{2r}}}{C_a} = \frac{171\cdot1}{150}, \qquad \alpha_{2_r} = 48\cdot76^\circ = \beta_{1_r}$$

Repeating these calculations at the tip ($r_t = 0\cdot2115m$, $U = 332\cdot2$ m/s) yields $\alpha_{1_t} = 39\cdot77^\circ = \beta_{2_t}$ and $\alpha_{2_t} = 54\cdot12^\circ = \beta_{1_t}$. The radial variations of the air angles are shown in Fig. 5.19, which shows the greatly reduced twist compared with the free vortex variation shown in Fig. 5.18. It will be recalled, however, that the radial equilibrium condition is not satisfied by the constant reaction design, and with the relatively low hub–tip ratio for this stage the constant reaction design is not likely to be the most efficient. It should be pointed out, also, that a small error is introduced by assuming a symmetrical velocity diagram based on the blade speeds at rotor inlet, because at hub and tip radii the blade speed is not constant throughout the stage due to the taper of the

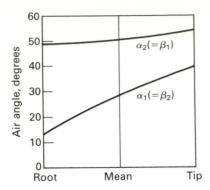

**FIG. 5.19 Radial variation of air angles: constant 50 per cent reaction, 3rd stage**

annulus. The reader should recognize that reaction is a useful concept in design but it does not matter if the actual value of reaction is not quite equal to the theoretical value.

To conclude this subsection, a third type of design will be examined, namely exponential blading with $\Lambda_m = 0.5$. The value of $R$ at the root section is given by $R = 0.1279/0.1697 = 0.754$. The reaction at the root is then given by

$$\Lambda_r = 1 + \left(1 - \frac{2}{0.754}\right)(1 - 0.50)$$

$$\Lambda_r = 0.174$$

This is slightly higher than the value obtained for the free vortex design (0.164), as would be expected from Fig. 5.12. With exponential blading $n = 0$, $C_{w_1} = a - (b/R)$ and $C_{w_2} = a + (b/R)$. Evaluating the constants $a$ and $b$,

$$a = U_m(1 - \Lambda_m) = 266.6(1 - 0.50) = 133.3 \text{ m/s}$$

$$b = \frac{c_p \Delta T}{2U_m \lambda} = \frac{1.005 \times 1000 \times 24}{2 \times 266.6 \times 0.88} = 51.4 \text{ m/s}$$

The following table gives the resulting whirl velocity distribution,

|  | Root | Mean | Tip |
|---|---|---|---|
| Inlet $R$ | 0.754 | 1.00 | 1.246 |
| $C_{w_1} = a - \dfrac{b}{R}$ (m/s) | 65.1 | 81.9 | 92.0 |
| Exit $R$ | 0.790 | 1.00 | 1.210 |
| $C_{w_2} = a + \dfrac{b}{R}$ (m/s) | 198.4 | 184.7 | 175.8 |

Setting $C_{a_{1m}} = 150$ m/s as before, from (5.26), at inlet

$$(C_{a_1})^2 - (C_{a_{1m}})^2 = -2\left[a^2 \ln R + \frac{ab}{R} - ab\right]$$

Hence

$$\left(\frac{C_{a_1}}{C_{a_{1m}}}\right)^2 = 1 - \frac{2}{C_{a_{1m}}^2}\left[a^2 \ln R + \frac{ab}{R} - ab\right]$$

Thus, at the root

$$\left(\frac{C_{a_1}}{C_{a_{1m}}}\right)^2 = 1\cdot247, \qquad C_{a_{1r}} = 167\cdot5 \text{ m/s}$$

at the tip $(C_{a_1}/C_{a_{1m}})^2 = 0\cdot773$, $C_{a_{1t}} = 131\cdot9$ m/s.
From equation (5.25), at exit from the rotor,

$$\left(\frac{C_{a_2}}{C_{a_{2m}}}\right)^2 = 1 - \frac{2}{C_{a_{2m}}^2}\left[a^2 \ln R - \frac{ab}{R} + ab\right]$$

giving

$$C_{a_{2r}} = 185\cdot8 \text{ m/s}$$

$$C_{a_{2t}} = 115\cdot5 \text{ m/s}$$

From the velocities calculated above, the rotor air angles can readily be calculated to give $\beta_{1r} = 39\cdot03°$, $\beta_{2r} = 2\cdot28°$, $\beta_{1t} = 61\cdot23°$ and $\beta_{2t} = 51\cdot79°$. The rotor air angles for the free vortex, constant reaction and exponential designs are compared in Fig. 5.20, both at inlet and exit from the rotor. It can be seen

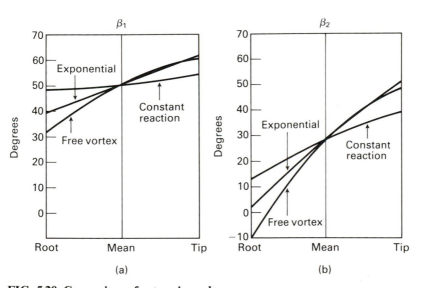

FIG. 5.20 **Comparison of rotor air angles**

that the free vortex exhibits the most marked twist over the blade span, with the constant reaction showing the least; the exponential design gives a compromise between the two. The fluid deflection, $\beta_1 - \beta_2$, for the three designs can be summarized as below.

|  | Root | Mean | Tip |
|---|---|---|---|
| Free vortex | 42·12 | 22·29 | 12·09 |
| Constant reaction | 35·73 | 22·29 | 14·35 |
| Exponential | 37·75 | 22·29 | 9·44 |

The aerodynamic loading at the root section of the free vortex is substantially higher than that for either of the other two designs.

It was pointed out earlier that the maximum relative inlet Mach numbers occur at the rotor blade tip and the stator blade root. Without detailing the calculations we may summarize the results as follows.

At rotor tip (i.e. at $r_t$):

|  | $C_{a_1}$ (m/s) | $C_{w_1}$ (m/s) | $V_1$ (m/s) | $T_1$ (K) | $M$ |
|---|---|---|---|---|---|
| Free vortex | 150·0 | 65·7 | 305·8 | 319·7 | 0·853 |
| Constant reaction | 150·0 | 82·5 | 291·3 | 314·0 | 0·820 |
| Exponential | 131·9 | 92·0 | 274·0 | 320·1 | 0·764 |

At stator tip (i.e. at $r_r$):

|  | $C_{a_2}$ (m/s) | $C_{w_2}$ (m/s) | $C_2$ (m/s) | $T_2$ (K) | $M$ |
|---|---|---|---|---|---|
| Free vortex | 150·0 | 233·7 | 277·7 | 318·6 | 0·776 |
| Constant reaction | 150·0 | 171·1 | 227·5 | 331·2 | 0·624 |
| Exponential | 185·8 | 198·4 | 271·8 | 320·2 | 0·758 |

Clearly the rotor tip value is the critical one, and it can be seen that the exponential design yields an appreciably lower Mach number. The importance of this will be discussed further in section 5.10. We have been considering the third stage here as an example, and the Mach numbers in the later stages will be lower because of the increased temperature and hence increased acoustic velocity.

The exponential design results in a substantial variation in axial velocity, both across the annulus and through the stage. We have based the annulus area on the premise of constant axial velocity, and this would have to be recalculated for the case of non-constant axial velocity. At any radius $r$, the rate of flow $\delta m$ through an element of width $\delta r$ is given by

$$\delta m = \rho 2\pi r \delta r C_a$$

and

$$m = 2\pi \int_{r_r}^{r_t} \rho r C_a \, \mathrm{d}r$$

The annulus area required to pass the known flow could be determined by numerical integration and would be slightly different to the value assumed for the previous cases of free vortex and constant reaction.

The three design methods considered all have advantages and the final choice of design would depend to a large extent on the design team's previous experience. The constant reaction design looks quite attractive, but it must be borne in mind that radial equilibrium has been ignored; ignoring radial equilibrium will result in flow velocities not being in agreement with the predicted air angles, leading to some loss in efficiency.

It is appropriate at this point to return to the question of the reduced loading in the first and last stages. The first stage is the most critical because of the high Mach number at the tip of the rotor; in addition, at certain flight conditions (e.g. yaw, high angle of attack or rapid turns of the aircraft) the inlet flow may become distorted with significant variations in axial velocity across the compressor annulus. By somewhat reducing the design temperature rise, and hence the aerodynamic loading, these problems can be partially alleviated. In the case of the last stage, the flow is delivered to the diffuser at entry to the combustion chamber; it would be desirable to have a purely axial velocity at exit, and certainly the swirl should be kept as low as possible. Once again, a slight reduction in the required temperature rise eases the aerodynamic design problem.

In this section we have been deciding on the air angles likely to lead to a satisfactory design of compressor. It is now necessary to discuss methods of obtaining the blade shapes which will lead to these air angles being achieved, and this will be done in the next section.

## 5.8   Blade design

Having determined the air angle distributions which will give the required stage work, it is now necessary to convert these into blade angle distributions from which the correct geometry of the blade forms may be determined.

The most obvious requirements of any particular blade row are: firstly, that it should turn the air through the required angle [$(\beta_1 - \beta_2)$ in the case of the rotor and $(\alpha_2 - \alpha_3)$ in the case of the stator]; and secondly that it should carry out its diffusing process with optimum efficiency, i.e. with a minimum loss of stagnation pressure. With regard to the first requirement, we shall see that due allowance must be made for the fact that the air will not leave a blade precisely in the direction indicated by the blade outlet angle. As far as the second requirement is concerned, certainly the air and blade angles at inlet to a blade row must be similar to minimize losses. But when choosing the blade angle, the designer must remember that the compressor has to operate over a wide range of speed and pressure ratio. The air angles have been calculated for the design speed and pressure ratio, and under different operating conditions both the fluid velocities and blade speed may change with resulting changes in the air angles. On the other hand, the blade angles, once chosen, are fixed. It follows that to obtain the best performance over a range of operating conditions it may not be the best policy to make the blade inlet angle equal to the design value of the relative air angle.

The number of variables involved in the geometry of a compressor blade row is so large that the design becomes to a certain extent dependent on the particular preference and previous experience of the designer. He will have at his disposal, however, correlated experimental results from wind tunnel tests on single blades or rows of blades. In the former case, the effects of adjacent blades in the row have to be accounted for by the application of empirical factors. The second type of data, from tests on rows or *cascades* of blades, is more widely used. Although it might appear desirable to test a cascade of blades in an annular form of wind tunnel, in an attempt to simulate conditions in an actual compressor, the blades are generally tested in the form of a straight cascade. An annular tunnel would not satisfactorily reproduce conditions in a real compressor and a considerable range of hub–tip ratios would have to be tested. By using a straight cascade, mechanical complication of the test rig is considerably reduced, and the two-dimensional flow conditions obtained in a tunnel of rectangular section greatly simplifies the interpretation of the test results.

In both Britain and the United States, much experimental research has been devoted to cascade tests on compressor blading. It is proposed to give a broad outline of this work and to show how the results obtained may be correlated in a fashion suitable for direct use by the compressor designer. Cascade tests result in two main items of information. These are (a) the angle through which the air is turned for a minimum loss, and (b) the corresponding profile drag coefficient from which the cascade efficiency may be estimated. When high velocities in the region of the sonic velocity are used, the tests also yield valuable information on compressibility effects. A typical cascade tunnel and the type of test result obtainable from it will now be described.

## Compressor blade cascade tunnel and typical test results

The tunnel consists essentially of an arrangement whereby a stream of air can be sucked or blown through a number of blades set out in the form of a

Cascade

Pitot tube

To manometer

Adjustable side walls

Slots for traversing instruments

Air supply

$\alpha_1$

$\alpha_1'$

Turn-table

**FIG. 5.21 Elevation and plan of simple cascade tunnel. (Traversing paths indicated by dotted lines)**

straight cascade (Fig. 5.21). Means are provided for traversing pressure and flow direction measuring instruments over two planes usually a distance of one blade chord upstream and downstream of the cascade. The height and length of the cascade are made as large as the available air supply will allow, in an attempt to eliminate interference effects due to the tunnel walls. Boundary layer suction on the walls is frequently applied to prevent contraction of the air stream as it passes through the tunnel.

The cascade is mounted on a turn-table so that its angular direction relative to the inflow duct can be set to any desired value. This device enables tests to be made on the cascade over a range of incidence angles of the air entering the cascade. In other more complex tunnels provision is also made for variation of the geometry of the blade row, such as the spacing between the blades and their setting angle, without removing the cascade from the tunnel. Pressure and velocity measurements are made by the usual form of L-shaped pitot and pitot-static tubes. Air directions are found by various types of instruments, the most common being the claw and cylindrical yawmeters illustrated in Fig. 5.22. The principle of operation is the same in both and consists of rotating the instrument about its axis until a balance of pressure from the two holes is

FIG. 5.22  Yawmeters: (a) cylindrical, (b) claw

obtained. The bisector of the angle between the two holes then gives the air direction.†

A cross-section of three blades forming a part of a typical cascade is shown in Fig. 5.23 which also includes details of the various angles, lengths and velocities associated with the cascade experiments. For any particular test, the

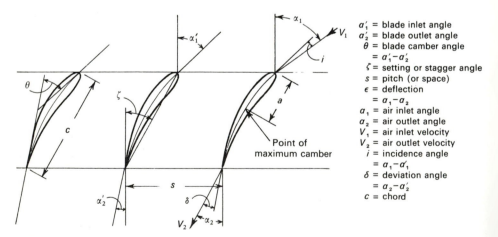

$a'_1$ = blade inlet angle
$a'_2$ = blade outlet angle
$\theta$ = blade camber angle
   $= a'_1 - a'_2$
$\zeta$ = setting or stagger angle
$s$ = pitch (or space)
$\epsilon$ = deflection
   $= a_1 - a_2$
$a_1$ = air inlet angle
$a_2$ = air outlet angle
$V_1$ = air inlet velocity
$V_2$ = air outlet velocity
$i$ = incidence angle
   $= a_1 - a'_1$
$\delta$ = deviation angle
   $= a_2 - a'_2$
$c$ = chord

FIG. 5.23  Cascade notation

† For a full description of cascade tunnel testing see Todd, K. W. Ref. (7) and Gostelow, J. P. (Ref. (8).

blade *camber* angle $\theta$, its *chord c* and the *pitch* (or *space*) *s* will be fixed and the blade inlet and outlet angles $\alpha_1'$ and $\alpha_2'$ determined by the chosen setting or *stagger* angle $\zeta$. The angle of *incidence*, $i$, is then fixed by the choice of a suitable air inlet angle $\alpha_1$ since $i = \alpha_1 - \alpha_1'$. This can be done by an appropriate setting of the turn-table on which the cascade is mounted. With the cascade in this position, the pressure and direction-measuring instruments are then traversed along the blade row in the upstream and downstream positions, and the test results plotted as in Fig. 5.24. This shows the variation of loss of

**FIG. 5.24 Variations of stagnation pressure loss and deflection for cascade at fixed incidence**

stagnation pressure, and *air deflection* $\varepsilon = \alpha_1 - \alpha_2$, covering two blades at the centre of the cascade. As the loss will be dependent on the magnitude of the velocity of the air entering the cascade, it is convenient to express it in a dimensionless form by dividing by the inlet dynamic head, i.e.

$$\text{loss} = \frac{p_{01} - p_{02}}{\frac{1}{2}\rho V_1^2} = \frac{w}{\frac{1}{2}\rho V_1^2} \tag{5.28}$$

This facilitates correlation of test results covering a range of values of $V_1$.

The curves of Fig. 5.24 can now be repeated for different values of incidence angle, and the whole set of results condensed to the form shown in Fig. 5.25 in which the mean loss $\bar{w}/\frac{1}{2}\rho V_1^2$ and mean deflection $\varepsilon$ are plotted against

FIG. 5.25 **Mean deflection and mean total-head pressure loss for cascade of fixed geometrical form**

incidence for a cascade of fixed geometrical form. These curves show that the mean loss remains fairly constant over a wide range of incidence, rising rapidly when the incidence has a large positive or negative value. At these extreme incidences the flow of air around the blades breaks down in a manner similar to the stalling of an isolated aerofoil. The mean deflection rises with increasing incidence, reaching a maximum value in the region of the positive stalling incidence.

Test results in the form given in Fig. 5.25 are obtained for a wide range of geometrical forms of the cascade, by varying the camber, pitch/chord ratio, etc. Reduction of the resulting data to a set of design curves is achieved by the following method.

From curves of the form given in Fig. 5.25, a single value of deflection which is most suitable for use with that particular form of cascade is selected. As the object of the cascade is to turn the air through as large an angle as possible with a minimum loss, the use of the maximum deflection is not possible because of the high loss due to stalling. Furthermore, remembering that in a compressor the blade will have to operate over a range of incidence under off-design conditions, the design incidence should be in the region of the flat portion of the loss curve. The practice has been to select a deflection which corresponds to a definite proportion of the stalling deflection. The proportion found to be most suitable is eight-tenths, so that the selected or *nominal deflection* $\varepsilon^* = 0.8\varepsilon_S$ where $\varepsilon_S$ is the stalling deflection. Difficulty is sometimes experienced in deciding on the exact position of the stall, so its position is standardized by assuming that the stall occurs when the loss has reached twice its minimum value.

Analysis of the values of the nominal deflection $\varepsilon^*$ determined from a large number of tests covering different forms of cascade, has shown that its value is

**FIG. 5.26 Design deflection curves**

mainly dependent on the pitch/chord ratio and air outlet angle $\alpha_2$. Its variation with change of other factors determining the geometrical form of the cascade, such as blade camber angle, is comparatively small. On this basis, the whole set of test results is reducible to the form shown in Fig. 5.26 in which the nominal deflection is plotted against air outlet angle with pitch/chord ratio as parameter. This set of master curves, as they may well be called, is of great value to the designer because, having fixed any two of the three variables involved, an appropriate value for the third may be determined. For instance, if the air inlet and outlet angles have been fixed by the air angle design, a suitable pitch/chord ratio can be read from the diagram.

As an example, consider the design of the third-stage rotor blade required for the free vortex design carried out in the previous section. At the mean radius of $0.1697$ m, where $\beta_1 = 50.92°$ and $\beta_2 = 28.63°$, $\varepsilon^* = \beta_1 - \beta_2 = 22.29°$ and from Fig. 5.26, with an air outlet angle of $28.63°$, $s/c = 0.9$.

Determination of the chord length will now depend on the pitch, which itself is clearly dependent on the number of blades in the row. When making a choice for this number, the *aspect ratio* of the blade, i.e. the ratio of length to chord, has to be considered because of its effect on secondary losses. This will be discussed in greater detail in the next section on stage performance. For the purpose of our example it will be assumed that an aspect ratio $h/c$ of about three will be suitable. The blade height can be obtained from the previously determined annulus dimensions as $0.0836$ m, so that the chord becomes

$$c = \frac{0.0836}{3} = 0.0279 \text{ m}$$

and the pitch is

$$s = 0.9 \times 0.0279 = 0.0251 \text{ m}$$

The number of blades $n$ is then given by

$$n = \frac{2\pi \times 0.1697}{0.0251} = 42.5$$

It is desirable to avoid numbers with common multiples for the blades in successive rows to reduce the likelihood of introducing resonant forcing frequencies. One method of doing this is to choose an even number for the stator blades and a prime number for the rotor blades. An appropriate number for the rotor blades in this stage would therefore be 43, and recalculation in the reverse order gives

$$s = 0.0248 \text{ m}, \qquad c = 0.0276 \text{ m}, \qquad \text{and} \qquad h/c = 3.03$$

A similar procedure could also be carried out for the stator blades.

The use of prime numbers for rotor blades is less common than used to be the case. This is a result of developments in mechanical design which have resulted in the ability to replace damaged rotor blades in the field without the need for re-balancing the rotor system. In the case of a fan for a high by-pass turbofan engine, for example, an even number of rotor blades is frequently used and airlines keep a stock of replacement blades in balanced pairs. In the event of a blade failure, the defective blade and the one diametrically opposite are replaced.

In the above example, the chord was determined simply from aerodynamic considerations. The chord chosen for the first stage, however, may well be determined by stringent requirements for resistance to Foreign Object Damage (FOD). This is often the case for aircraft engines, and it may result in a somewhat lower than optimum aspect ratio.

One further item of information is necessary before the design of the blade forms at this radius can be completed. Whereas the blade inlet angle $\alpha_1'$ will be known from the air inlet angle and chosen incidence (usually taken as zero so that $\alpha_1' = \alpha_1$), the blade outlet angle $\alpha_2'$ cannot be determined from the air outlet angle $\alpha_2$ until the deviation angle $\delta = \alpha_2 - \alpha_2'$ has been determined. Ideally, the mean direction of the air leaving the cascade would be that of the outlet angle of the blades, but in practice it is found that there is a deviation which is due to the reluctance of the air to turn through the full angle required by the shape of the blades. This will be seen from Fig. 5.23. An analysis of the relation between the air and blade outlet angles from cascade tests shows that their difference is mainly dependent on the blade camber and the pitch/chord ratio. It is also dependent on the shape of the camber-line of the blade section and on the air outlet angle itself. The whole of this may be summed up in the following empirical rule for deviation:

$$\delta = m\theta \sqrt{(s/c)} \tag{5.29}$$

where

$$m = 0.23 \left(\frac{2a}{c}\right)^2 + 0.1 \left(\frac{\alpha_2}{50}\right)$$

$a$ is the distance of the point of maximum camber from the leading edge of the blade, which feature is illustrated in Fig. 5.23, and $\alpha_2$ is in degrees. Frequently a circular arc camber-line is chosen so that $2a/c = 1$, thereby simplifying the formula for $m$, but its general form as given above embraces all shapes including a parabolic arc which is sometimes used. (For inlet guide vanes, which are essentially nozzle vanes giving accelerating flow, the power of $s/c$ in equation (5.29) is taken as unity instead of $0.5$ and $m$ is given a constant value of $0.19$.)

## Construction of blade shape

On the assumption of a *circular arc* camber-line, the deviation in the current example will be

$$\delta = \left[0.23 + 0.1 \times \frac{28.63}{50}\right] \sqrt{(0.9)}\theta = 0.273\theta$$

With this information it is now possible to fix the main geometrical parameters of the rotor blade row of our example. The procedure is as follows.

Since $\theta = \alpha_1' - \alpha_2'$ and $\alpha_2' = \alpha_2 - \delta$

$$\theta = \alpha_1' - \alpha_2 + \delta$$
$$= \alpha_1' - \alpha_2 + 0.273\theta$$
$$0.727\theta = \alpha_1' - \alpha_2$$
$$= 50.92 - 28.63$$

(assuming zero incidence, $\alpha_1' = \alpha_1$)

It follows that $\theta = 30.64°$ and $\alpha_2' = \alpha_1' - \theta = 20.28°$. The deviation angle is $8.63°$.

The position of the blade chord can be fixed relative to the axial direction by the stagger angle $\zeta$ given by

$$\zeta = \alpha_1' - \frac{\theta}{2}$$
$$= 50.92 - (30.64/2)$$
$$= 35.60°$$

Referring to Fig. 5.27, the chord line AB is drawn $0.0276$ m long at $35.60°$ to the axial direction. The lines AC and BD at angles $\alpha_1'$ and $\alpha_2'$ are then added

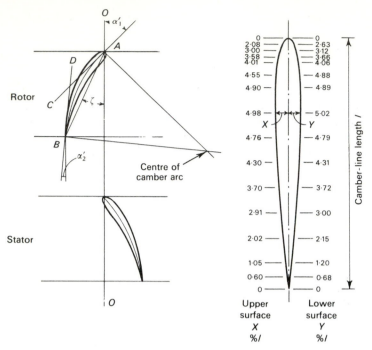

**FIG. 5.27 Blade design and base profile**

and a circular arc constructed tangential to these lines and having AB as chord. This arc will now be the camber-line of the blade around which an aerofoil section can be built up.

The manner of specifying the base profile is shown in Fig. 5.27, ordinates being given at definite positions along the camber-line. The RAF 27 profile and the so-called 'C series' of profiles have been widely used in British practice, and the NACA series in America. For moderately loaded stages, in which the velocities are well removed from sonic values, small variations in form are found to have little effect on the final performance of the compressor. Further details of base profiles, together with the geometrical relations necessary when parabolic camber-lines are employed, can be found in Ref. (9). For stages operating with relative Mach numbers in the transonic range, it has been found that double circular arc blade sections offer the best performance.

The method outlined could now be applied to a selected number of points along the blade length, bearing in mind that having once fixed the pitch at the mean diameter by the choice of a particular number of blades, the pitch at all other radii is determined. As the $s/c$ ratio is derived from the air angles, the length of the chord of the blade at any particular radius will be determined from the pitch. By this means a complete picture of the blade can be built up. The blade stresses can then be accurately assessed.

Now that the reader understands the basis for selecting the pitch/chord ratio, it is appropriate to check the diffusion factor for a couple of stages. Recalling for convenience equation (5.7),

$$D = 1 - \frac{V_2}{V_1} + \frac{\Delta C_w}{2V_1} \frac{s}{c}$$

the velocity terms are all available from the calculated velocity triangles. Considering the free vortex design of the third stage, at the tip section $C_a = 150$ m/s, $\beta_1 = 60 \cdot 63°$, $\beta_2 = 48 \cdot 54°$, $\beta_1 - \beta_2 = 12 \cdot 09°$ and $\Delta C_w = 87$ m/s. From Fig. 5.26 a suitable value of $s/c$ would be about $1 \cdot 1$ and the values of $V_2$ and $V_1$ can be readily calculated from $C_a/\cos \beta$ to be $226 \cdot 5$ and $305 \cdot 8$ m/s respectively. Hence

$$D = 1 - \frac{226 \cdot 5}{305 \cdot 8} + \frac{87 \times 1 \cdot 1}{2 \times 305 \cdot 8} = 0 \cdot 42$$

Referring back to Fig. 5.8, it can be seen that this value is quite satisfactory. Repeating the calculations at the root results in a slightly higher value of $0 \cdot 45$, primarily due to the increase in $\Delta C_w$. It is worth noting that the de Haller numbers for the third stage were close to the limit of $0 \cdot 72$, and the diffusion factors are also near the point at which losses increase rapidly. At the tip of the first-stage rotor, the diffusion factor can be calculated to be about $0 \cdot 23$, and this again correlates well with the light aerodynamic loading of this stage; the transonic relative Mach number, however, leads to further losses which will be discussed in section 5.10. The experienced compressor designer will have a good idea of the expected pitch/chord ratios at the outset of the design process, and the diffusion factor is an excellent preliminary check on aerodynamic loading.

## 5.9   Calculation of stage performance

After completion of the stage design, it will now be necessary to check over the performance, particularly in regard to the efficiency which for a given work input will completely govern the final pressure ratio. This efficiency is dependent on the total drag coefficient for each of the blade rows comprising the stage, and in order to evaluate these quantities it will be necessary to revert to the loss measurements in cascade tests. From the measured values of mean loss $\bar{w}$, two coefficients can be obtained. These are the lift and profile drag coefficients $C_L$ and $C_{Dp}$, formulae for which may be obtained as follows.

Referring to the diagram of forces acting on the cascade as shown in Fig. 5.28, the static pressure rise across the blades is given by

$$\Delta p = p_2 - p_1$$

$$= (p_{02} - \tfrac{1}{2}\rho V_2^2) - (p_{01} - \tfrac{1}{2}\rho V_1^2)$$

the incompressible flow formula being used because the change of density is

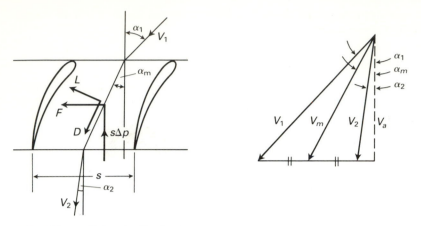

**FIG. 5.28 Applied and effective forces acting on cascade**

negligible. Hence, still using cascade notation for velocities and angles,

$$\Delta p = \tfrac{1}{2}(V_1^2 - V_2^2) - \bar{w}$$
$$= \tfrac{1}{2}\rho V_a^2(\tan^2 \alpha_1 - \tan^2 \alpha_2) - \bar{w} \qquad (5.30)$$

the axial velocity $V_a$ being assumed the same at inlet and outlet.

The axial force *per unit length* of each blade is $s\Delta p$ and, from consideration of momentum changes, the force acting *along* the cascade per unit length is given by

$$F = s\rho V_a \times \text{change in velocity component along cascade}$$
$$= s\rho V_a^2(\tan \alpha_1 - \tan \alpha_2) \qquad (5.31)$$

The coefficients $C_L$ and $C_{Dp}$ are based on a vector mean velocity $V_m$ defined by the velocity triangles in Fig. 5.28. Thus

$$V_m = V_a \sec \alpha_m$$

where $\alpha_m$ is given by

$$\tan \alpha_m = [\tfrac{1}{2}(V_a \tan \alpha_1 - V_a \tan \alpha_2) + V_a \tan \alpha_2]/V_a = \tfrac{1}{2}(\tan \alpha_1 + \tan \alpha_2)$$

If $D$ and $L$ are the drag and lift forces along and perpendicular to the direction of the vector mean velocity, then resolving along the vector mean gives

$$D = \tfrac{1}{2}\rho V_m^2 c C_{Dp} \quad \text{(by definition of } C_{Dp}\text{)}$$
$$= F \sin \alpha_m - s\Delta p \cos \alpha_m$$

Hence from equations (5.30) and (5.31),

$$\tfrac{1}{2}\rho V_m^2 c C_{Dp} = s\rho V_a^2(\tan \alpha_1 - \tan \alpha_2) \sin \alpha_m$$
$$- \tfrac{1}{2}\rho V_a^2 s(\tan^2 \alpha_1 - \tan^2 \alpha_2) \cos \alpha_m + \bar{w}s \cos \alpha_m$$

Since $\tan^2 \alpha_1 - \tan^2 \alpha_2 = (\tan \alpha_1 - \tan \alpha_2)(\tan \alpha_1 + \tan \alpha_2)$

$$= 2(\tan \alpha_1 - \tan \alpha_2)\tan \alpha_m$$

the first two terms in the expressions for $C_{Dp}$ are equal and the equation reduces to

$$C_{Dp} = \left(\frac{s}{c}\right)\left(\frac{\bar{w}}{\frac{1}{2}\rho}\right)\left(\frac{\cos \alpha_m}{V_m^2}\right)$$

$$= \left(\frac{s}{c}\right)\left(\frac{\bar{w}}{\frac{1}{2}\rho}\right)\left(\frac{\cos^3 \alpha_m}{V_a^2}\right)$$

i.e.

$$C_{Dp} = \left(\frac{s}{c}\right)\left(\frac{\bar{w}}{\frac{1}{2}\rho V_1^2}\right)\left(\frac{\cos^3 \alpha_m}{\cos^2 \alpha_1}\right) \tag{5.32}$$

Also, resolving perpendicularly to the vector mean,

$$L = \tfrac{1}{2}\rho V_m^2 cC_L \text{ (by definition of } C_L)$$

$$= F \cos \alpha_m + s\Delta p \sin \alpha_m$$

Therefore

$$\tfrac{1}{2}\rho V_m^2 cC_L = s\rho V_a^2(\tan \alpha_1 - \tan \alpha_2) \cos \alpha_m$$

$$+ \tfrac{1}{2}\rho V_a^2 s(\tan^2 \alpha_1 - \tan^2 \alpha_2) \sin \alpha_m - \bar{w}s \sin \alpha_m$$

which on reduction finally gives

$$C_L = 2(s/c)(\tan \alpha_1 - \tan \alpha_2) \cos \alpha_m - C_{Dp} \tan \alpha_m \tag{5.33}$$

Using these formulae the values of $C_L$ and $C_{Dp}$ may be calculated from the data given by the curves of Fig. 5.25. Since $\alpha_1'$ is known from the geometry of the blade row, the following data may be found for any incidence angle $i$:

$$\alpha_1 = \alpha_1' + i$$

$$\alpha_2 = \alpha_1 - \varepsilon^*$$

$$\alpha_m = \tan^{-1}[\tfrac{1}{2}(\tan \alpha_1 + \tan \alpha_2)]$$

Then by using values of $\bar{w}/\tfrac{1}{2}\rho V_1^2$ read from the curve and the known value of $s/c$ for the cascade, $C_{Dp}$ and $C_L$ may be calculated from equations (5.23) and (5.33) and plotted against incidence as in Fig. 5.29.

Because the value of the term $C_{Dp} \tan \alpha_m$ in equation (5.33) is negligibly small, it is usual to use the more convenient 'theoretical' value of $C_L$ given by

$$C_L = 2(s/c)(\tan \alpha_1 - \tan \alpha_2)\cos \alpha_m \tag{5.34}$$

in which the effect of profile drag is ignored. Using this formula, curves of $C_L$ can be plotted for nominal (or design) conditions to correspond with the curves of deflection given in Fig. 5.26. These curves, which are again plotted

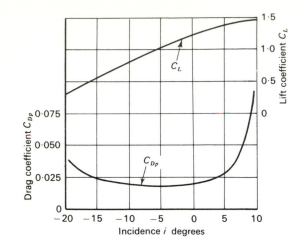

**FIG. 5.29  Lift and drag coefficient for cascade of fixed geometrical form**

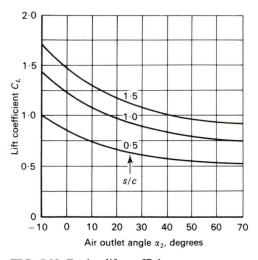

**FIG. 5.30  Design lift coefficients**

against air outlet angle $\alpha_2$ for fixed values of pitch/chord ratio $s/c$, are given in Fig. 5.30.

Before these coefficients can be applied to the blade rows of the compressor stage, two additional factors must be taken into account. These are the additional drag effects due to the walls of the compressor annulus, and the secondary loss due to trailing vortices and tip clearance. The flow effects which give rise to these losses are illustrated in Fig. 5.31. Analysis of compressor performance figures have shown that the secondary loss is of major importance and that its magnitude is of the same order as that incurred by the

(a) Annulus drag

(b) Secondary losses

**FIG. 5.31 Three-dimensional flow effects in compressor annulus**

profile drag of the blades. It is greatly influenced by tip clearance which should consequently be kept as small as possible—in the region of 1–2 per cent of the blade height. For typical axial compressor designs, the following empirical formula for the additional drag coefficient arising from secondary losses has been derived,

$$C_{DS} = 0.018C_L^2 \qquad (5.35)$$

where $C_L$ is the lift coefficient as given by equation (5.34) and the curves of Fig. 5.30.

The loss due to annulus drag is naturally dependent on the relative proportions of the blade row, its influence increasing as the blades become shorter relative to their chord length. It has been found convenient to relate the drag coefficient, resulting from this loss, to the blade row dimensions by the empirical formula,

$$C_{DA} = 0.020(s/h) \qquad (5.36)$$

where $s$ and $h$ are the pitch and height of the blades respectively. Therefore, on inclusion of these factors, an overall drag coefficient can be determined which is given by

$$C_D = C_{Dp} + C_{DA} + C_{DS} \qquad (5.37)$$

The argument used in deriving equation (5.32) for the profile drag coefficient in the case of a straight cascade will apply equally well to the annular case if $C_D$ is substituted for $C_{Dp}$. Hence for the annular case

$$C_D = \left(\frac{s}{c}\right)\left(\frac{\bar{w}}{\frac{1}{2}\rho V_1^2}\right)\left(\frac{\cos^3 \alpha_m}{\cos^2 \alpha_1}\right) \qquad (5.38)$$

This enables the loss coefficient $\bar{w}/\frac{1}{2}\rho V_1^2$ for the blade row to be determined. The *theoretical* static pressure rise through the blade row is found by putting the loss $\bar{w}$ equal to zero in equation (5.30), which gives

$$\Delta p_{th} = \tfrac{1}{2}\rho V_a^2(\tan^2 \alpha_1 - \tan^2 \alpha_2)$$
$$= \tfrac{1}{2}\rho V_a^2(\sec^2 \alpha_1 - \sec^2 \alpha_2)$$

Therefore

$$\frac{\Delta p_{th}}{\tfrac{1}{2}\rho V_a^2 \sec^2 \alpha_1} = 1 - \frac{\sec^2 \alpha_2}{\sec^2 \alpha_1}$$

so that the theoretical pressure rise in terms of the inlet dynamic head and cascade air angles becomes

$$\frac{\Delta p_{th}}{\tfrac{1}{2}\rho V_1^2} = 1 - \frac{\cos^2 \alpha_1}{\cos^2 \alpha_2}$$

The efficiency of the blade row, $\eta_b$, which is defined as the ratio of the actual pressure rise to the theoretical pressure rise, can then be found from $\eta_b = (\Delta p_{th} - \bar{w})/\Delta p_{th}$. Or in non-dimensional terms,

$$\eta_b = 1 - \frac{\bar{w}/\tfrac{1}{2}\rho V_1^2}{\Delta p_{th}/\tfrac{1}{2}\rho V_1^2} \tag{5.39}$$

When dealing with cascade data, efficiency is evaluated from pressure rises, whereas compressor stage efficiency is defined in terms of temperature rises. Furthermore, the stage efficiency encompasses both a rotor and a stator row. It is therefore not obvious how the efficiency of a blade row obtained from cascade tests can be related to the stage efficiency. Initially we will consider this problem in the context of a stage designed for 50 per cent reaction at the mean diameter. In such a case, because of the symmetrical blading, $\eta_b$ [evaluated from equation (5.39) for conditions obtaining at the mean diameter] will be virtually the same for both the rotor and stator blade row.† What we shall now proceed to show is that this value of $\eta_b$ can be regarded as being equal to the isentropic efficiency of the whole stage, $\eta_S$.

If $p_1$ and $p_2$ are the static pressures at inlet to and outlet from the rotor,

$$\eta_b = \frac{p_2 - p_1}{p_2' - p_1}$$

where $p_2'$ denotes the ideal pressure at outlet corresponding to no losses. If the stage efficiency $\eta_S$ is defined as the ratio of the isentropic static temperature rise to the actual static temperature rise,

$$\frac{p_2}{p_1} = \left[ 1 + \frac{\eta_S \Delta T_S}{2T_1} \right]^{\gamma/(\gamma - 1)}$$

† Owing to slight differences in blade pitch and height between the rotor and stator rows of a stage, $C_{DA}$ from equation (5.36) may differ marginally. But as the following numerical example will show, $C_{DA}$ is only a small proportion of $C_D$ and this difference will have little effect on $\eta_b$.

because $\Delta T_S / 2$ will be the temperature rise in the rotor for 50 per cent reaction. Also

$$\frac{p_2'}{p_1} = \left[1 + \frac{\Delta T_S}{2T_1}\right]^{\gamma/(\gamma-1)}$$

Therefore

$$\eta_b = \left[\frac{p_2}{p_1} - 1\right] \bigg/ \left[\frac{p_2'}{p_1} - 1\right]$$

$$= \left\{\left[1 + \frac{\eta_S \Delta T_S}{2T_1}\right]^{\gamma/(\gamma-1)} - 1\right\} \bigg/ \left\{\left[1 + \frac{\Delta T_S}{2T_1}\right]^{\gamma/(\gamma-1)} - 1\right\}$$

After expanding and neglecting second-order terms, this reduces to

$$\eta_b = \eta_S \left[1 - \frac{1}{\gamma - 1} \times \frac{\Delta T_S}{4T_1}(1 - \eta_S)\right]$$

But $\Delta T_S$ is of the order 20 K and $T_1$ about 300 K, so that the second term in the bracket is negligible and

$$\eta_b = \eta_S$$

For cases other than 50 per cent reaction at the mean diameter, an approximate stage efficiency can be deduced by taking the arithmetic mean of the efficiencies of the two blade rows, i.e.

$$\eta_S = \tfrac{1}{2}(\eta_{b\,\text{rotor}} + \eta_{b\,\text{stator}})$$

If the degree of reaction is far removed from the 50 per cent condition, then a more accurate expression for the stage efficiency may be taken as

$$\eta_S = \Lambda \eta_{b\,\text{rotor}} + (1 - \Lambda)\eta_{b\,\text{stator}}$$

where $\Lambda$ is the degree of reaction as given in section 5.5.

## EXAMPLE

We shall again consider the third stage of the compressor used as an example of compressor design, for which the degree of reaction was 50 per cent at the mean diameter. From the blade design given in the previous section, and remembering that $\beta_1$ and $\beta_2$ are the air inlet and outlet angles $\alpha_1$ and $\alpha_2$ respectively in cascade terminology,

$$\tan \alpha_m = \tfrac{1}{2}(\tan \alpha_1 + \tan \alpha_2) = \tfrac{1}{2}(\tan 50 \cdot 92 + \tan 28 \cdot 63) = 0 \cdot 889$$

whence

$$\alpha_m = 41 \cdot 63°$$

From Fig. 5.30, $C_L$ at $s/c = 0 \cdot 9$ and $\alpha_2 = 28 \cdot 63$ is equal to $0 \cdot 875$. Hence from equation (5.33)

$$C_{DS} = 0.018(0.875)^2 = 0.0138$$

Recalling that $s = 0.0248$ m and $h = 0.0836$ m we have from equation (5.36)

$$C_{DA} = \frac{0.020 \times 0.0248}{0.0836} = 0.0059$$

From Fig. 5.29, at zero incidence $C_{DP} = 0.018$; hence the total drag coefficient is

$$C_D = C_{DP} + C_{DA} + C_{DS} = 0.018 + 0.0059 + 0.0138 = 0.0377$$

Therefore, from equation (5.38)

$$\frac{\bar{w}}{\frac{1}{2}\rho V_1^2} = C_D \bigg/ \left[ \left(\frac{s}{c}\right) \frac{\cos^3 \alpha_m}{\cos^2 \alpha_1} \right] = \frac{0.0377 \cos^2 50.92°}{0.9 \cos^3 41.63°} = 0.0399$$

We also have

$$\frac{\Delta p_{th}}{\frac{1}{2}\rho V_1^2} = 1 - \frac{\cos^2 \alpha_1}{\cos^2 \alpha_2} = 1 - \frac{\cos^2 50.92°}{\cos^2 28.63°} = 0.4842$$

Therefore, from equation (5.39)

$$\eta_b = 1.0 - \frac{0.0399}{0.4842} = 0.918$$

The efficiency of the stator row will be similar, and hence we may take

$$\eta_S = 0.92$$

The static temperature at entry to the stage can be found to be 318·5 K, and with a stage temperature rise of 24 K, the *static* pressure ratio of the stage is found to be

$$R_S = \left[ 1 + \frac{\eta_S \Delta T_S}{T_1} \right]^{\gamma/(\gamma-1)} = \left[ 1 + \frac{0.92 \times 24}{318.5} \right]^{3.5} = 1.264$$

We have now seen how the design point performance of a stage may be estimated. It will have been noticed that the results are in terms of an isentropic stage efficiency based on *static* temperatures and a stage pressure ratio based on *static* pressures. To the compressor designer, however, it is the efficiency based on *stagnation* temperature rise and the *stagnation* pressure ratio which are of interest. Making use of the relations $p_0/p = (T_0/T)^{\gamma/(\gamma-1)}$ and $T_0 = T + C^2/2c_p$ it is of course possible to transform a static pressure ratio into a stagnation pressure ratio when the inlet and outlet velocities are known. But for the common case where these velocities are equal (i.e. $C_3 = C_1$) there is no need for this elaboration. $\Delta T_{0S} = \Delta T_S$, and it is easy to show as follows that $\eta_S$ is virtually the same whether based on static or stagnation temperatures.

**FIG. 5.32**

Then merely by substituting $T_{01}$ for $T_1$ in the foregoing equation for $R_S$ the stagnation pressure ratio is obtained directly.

Referring to Fig. 5.32, where the stage inlet and outlet states are denoted by 1 and 3 (static) and 01 and 03 (stagnation), we have

$$\eta_S = \frac{T'_3 - T_1}{T_3 - T_1} = \frac{T_3 - T_1 - x}{T_3 - T_1} = 1 - \frac{x}{\Delta T_S}$$

And when based on stagnation temperatures,

$$\eta_S = \frac{T'_{03} - T_{01}}{T_{03} - T_{01}} = 1 - \frac{y}{\Delta T_{0S}}$$

With $C_3 = C_1$, $\Delta T_{0S} = \Delta T_S$. Furthermore, since the pressure ratio per stage is small the constant $p_{03}$ and $p_3$ lines are virtually parallel between 3' and 3 so that $y \approx x$. It follows that $\eta_S$ has the same value on either basis. For the stage of our example, $T_{01}$ was 333 K, and hence the stagnation pressure ratio is

$$R_S = \left[1 + \frac{0 \cdot 92 \times 20}{333}\right]^{3 \cdot 5} = 1 \cdot 252$$

It will be remembered that it was necessary to make use of two assumed efficiencies at the start of the design process: a polytropic efficiency for the compressor as a whole, and a stage efficiency. As a first approximation these were taken to be equal and a value of 0·90 was assumed. The estimated value of $\eta_S$ for the third stage, i.e. 0·92, is in sufficient agreement, bearing in mind the uncertainties in predicting the secondary and annulus losses. If similar agreement were to be obtained for all the stages, we might conclude that the design had been conservative and that the compressor should have no difficulty in achieving the specified performance. To obtain an estimate of the overall efficiency it would be necessary to repeat the foregoing for all the stages. The product of the stage pressure ratios would then yield the overall pressure ratio from which the overall isontropic temperature rise, and hence the overall efficiency, could be calculated.

Before continuing, it may be helpful to summarize the main steps in the design procedure described in the previous sections. Having made appropriate assumptions about the efficiency, tip speed, axial velocity, and so on, it was possible to size the annulus at inlet and outlet of the compressor and calculate the air angles required for each stage at the mean diameter. A choice was then made of a suitable vortex theory to enable the air angles to be calculated at various radii from root to tip. Throughout this work it was necessary to ensure that limitations on blade stresses, rates of diffusion and Mach number were not exceeded. Cascade test data were used to determine a blade geometry which would give these air angles, and also the lift and drag coefficients for a two-dimensional row of blades of this form. Finally, empirical correction factors were evaluated to enable these coefficients to be applied to the annular row at the mean diameter so that the stage efficiency and pressure ratio could be estimated.

## 5.10   Compressibility effects

Over a long period of time, the development of high-performance gas turbines has led to the use of much higher flow rates per unit frontal area and blade tip speeds, resulting in higher velocities and Mach numbers in compressors. While early units had to be designed with subsonic velocities throughout, Mach numbers exceeding unity are now found in the compressors of industrial gas turbines and Mach numbers as high as 1·5 are used in the design of fans for turbofans of high by-pass ratio. It is not within the scope of this book to cover transonic design in detail, and the treatment will be confined to a brief introduction and provision of some key references; again, it should be realized that much of the relevant information is of a proprietary nature and is not available in the open literature.

High-speed cascade testing is required to provide experimental data on compressibility effects, and in particular to determine the values of the Mach numbers, corresponding to entry velocities relative to the blades, which bring about poor cascade performance. The first high velocity of interest is that corresponding to what is called the 'critical' Mach number $M_c$; at entry velocities lower than this, the performance of the cascade differs very little from that at low speeds. Above this velocity, the losses begin to show a marked increase until a point is reached where the losses completely cancel the pressure rise and the blade ceases to be any use as a diffuser. The corresponding Mach number is then referred to as the 'maximum' value $M_m$. For a typical subsonic compressor cascade at zero incidence, the values of these Mach numbers are in the region of 0·7 and 0·85 respectively. A typical variation of fluid deflection and pressure loss for subsonic blading was shown in Fig. 5.25, representing flow at low Mach numbers. As the Mach number increases, two important effects take place: first, the overall level of losses increase substantially, and second, the range of incidence for which losses are

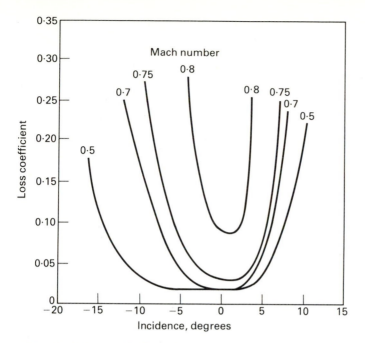

FIG. 5.33 Effect of Mach number on losses

acceptable is drastically reduced. This means that off-design performance of the compressor may be seriously affected. Figure 5.33 shows test results for a subsonic blade section over a range of Mach number from 0·5 to 0·8; the usable range of incidence at M 0·8 can be seen to be very narrow and clearly compressor blading of this type could not be used at Mach numbers approaching or exceeding unity. It should be pointed out that for compressible flow the denominator in the loss coefficient is $(p_{01} - p_1)$ rather than $\rho V_1^2/2$.

Compressibility effects will be most important at the front of the compressor where the inlet temperature, and hence the acoustic velocity, are lowest. The Mach number corresponding to the velocity relative to the tip of the rotor is the highest encountered and is important both from the viewpoint of shock losses and noise. The stator Mach number is generally highest at the hub radius because of the increased whirl velocity normally required to give constant work input at all radii in the rotor. The rotor and stator Mach numbers corresponding to the first stage of the compressor designed in section 5.7 are shown in Fig. 5.34, which shows that the flow relative to the rotor is supersonic over a considerable length of the blade. The stator Mach numbers can be seen to be significantly lower, this being due to the fact that they are unaffected by the blade speed.

Analysis of a large amount of compressor tests by NACA [Refs. (10, 11)] showed that losses for subsonic conditions correlated well on the basis of

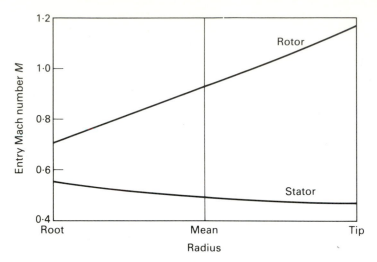

FIG. 5.34 **Variation of entry Mach numbers**

diffusion factor, as shown in Fig. 5.8, but were significantly higher for transonic conditions. It was deduced that this increase in loss must be due to shock losses, but it was also found that the spacing of the blades had a considerable effect; a reduction in solidity (i.e. an increase in pitch/chord ratio) caused a rapid increase in loss. A simple method for predicting the loss was developed and can be explained with reference to Fig. 5.35.

A pair of double circular arc (DCA) blades are shown, with a supersonic velocity entering in a direction aligned with the leading edge. The supersonic expansion along the uncovered portion of the suction side can be analysed by means of the Prandtl–Meyer relations discussed in Appendix A.8, and the

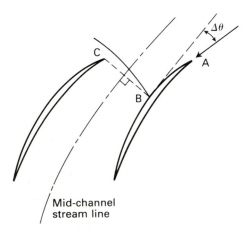

FIG. 5.35 **Shock loss model**

Mach number will increase as the flow progresses along the suction surface. A shock structure is assumed in which the shock stands near the entrance to the blade passage, striking the suction surface at B, extending in front of the blade at C and then bending back similar to a bow wave. It is then assumed that the loss across the shock can be approximated by the normal shock loss taken for the average of the Mach numbers at A and B. At A the Mach number is taken to be the inlet relative Mach number. The value at B can be calculated from the inlet Mach number and the angle of supersonic turning to the point B, where the flow is tangential to the surface, using the Prandtl–Meyer relations. The location of the point B is taken to be at the intersection of the suction surface and a line drawn normal to the mean passage camber-line and through the leading edge of the next blade. The requirement, then, is to establish the supersonic turning from inlet to point B and this clearly depends on the blade spacing; moving the blades further apart (reducing solidity) results in point B moving back, increasing both the turning and Mach number, resulting in an increased shock loss. Figure 5.36 shows the relationship between inlet Mach number, supersonic turning and shock loss. If we consider the tip of our first-stage rotor, the relative Mach number was 1·165 and the rotor air angles were 67·11° at inlet and 62·69° at exit. The total turning was therefore 4·42°, and the supersonic turning to point B would depend on the blade spacing; making a reasonable assumption of 2°, from Fig. 5.36 it is found that $M_B = 1·22$ and $w = 0·015$. The additional loss due to shocks is significant and would increase rapidly with increased supersonic turning; this is another reason for the designer to select a slightly reduced temperature rise in the first stage.

The fans of high by-pass turbofans require a large diameter, and even though they run at a lower rotational speed than a conventional jet engine, the tip speeds are significantly higher; typical Mach numbers at the tip would be of the order of 1·4–1·6 and DCA blading is not satisfactory. Blade sections have been specially developed to meet this high Mach number requirement and are not based on aerofoil sections. To keep shock losses to an acceptable level, it is necessary to provide supersonic diffusion to a Mach number of about 1·2 prior to the normal shock at entry to the blade passage. It is shown in Appendix A.2 that supersonic diffusion requires a *decrease* in flow area. This may be accomplished in two ways, either by decreasing the annulus area or by making the suction side slightly concave. Both methods may be used together. A typical blade profile is shown in Fig. 5.37, where it can be seen that the amount of turning is very small. Until recently it was thought that shock losses explained the decrease in efficiency with Mach number which has been observed, but it is now realized that the shocks only account for part of the additional losses. There are also significant losses resulting from shock–boundary layer interaction and the net effect appears to be a magnification of the viscous losses. Kerrebrock [Ref. (12)] gives a comprehensive review of the aerodynamic problems of transonic compressors and fans.

Fan blades tend to be long and flexible, and most require part-span dampers to prevent excessive or torsional movement. These are usually

**FIG. 5.36  Shock loss coefficient**

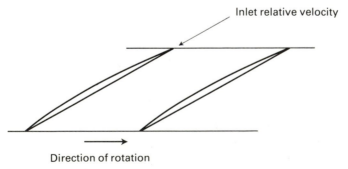

**FIG. 5.37  Blade section for supersonic flow**

positioned at about 65 per cent of the span measured from the root, and it will be appreciated that they are located in a region of high Mach number and lead to considerable local decreases in efficiency. Some manufacturers can now design fan blades without dampers, giving a considerable improvement in aerodynamic performance; blades without dampers normally have significantly wider chords and this helps to meet bird strike certification requirements. Figure 5.38(a) shows the effect of a seagull strike on a large turbofan during take-off and gives some idea of the impact forces involved. The blade shown has two part-span dampers, but later versions have only one. Figure 5.38(b) shows the shape of the tip section and the use of a concave portion on the suction side can be observed.

FIG. 5.38(a)

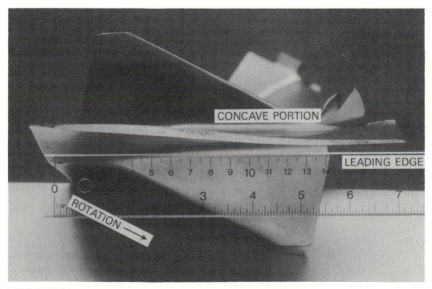

FIG. 5.38(b)

## 5.11   Off-design performance

Attention so far has focused on the aerothermodynamic design of axial flow compressors to meet a particular design point of mass flow, pressure ratio and efficiency. It must be realized at the outset, however, that any compressor will be required to operate at conditions far removed from the design point including engine starting, idling, reduced power, maximum power, acceleration and deceleration. Thus it is clear that the compressor must be capable of satisfactory operation over a wide range of rotational speeds and inlet conditions. With the compressor blading and annulus dimensions chosen to satisfy the design point condition, it is obvious that these will not be correct for conditions far removed from design. Previous sections showed that compressor blading had a limited range of incidence before losses became unacceptably high, resulting in low compressor efficiency. What is even more important, however, is that the problems of blade stalling may lead to surging of the whole compressor, preventing operation of the engine at particular conditions; this may lead to severe engine damage or, in the case of aircraft, a critical safety hazard.

An overall picture of compressor off-design performance can be built up from consideration of the behaviour of individual stages and the interaction between a series of stages. Recalling equation (5.5), the temperature rise in a stage is given by

$$\Delta T_{0S} = \frac{UC_a}{c_p} (\tan \beta_1 - \tan \beta_2)$$

which is expressed in terms of the rotor air angles. This can be recast to give

$$\Delta T_{0S} = \frac{U}{c_p} [U - C_a(\tan \alpha_1 + \tan \beta_2)]$$

The angle $\alpha_1$ is the outlet air angle from the preceding stator, $\beta_2$ is the rotor outlet air angle, and these can be considered essentially constant, being determined by the blading geometry; $\beta_1$, on the other hand, will vary widely as $C_a$ and $U$ change. Dividing the previous equation throughout by $U^2$ and rearranging we get

$$\frac{c_p\Delta T_{0S}}{U^2} = 1 - \frac{C_a}{U} (\tan \alpha_1 + \tan \beta_2)$$

The term $C_a/U$ is known as the *flow coefficient* ($\phi$) and $c_p\Delta T_{0S}/U^2$) as the *temperature coefficient* ($\psi$). With the stage operating at the design value of $\phi$ the incidence will be at its design value and a high efficiency will be achieved. With the assumption that $\alpha_1$ and $\beta_2$ are constant,

$$\psi = 1 - \phi k$$

where $k = \tan \alpha_1 + \tan \beta_2$. This relationship is shown by the dotted straight

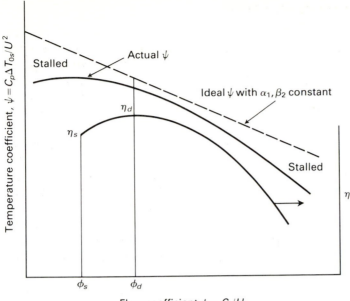

**FIG. 5.39 Stage characteristic**

line in Fig. 5.39. It is obvious that the temperature coefficient increases as the flow coefficient decreases and reaches a value of 1 when $\phi = 0$. The temperature coefficient can be interpreted from the shape of the velocity diagram. Recalling that

$$\Delta T_{0s} = \frac{U \Delta C_w}{c_p}$$

it immediately follows that

$$\frac{c_p \Delta T_{0s}}{U^2} = \frac{\Delta C_w}{U}$$

Thus, if $\psi = 1$, $\Delta C_w = U$ and from earlier examination of velocity diagrams it will be recognized that this results in excessive diffusion in the blade passage and efficiency will decrease. For satisfactory operation, $\Delta C_w/U$, and hence $\psi$, should be around 0·3–0·4.

The pressure rise across the stage, $\Delta p_{0s}$, is determined by the temperature rise and the isentropic efficiency of the stage.

$$\frac{p_{01} + \Delta p_{0s}}{p_{01}} = 1 + \frac{\Delta p_{0s}}{p_{01}} = \left(1 + \frac{\eta_s \Delta T_{0s}}{T_{01}}\right)^{\gamma/(\gamma - 1)}$$

Expanding by means of the Binomial Theorem, assuming that $\Delta T_{0s} \ll T_{01}$, we have

$$\frac{\Delta p_{os}}{p_{01}} = \frac{\gamma}{\gamma - 1} \eta_s \frac{\Delta T_{os}}{T_{01}}$$

Making use of the Equation of State and the relationship $(\gamma - 1)/\gamma = R/c_p$, it follows that

$$\frac{\Delta p_{os}}{\rho_{01}} = \eta_s c_p \Delta T_{os}$$

and hence

$$\frac{\Delta p_{os}}{\rho_{01} U^2} = \eta_s \frac{c_p \Delta T_{os}}{U^2}$$

The term $\Delta p_{os}/\rho_{01} U^2$ is referred to as the *pressure coefficient*, which can be seen to be the product of the stage efficiency and temperature coefficient. Thus the overall performance of a stage can be expressed in terms of flow coefficient, temperature coefficient and either a pressure coefficient or the stage efficiency. The resulting form of the stage characteristic is shown in Fig. 5.39; in practice, $\alpha_1$ and $\beta_2$ will not remain constant due to increased deviation as conditions change from the design point. In regions of blade stalling, both at positive and negative incidence, there will be considerable divergence giving the shape shown. Choking will occur at a high value of flow coefficient, leading to a very rapid drop in efficiency and placing an upper limit on the flow which can be passed at a given blade speed. Stage characteristics may be obtained from single-stage tests, by analysis of interstage data on a complete compressor or by prediction using cascade data. In practice, not all the constant speed lines would collapse into a single curve as shown, but for the following brief discussion a single line characteristic will be assumed.

The $\psi - \phi$ curve in Fig. 5.39 is drawn for the case where the efficiency is a maximum at the design flow coefficient, $\phi_d$. Moving away from $\phi_d$ results in a change in incidence and increased losses. Reducing $\phi$ results in increased positive incidence and stall at $\phi_s$; increasing $\phi$ eventually results in choking of the stage and a severe drop in efficiency. It is essential that all individual stages of a compressor operate in the region of high efficiency without encountering either stall or choke at normal operating conditions; at conditions far removed from design it may not be possible to achieve this without remedial action involving changes in compressor geometry. The difficulties involved in achieving correct matching of the stages can be understood by considering the operation of several identical stages in series. This procedure is known as 'stage stacking' and is an invaluable tool to the aerodynamacist who is concerned with the overall performance of a compressor and determining the reasons for sub-standard performance; stage stacking can be used to produce overall compressor characteristics, showing the regions in which individual stages stall. In this elementary treatment we will only touch on the use of this

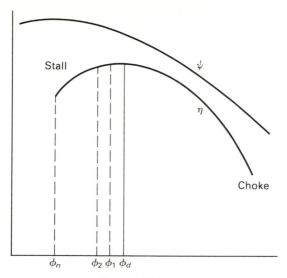

FIG. 5.40  **Effect of reduction in $\phi$ at inlet**

procedure. An idealized stage characteristic is shown in Fig. 5.40 and we will consider that at the design point all stages operate at $\phi_d$. If the compressor should be subjected to a decrease in mass flow, this will result in a reduction in the flow coefficient entering the first stage to $\phi_1$, which will result in an increase in pressure ratio causing the density at entry to the second stage to be increased. The axial velocity at entry to the second stage is determined by the Equation of Continuity and both effects combine to give a further decrease in flow coefficient for that stage to $\phi_2$. This effect is propagated through the compressor and eventually some stage will stall at $\phi_n$. Increasing the flow coefficient has the opposite effect and will drive some stage into choke.

## 5.12  Axial compressor characteristics

The characteristic curves of an axial compressor take a form similar to those of the centrifugal type, being plotted on the same non-dimensional basis, i.e. pressure ratio $p_{02}/p_{01}$ and isentropic efficiency $\eta_c$ against the non-dimensional mass flow $m\sqrt{T_{01}}/p_{01}$ for fixed values of the non-dimensional speed $N/\sqrt{T_{01}}$.

A typical set of such curves is shown in Fig. 5.41 and in comparison with Fig. 4.10 it will be observed that the characteristics for fixed values of $N/\sqrt{T_{01}}$ cover a much narrower range of mass flow than in the case of the centrifugal compressor. At high rotational speeds the constant speed lines become very steep and ultimately may be vertical. The same limitations occur at either end

FIG. 5.41  Axial compressor characteristics

of the $N/\sqrt{T_{01}}$ lines due to surging and choking. The surge points, however, are normally reached before the curves reach a maximum value and, because of this, the design operating point, which is always near the peak of the characteristic, is also very near the surge line. Consequently the range of stable operation of axial compressors is narrow and gas turbine plant incorporating this type of compressor call for great care in matching the individual components if instability is to be avoided at operating conditions removed from the design point. Chapter 8 deals with this aspect in detail. The mechanism of surging in axial compressors is complex and is still not yet fully understood. It is often difficult to distinguish between surging and stalling and one phenomenon may easily lead to the other. The phenomenon of rotating stall referred to in section 4.6 may also be encountered: it can lead to loss of performance and severe blade vibration without actually causing surge. A comprehensive review of compressor stall phenomena is given by Greitzer in Ref. (13).

The complete compressor characteristic can be obtained only if the compressor is driven by an external power source; it will be shown in Chapter 8 that the running range of the compressor is severely restricted when it is combined with a combustion chamber and turbine. The driving unit must be capable of variable speed operation, with the speed continuously variable and closely controlled. Compressor test rigs in the past have been driven by electric motors, steam turbines or gas turbines. One of the major problems of testing compressors is the very large power requirement. As an example, at take-off condition the compressor of an Olympus 593 in Concorde absorbs about 75 MW, with approximately 25 MW going to the LP compressor and 50 MW to the HP compressor. A typical fan stage from a large turbofan will compress about 700 kg/s through a pressure ratio of about 1·5, requiring about 30 MW. Two methods of reducing the total power requirement are to throttle the intake to the compressor or to test a scaled-down model. Throttling the intake reduces the density and hence the mass flow, thereby reducing the power input; a major problem that occurs, however, is the large drop in Reynolds number with reduced density. The Reynolds number may well be reduced by a factor of 3 or 4, which has the effect of increasing the relative magnitude of the viscous losses giving unrealistically low values of compressor efficiency. Model testing of single-stage fans is more common than throttled intake testing. Another problem with compressor testing is that the operating environment of the compressor in the test rig is not the same as that encountered in the engine, due to causes such as changes in tip clearance as a result of different casing temperatures and shaft expansions. Some manufacturers prefer to carry out compressor tests on a gas generator rig, using the actual combustion system and turbine, but this requires the use of a variable exhaust nozzle to cover a reasonable range of the compressor characteristic; the compressor, however, is exposed to realistic engine operating conditions. Compressor testing is both complex and expensive, but is absolutely essential for the development of a high-performance engine. A typical compressor development programme for an industrial gas turbine is described in Ref. (15).

## Some deductions from the compressor characteristics

The axial flow compressor consists of a series of stages, each of which has its

own characteristic: stage characteristics are similar to the overall
characteristic but have much lower pressure ratios. The mass flow through the
compressor is limited by choking in the various stages and under some
conditions this will occur in the early stages and under others in the rear
stages.

We have noted that if an axial flow compressor is designed for a constant
axial velocity through all stages, the annulus area must progressively decrease
as the flow proceeds through the compressor, because of the increasing
density. The annulus area required for each stage will be determined for the
design condition and clearly at any other operating condition the fixed area
will result in a variation of axial velocity through the compressor. When the
compressor is run at a speed lower than design, the temperature rise and
pressure ratio will be lower than the design value. The effect of the reduction in
density will be to increase the axial velocity in the rear stages, where choking
will eventually occur and limit the mass flow. Thus at low speeds the mass flow
will be determined by choking of the rear stages, as indicated in Fig. 5.42. As

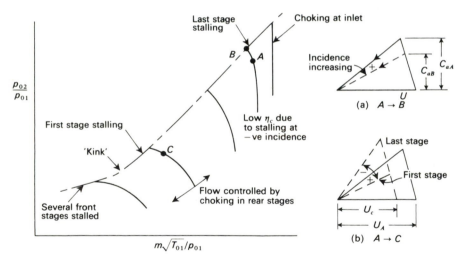

**FIG. 5.42 Phenomena at off-design operation**

the speed is increased, the density in the stages is increased to the design value
and the rear stages of the compressor can pass all the flow provided by the
early stages. Eventually, however, choking will occur at the inlet; the vertical
constant speed line in Fig. 5.42 is due to choking at the inlet of the compressor.

When the compressor is operating at its design condition, all stages are
operating at the correct value of $C_a/U$ and hence at the correct incidence. If we
consider moving from design point $A$ to point $B$ on the surge line at the design
speed it can be seen that the density at the compressor exit will be increased
due to the increase in delivery pressure, but the mass flow is slightly reduced.

Both of these effects will reduce the axial velocity in the last stage and hence increase the incidence as shown by the velocity triangle (a) in Fig. 5.42. A relatively small increase in incidence will cause the rotor blade to stall, and it is thought that surge at high speeds is caused by stalling of the last stage.

When speed is reduced from $A$ to $C$, the mass flow generally falls off more rapidly than the speed, and the effect is to decrease the axial velocity at inlet and cause the incidence on the first-stage blade to increase as shown in Fig. 5.42(b). The axial velocity in the latter stages, however, is increased because of the lower pressure and density, so causing the incidence to decrease as shown. Thus at low speeds, surging is probably due to stalling of the first stage. It is possible for axial compressors to operate with several of the early stages stalled, and this is thought to account for the 'kink' in the surge line which is often encountered with high-performance compressors (see Fig. 5.42). A detailed discussion of the relation between stage stalling and the surge line using stage stacking techniques is given in Ref. (14).

At conditions far removed from surge the density will be much lower than required and the resulting high axial velocities will give a large decrease in incidence, which will eventually result in stalling at negative incidences. The efficiency will be very low in these regions.

### The twin-spool compressor

It will be evident that as the design pressure ratio is increased, the difference in density between design and off-design conditions will be increased, and the probability of blades stalling due to incorrect axial velocities will be much higher. The effects of increased axial velocity towards the rear of the compressor can be alleviated by means of *blow-off*, where air is discharged from the compressor at some intermediate stage through a valve to reduce the mass flow through the later stages. Blow-off is wasteful, but is sometimes necessary to prevent the engine running line intersecting the surge line; this will be discussed in Chapter 8. A more satisfactory solution is to use a twin-spool compressor.

We have seen that reduction of compressor speed from the design value will cause an increase of incidence in the first stage and a decrease of incidence in the last stage; clearly the effect will increase with pressure ratio. The incidence could be maintained at the design value by increasing the speed of the last stage and decreasing the speed of the first stage, as indicated in Fig. 5.43. These conflicting requirements can be met by splitting the compressor into two (or more) sections, each being driven by a separate turbine, as shown in Fig. 1.7 for example. In the common twin-spool configuration the LP compressor is driven by the LP turbine and the HP compressor by the HP turbine. The speeds of the two spools are mechanically independent but a strong aerodynamic coupling exists which has the desired effect on the relative

**FIG. 5.43**

speeds when the gas turbine is operating off the design point. This will be discussed in Chapter 9.


## The variable geometry compressor

An alternative approach to satisfying the off-design performance of high-pressure ratio compressors is to use several rows of *variable stators* at the front of the compressor, permitting pressure ratios of over 16:1 to be achieved in a single spool. If the stators are rotated away from the axial direction, increasing $\alpha_1$ in Fig. 5.4, the effect is to decrease the axial velocity and mass flow for a given rotational speed. This delays stalling of the first few stages and choking in the last stages, at low rotational speeds. The earlier discussion on stage characteristics showed that ideally

$$\psi = 1 - k\phi$$

where $k = \tan \alpha_1 + \tan \beta_2$. Using variable stators, it is possible to increase $\alpha_1$ with $\beta_2$ remaining constant. The effect is to decrease the temperature coefficient for a given flow coefficient; the pressure coefficient will also be reduced because of the reduction in temperature coefficient. The stage characteristic will be shifted to the left as shown in Fig. 5.44 because of the reduction in mass flow. The stage stacking techniques mentioned in section 5.11 can be used to predict the effects of incorporating variable geometry on the overall compressor characteristics. One of the major benefits is an improvement in the surge margin at low engine speeds, which is particularly important during starting and operation at idle conditions: these topics will be discussed further in Chapter 8.

The compressor of the Ruston Tornado, shown in Fig. 1.13 and described in Ref. (15), uses variable IGVs followed by four rows of variable stators to achieve a pressure ratio of 12:1. The design speed of the compressor is 11 085 rev/min and at speeds below 10 000 rev/min the stators start to close progressively until they are in the fully closed position for all speeds below 8000 rev/min: the IGVs are rotated through 35° and the stators through 32°, 25°, 25° and 10° from the first to the fourth stage. While the effect of variable stators can be predicted with reasonable accuracy, it is necessary to optimize the settings of each row by rig-testing the compressor.

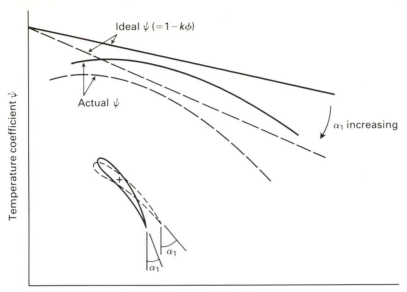

FIG. 5.44 **Effect of variable stators**

## 5.13  Closure

It must be emphasized that the foregoing treatment of axial compressor theory and design has been kept at an elementary level to serve as an introduction to what has now become an extremely complex field of study. Since the early pioneering work of NGTE and NACA in establishing rational design techniques based on empirical data, much effort has gone into the development of sophisticated methods making use of digital computers. While digital computers have been used in compressor development for many years, the enormous improvements in computing speed and storage capacity in recent years have had a major impact on turbomachinery design; this effort is truly international in scope. Two of these methods are referred to as 'streamline curvature' and 'matrix through flow' techniques. These attempt to determine the flow pattern in the so-called 'meridional plane' which is the plane containing the axial and radial directions. This is in contrast to earlier methods which were essentially concerned with flow patterns in the peripheral–axial planes at various positions along the blade span. For a summary of these techniques the reader is referred to Ref. (16) which compares them from the user's point of view. A third approach involving advanced numerical analysis is the so-called 'time marching' method, in which an initial estimate of the flow pattern is made and unsteady aerodynamic theory is used to predict succeeding solutions until equilibrium is finally achieved, Ref. (17). Reviews of computational methods and their application are given in Refs (18) and (19). Modern methods of numerical analysis have resulted in developments such as 'controlled diffusion' and 'end bend' blading; the first of these makes use of supercritical airfoil sections developed from high subsonic wing sections and the second alters the shape of the blades at both inner and outer annulus walls to allow for the decreased axial velocity in the boundary layer. Computational methods will play an ever-increasing role in compressor design in the future, but it is unlikely that empiricism or experimental tests will ever be completely displaced.

Finally, it should be recognized that the very limited number of references supplied are intended to be central to the foundation of compressor design. The literature is expanding very rapidly and those actively involved in compressor design have to follow this closely, making it all too easy to forget the source of this technology. The books by Horlock [Ref. (5)], updated in 1982, and Gostelow [Ref. (8)] both contain extensive bibliographies.

## NOMENCLATURE

For velocity triangle notation ($U$, $C$, $V$, $\alpha$, $\beta$) see Fig. 5.4
For cascade notation ($\alpha'$, $\delta$, $\varepsilon$, $\zeta$, $\theta$, $i$, $s$, $c$) see Fig. 5.23

$C_D$   overall drag coefficient
$C_{DA}$ annulus drag coefficient
$C_{Dp}$ profile drag coefficient
$C_{DS}$ secondary loss coefficient
$C_L$   lift coefficient
$D$    diffusion factor
$h$    blade height, specific enthalpy
$n$    number of blades
$N$   number of stages
$r$    radius
$R$   $r/r_m$ as well as pressure ratio and gas constant
$w$   stagnation pressure loss
$h/c$ aspect ratio
$s/c$ pitch/chord ratio
$\lambda$    work-done factor
$\Lambda$   degree of reaction ($\Delta T_{\text{rotor}}/\Delta T_{\text{stage}}$)
$\phi$    flow coefficient ($C_a/U$)
$\psi$    temperature coefficient ($c_p \Delta T_{0s}/U^2$)

*Suffixes*

$a, w$ axial, whirl, component
$b$    blade row
$m$   mean, vector mean
$s$    stage
$r$    root radius, radial component
$t$    tip radius

# 6

# *Combustion systems*

The combustion system is not yet so amenable to theoretical treatment as the other components of a gas turbine. Although sufficient experience has been gained to enable a new design of combustion system to be undertaken with at least a qualitative idea of its probable performance, the design can only be perfected by testing the component independently and modifying it in the light of test results. The main purpose of this chapter is to show how the problem of design is basically one of reaching the best compromise between a number of conflicting requirements. Since the most common fuels for gas turbines are liquid petroleum distillates and natural gas, it is the form of combustion system appropriate to such fuels which will be described: a brief mention of the special problems associated with the use of residual fuel oil and solid fuel will conclude the chapter.

## 6.1   Form of combustion system

For gas turbines operating on the closed cycle wherein the air is used repeatedly, an air-tight heat-exchanger must be used to impart heat to the working fluid. This is not a very compact method and in the open cycle gas turbine, where a fresh supply of air is inhaled continuously, the direct method of burning fuel in the working stream itself is always used; it is this type of combustion system which is described in this chapter. Because the combustion is continuous, an electric spark is only required for initiating the process, and thereafter the flame must be self-sustaining or 'self-piloting'.

Frequently the air leaving the compressor is split into a number of separate streams, each supplying a *tubular* combustion chamber. These chambers are spaced around the shaft connecting the compressor and turbine as shown in Fig. 6.1(a), each chamber having its own fuel jet fed from a common supply line. This type of layout is particularly suitable for gas turbines using centrifugal compressors where the air stream is already divided by the diffuser vanes, and was widely used in early aircraft engines. Single tubular chambers are also commonly used for small gas turbines.

A single *annulur* combustion chamber surrounding the rotor shaft, Fig. 6.1(b), would seem to be more suitable for use with axial compressors. Maximum use is made of the space between the compressor and turbine, so that the annular system should have a lower pressure loss and result in an engine of minimum diameter. The disadvantages are as follows. Firstly, although a large number of fuel jets can be employed, it is more difficult to obtain an even fuel–air distribution and an even outlet temperature distribution: the natural symmetry of the tubular chamber is very helpful in this respect. Secondly, the annular chamber is inevitably weaker structurally

(a) Cylindrical combustion chambers

(b) Annular combustion chamber

(c)  Combustion chamber for industrial gas turbine

**FIG. 6.1  Three types of combustion system**

and it is difficult to avoid buckling of the hot flame-tube walls: this presents problems in engines of large diameter. Thirdly, most of the development work must be carried out on the complete chamber, requiring a test facility capable of supplying the full engine air mass flow, whereas with the multi-chamber layout all the work can be done on one tubular chamber using its appropriate fraction of the total mass flow.

These disadvantages led designers to the idea of a *tubo-annular* (or *cannular*) system which is a combination of (*a*) and (*b*). This consists of tubular flame-tubes (or 'cans') uniformly spaced around an annular casing. The third disadvantage mentioned above is only partially mitigated because although initial development can be undertaken on a sector of the annulus containing a single flame-tube, the side-walls of the sector prevent it from being truly representative of the complete chamber. Tubo-annular chambers have been widely used for large aircraft engines.

The space occupied by the combustion system of an industrial gas turbine is usually only of secondary importance, and the combustion may be carried out in one or two large cylindrical chambers feeding the turbine via a scroll or volute. The chambers may be situated at one end of the heat-exchanger when this component is included in the cycle, Fig. 6.1(c). Large combustion chambers might enable adjustable valves or dampers to be incorporated for regulating the air flowing to different points in the combustion process. The value of this will be discussed in section 6.3.

For the remainder of this chapter we shall concentrate mainly on the way in which combustion is arranged to take place inside a flame-tube and need not be concerned with the overall configuration of the combustion system.

## 6.2  Some important factors affecting combustion chamber design

Little of the experience gained from the design of commercial heating plant, such as oil-fired boilers, is of use to the designer of combustion systems for gas turbines. A number of requirements peculiar to the gas turbine makes the problem quite different, and these may be summarized as follows.

(*a*)  The temperature level of the gases after combustion must be comparatively low to suit the highly stressed turbine materials.

(*b*)  At the end of the combustion space the temperature distribution must be of known form if a high turbine performance is to be realized and the blades are not to suffer from local overheating. It need not be uniform, but might with advantage increase with radius over the turbine annulus because the blade stresses decrease from root to tip.

(*c*)  Combustion must be maintained in a stream of air moving with a high velocity in the region of 30–60 m/s, and stable operation is required over a wide range of air/fuel ratio from full load to idling conditions. The

even  fuel-air distr

air/fuel ratio might vary from about 60:1 to 120:1 for simple gas turbines and from 100:1 to 200:1 if a heat-exchanger is employed. The high dilution implied by these figures is necessary to satisfy the first requirement.

(d)  The formation of carbon deposits ('coking') must be avoided, particularly the hard brittle variety. Small particles carried into the turbine in the high-velocity gas stream can erode the blades; furthermore, aerodynamically excited vibration in the combustion chamber might cause sizeable pieces of carbon to break free resulting in even worse damage to the turbine.

(e)  In aircraft gas turbines, combustion must also be stable over a wide range of chamber pressure because this parameter changes with altitude and forward speed. Furthermore, there are such special problems as the capability of relighting with the compressor wind-milling at high altitude, and the need to avoid a smoky exhaust.

Probably the only feature of the gas turbine which eases the combustion designer's problem is the peculiar interdependence of compressor delivery air density and mass flow which leads to the velocity of the air at entry to the combustion system being reasonably constant over the operating range.

For aircraft applications there are the additional limitations of small space and low weight, which are, however, slightly offset by somewhat shorter endurance requirements. Aircraft engine combustion chambers are normally constructed of light-gauge heat-resisting alloy sheet (approx. 0·8 mm thick), but are only expected to have a life of some 10 000 hours. Combustion chambers for industrial gas turbine plant may be constructed on much sturdier lines but, on the other hand, a life of about 100 000 hours is required. Refractory linings may eventually be used in heavy chambers, although the remarks made in (d) about the effects of hard carbon deposits breaking free apply with even greater force to refractory material. Research on possible ceramics and ceramic coatings is surveyed in Ref. (1).

We have seen that the gas turbine cycle is very sensitive to component inefficiencies, and it is of the utmost importance that the aforementioned requirements should be met without sacrificing combustion efficiency. That is, it is essential that over most of the operating range all the fuel injected should be completely burnt and the full calorific value realized. Also, because any pressure drop between inlet and outlet of the combustion chamber leads to both an increase in s.f.c. and reduction in specific power output, it is essential to keep the pressure loss to a minimum. It will be appreciated from the discussion which follows, that the smaller the space available for combustion and hence the shorter the time available for the necessary chemical reactions, the more difficult it is to meet all the requirements and still obtain a high combustion efficiency with low pressure loss. Clearly in this respect the designer of combustion systems for industrial gas turbines has an easier task than his counterpart in the aircraft field.

## 6.3   The combustion process

Combustion of a liquid fuel involves the mixing of a fine spray of droplets with air, vaporization of the droplets, the breaking down of heavy hydrocarbons into lighter fractions, the intimate mixing of molecules of these hydrocarbons with oxygen molecules, and finally the chemical reactions themselves. A high temperature, such as is provided by the combustion of an approximately stoichiometric mixture, is necessary if all these processes are to occur sufficiently rapidly for combustion in a moving air stream to be completed in a small space. It should be clear that the combustion of a gaseous fuel presents fewer problems, but much of what follows is still applicable.

Since the overall air/fuel ratio is in the region of 100:1, while the stoichiometric ratio is approximately 15:1, the first essential is that the air should be introduced in stages. Three such stages can be distinguished. About 15–20 per cent of the air is introduced around the jet of fuel in the *primary zone* to provide the necessary high temperature for rapid combustion. Some 30 per cent of the total air is then introduced through holes in the flame-tube in the *secondary zone* to complete the combustion. For high combustion efficiency, this air must be injected carefully at the right points in the process, to avoid chilling the flame locally and drastically reducing the reaction rate in that neighbourhood. Finally, in the *tertiary* or *dilution zone* the remaining air is mixed with the products of combustion to cool them down to the temperature required at inlet to the turbine. Sufficient turbulence must be promoted so that the hot and cold streams are thoroughly mixed to give the desired outlet temperature distribution, with no hot streaks which would damage the turbine blades.

The zonal method of introducing the air cannot by itself give a self-piloting flame in an air stream which is moving an order of magnitude faster than the flame speed in a burning mixture. The second essential feature is therefore a recirculating flow pattern which directs some of the burning mixture in the primary zone back on to the incoming fuel and air. One way of achieving this is shown in Fig. 6.2, which is typical of British practice. The fuel is injected in the same direction as the air stream, and the primary air is introduced through twisted radial vanes, known as *swirl vanes*, so that the resulting

**FIG. 6.2 Combustion chamber with swirl vanes**

vortex motion will induce a region of low pressure along the axis of the chamber. This vortex motion is sometimes enhanced by injecting the secondary air through short tangential chutes in the flame-tube, instead of through plain holes as in the figure. The net result is that the burning gases tend to flow towards the region of low pressure, and some portion of them is swept round towards the jet of fuel as indicated by the arrows.

Many other solutions to the problem of obtaining a stable flame are possible. One American practice is to dispense with the swirl vanes and achieve the recirculation by a careful positioning of holes in the flame-tube downstream of a hemispherical baffle as shown in Fig. 6.3(a). Figure 6.3(b) shows a possible solution using upstream injection which gives good mixing of the fuel and primary air. It is difficult to avoid overheating the fuel injector, however, and upstream injection is employed more for after-burners (or 'reheat') in the jet-pipe of aircraft engines than in main combustion systems. After-burners operate only for short periods of thrust-boosting. Finally,

(a)                                  (b)                                  (c)

**FIG. 6.3 Methods of flame stabilization**

Fig. 6.3(c) illustrates a vaporizer system wherein the fuel is injected at low pressure into walking-stick shaped tubes placed in the primary zone. A rich mixture of fuel vapour and air issues from the vaporizer tubes in the upstream direction to mix with the remaining primary air passing through holes in a baffle around the fuel supply pipes. The fuel system is much simpler, and the difficulty of arranging for an adequate distribution of fine droplets over the whole operating range of fuel flow [see under heading 'Fuel injection', p. 176] is overcome. The problem in this case is to avoid local 'cracking' of the fuel in the vaporizer tubes with the formation of deposits of low thermal conductivity leading to overheating and burn-out. Vaporizer schemes are receiving increasing attention, particularly for annular chambers wherein it is inherently more difficult to obtain a satisfactory fuel–air distribution with sprays of droplets from high-pressure injectors. They have been employed in several successful aircraft engines.

Having described the way in which the combustion process is accomplished, it is now possible to see how incomplete combustion and pressure losses arise. When not due simply to poor fuel injector design leading to fuel droplets being carried along the flame-tube wall, incomplete combustion may be caused by local chilling of the flame at points of secondary air entry. This can easily reduce the reaction rate to the point where some of the products into which the fuel has decomposed are left in their partially burnt state, and the temperature at the downstream end of the chamber is normally below that at which the burning of these products can be expected to take place. Since the lighter hydrocarbons into which the fuel has decomposed have a higher ignition temperature than the original fuel, it is clearly difficult to prevent some chilling from taking place, particularly if space is limited and the secondary air cannot be introduced gradually enough. If devices are used to increase large-scale turbulence and so distribute the secondary air more uniformly throughout the burning gases, the combustion efficiency will be improved but at the expense of increased pressure loss. A satisfactory compromise must somehow be reached.

Although a high combustion efficiency may be achieved over the important part of the operating range, the performance of a combustion chamber of fixed proportions will inevitably deteriorate as the air/fuel ratio, air mass flow and chamber pressure depart markedly from design point conditions. The value of adjustable valves which might be incorporated in large combustion chambers for industrial gas turbines, Fig. 6.1(c), should now be clear. Such valves might take the form of plate or cylindrical shutters which could be rotated relative to the holes in the 'cold' end of the flame tube. These would enable the proportions of primary and secondary air to be adjusted as the operating conditions vary, so helping to maintain a high combustion efficiency when the gas turbine is operating at part load. They would also enable different grades of fuel, which to give the best results might require different values of air/fuel ratio in the primary zone, to be used without modification to the combustion chamber.

Combustion chamber pressure loss is due to two distinct causes: (i) skin friction and turbulence and (ii) the rise in temperature due to combustion. The stagnation pressure drop associated with the latter, often called the *fundamental loss*, arises because an increase in temperature implies a decrease in density and consequently an increase in velocity and momentum of the stream. A pressure force $(\Delta p \times A)$ must be present to impart the increase in momentum. One of the standard idealized cases considered in gas dynamics is that of a heated gas stream flowing without friction in a duct of constant cross-sectional area. The stagnation pressure drop in this situation, for any given temperature rise, can be predicted with the aid of the Rayleigh-line functions (see Appendix A.4). When the velocity is low and the fluid flow can be treated as incompressible (in the sense that although $\rho$ is a function of $T$ it is independent of $p$), a simple equation for the pressure drop can be found as follows.

The momentum equation for one-dimensional frictionless flow in a duct of constant cross-sectional area $A$ is

$$A(p_2 - p_1) + m(C_2 - C_1) = 0$$

For incompressible flow the stagnation pressure $p_0$ is simply $(p + \rho C^2/2)$, and

$$p_{02} - p_{01} = (p_2 - p_1) + \tfrac{1}{2}(\rho_2 C_2^2 - \rho_1 C_1^2)$$

Combining these equations, remembering that $m = \rho_1 A C_1 = \rho_2 A C_2$,

$$p_{02} - p_{01} = -(\rho_2 C_2^2 - \rho_1 C_1^2) + \tfrac{1}{2}(\rho_2 C_2^2 - \rho_1 C_1^2)$$
$$= -\tfrac{1}{2}(\rho_2 C_2^2 - \rho_1 C_1^2)$$

The stagnation pressure loss as a fraction of the inlet dynamic head then becomes

$$\frac{p_{01} - p_{02}}{\rho_1 C_1^2/2} = \left(\frac{\rho_2 C_2^2}{\rho_1 C_1^2} - 1\right) = \left(\frac{\rho_1}{\rho_2} - 1\right)$$

Finally, since $\rho \propto 1/T$ for incompressible flow,

$$\frac{p_{01} - p_{02}}{\rho_1 C_1^2/2} = \left(\frac{T_2}{T_1} - 1\right)$$

This will be seen from the Appendix to be the same as the compressible flow value of $(p_{01} - p_{02})/(p_{01} - p_1)$ in the limiting case of zero inlet Mach number. At this condition $T_2/T_1 = T_{02}/T_{01}$.

Although the assumptions of incompressible flow and constant cross-sectional area are not quite true for a combustion chamber, the result is sufficiently accurate to provide us with the order of magnitude of the fundamental loss. Thus, since the outlet/inlet temperature ratio is in the region of 2–3, it is clear that the fundamental loss is only about 1–2 inlet dynamic heads. The pressure loss due to friction is found to be very much higher—of the order of 20 inlet dynamic heads. When measured by pitot traverses at inlet and outlet with no combustion taking place, it is known as the *cold loss*. That the friction loss is so high, is due to the need for large-scale turbulence. Turbulence of this kind is created by the devices used to stabilize the flame, e.g. the swirl vanes in Fig. 6.2. In addition, there is the turbulence induced by the jets of secondary and dilution air. The need for good mixing of the secondary air with the burning gases to avoid chilling has been emphasized. Similarly, good mixing of the dilution air to avoid hot streaks in the turbine is essential. In general, the more effective the mixing the higher the pressure loss. Here again a compromise must be reached: this time between uniformity of outlet temperature distribution and low pressure loss.

Usually it is found that adequate mixing is obtained merely by injecting air through circular or elongated holes in the flame-tube. Sufficient penetration of the cold air jets into the hot stream is achieved as a result of the cold air having the greater density. The pressure loss produced by such a mixing process is associated with the change in momentum of the streams before and after mixing. In aircraft gas turbines the duct between combustion chamber outlet and turbine inlet is very short, and the compromise reached between good temperature distribution and low pressure loss is normally such that the temperature non-uniformity is up to $\pm 10$ per cent of the mean value. The length of duct is often greater in an industrial gas turbine and the temperature distribution at the turbine inlet may be more uniform, although at the expense of increased pressure drop due to skin friction in the ducting. The paper by Lefebvre and Norster in Ref. (5) outlines a method of proportioning a tubular combustion chamber to give the most effective mixing for a given pressure loss. Making use of empirical data from mixing experiments, such as dilution hole discharge coefficients, the authors show how to estimate the optimum ratio of flame-tube to casing diameter, and the optimum pitch/diameter ratio and number of dilution holes.

## 6.4   Combustion chamber performance

The main factors of importance in assessing combustion chamber performance are (a) pressure loss, (b) combustion efficiency, (c) outlet temperature distribution, (d) stability limits and (e) combustion intensity. We need say no more of (c), but (a) and (b) require further comment, and (d) and (e) have not yet received attention.

### Pressure loss

We have seen in section 6.3 that the overall stagnation pressure loss can be regarded as the sum of the fundamental loss (a small component which is a function of $T_{02}/T_{01}$) and the frictional loss. Our knowledge of friction in ordinary turbulent pipe flow at high Reynolds number would suggest that when the pressure loss is expressed non-dimensionally in terms of the dynamic head it will not vary much over the range of Reynolds number under which combustion systems operate. Experiments have shown, in fact, that the overall pressure loss can often be expressed adequately by an equation of the form

$$\text{pressure loss factor (P.L.F.)} = \frac{\Delta p_0}{m^2/2\rho_1 A_m^2} = K_1 + K_2\left(\frac{T_{02}}{T_{01}} - 1\right) \qquad (6.1)$$

Note that rather than $\rho_1 C_1^2/2$, a conventional dynamic head is used based on a velocity calculated from the inlet density, air mass flow $m$, and maximum cross-sectional area $A_m$ of the chamber. This velocity—sometimes known as the reference velocity—is more representative of conditions in the chamber, and the convention is useful when comparing results from chambers of different shape. Equation (6.1) is illustrated in Fig. 6.4. If $K_1$ and $K_2$ are determined from a combustion chamber on a test rig from a cold run and a hot run, then equation (6.1) enables the pressure loss to be estimated when the chamber is operating as part of a gas turbine over a wide range of conditions of mass flow, pressure ratio and fuel input.

   To give an idea of relative orders of magnitude, typical values of P.L.F. at design operating conditions for tubular, tubo-annular and annular combustion chambers are 35, 25 and 18 respectively. There are two points which must be remembered when considering pressure loss data. Firstly, the velocity of the air leaving the last stage of an axial compressor is quite high—say 150 m/s —and some form of diffusing section is introduced between the compressor and combustion chamber to reduce the velocity to about 60 m/s. It is a matter of convention, depending upon the layout of the gas turbine, as to how much of the stagnation pressure loss in this diffuser is included in the P.L.F. of the combustion system. In other words, it depends on where the compressor is deemed to end and the combustion chamber begin.

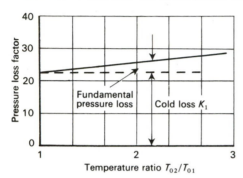

FIG. 6.4 **Variation of pressure loss factor**

Secondly, it should be appreciated from Chapters 2 and 3 that from the point of view of cycle performance calculations it is $\Delta p_0$ as a fraction of the compressor delivery pressure ($p_{01}$ in the notation of this chapter) which is the useful parameter. This is related to the P.L.F. as follows.

$$\frac{\Delta p_0}{p_{01}} = \frac{\Delta p_0}{m^2/2\rho_1 A_m^2} \times \frac{m^2/2\rho_1 A_m^2}{p_{01}} = \text{P.L.F.} \times \frac{R}{2}\left(\frac{m\sqrt{T_{01}}}{A_m p_{01}}\right)^2 \qquad (6.2)$$

where the difference between $\rho_1$ and $\rho_{01}$ has been ignored because the velocity is low. By combining equations (6.1) and (6.2) it can be seen that $\Delta p_0/p_{01}$ can be expressed as a function of non-dimensional mass flow at entry to the combustion chamber and combustion temperature ratio: such a relation is useful when predicting pressure losses at conditions other than design, as discussed in Chapter 8. Consider now the two extreme cases of tubular and annular designs. If the values of $\Delta p_0/p_{01}$ are to be similar, it follows from equation (6.2) and the values of P.L.F. given above that the chamber cross-sectional area per unit mass flow ($A_m/m$) can be smaller for the annular design. For aircraft engines, where space and weight are vital, the value of $A_m/m$ is normally chosen to yield a value of $\Delta p_0/p_{01}$ between 4 and 7 per cent. For industrial gas turbine chambers, $A_m/m$ is usually such that $\Delta p_0/p_{01}$ is little more than 2 per cent.

## Combustion efficiency

The efficiency of a combustion process may be found from a chemical analysis of the combustion products. Knowing the air/fuel ratio used and the proportion of incompletely burnt constituents, it is possible to calculate the ratio of the actual energy released to the theoretical quantity available. This approach via chemical analysis is not easy, because not only is it difficult to obtain truly representative samples from the high velocity stream, but also, owing to the high air/fuel ratios employed in gas turbines, the unburnt constituents to be measured are a very small proportion of the whole sample. Ordinary gas analysis apparatus, such as the Orsat, is not adequate and much more elaborate techniques have had to be developed—see Ref. (2).

If an overall combustion efficiency is all that is required, however, and not an investigation of the state of the combustion process at different stages, it is easier to conduct development work on a test rig on the basis of the combustion efficiency which was defined in Chapter 2, namely

$$\eta_b = \frac{\text{theoretical } f \text{ for actual } \Delta T}{\text{actual } f \text{ for actual } \Delta T}$$

For this purpose, the only measurements required are those necessary for determining the fuel/air ratio and the mean stagnation temperatures at inlet and outlet of the chamber. The theoretical $f$ can be obtained from curves such as those in Fig. 2.15.

It is worth describing how the mean stagnation temperature may be measured: there are two aspects, associated with the adjectives 'mean' and 'stagnation'. Firstly, it should be realized from the discussion under heading 'Fuel/air ratio, combustion efficiency and cycle efficiency', p. 52, that the expression for $\eta_b$ arises from the energy equation which consists of such terms as $mc_p T_0$. Since in practice there is always a variation in velocity as well as temperature over the cross-section, it is necessary to determine, not the ordinary arithmetic mean of a number of temperature readings, but what is known as the 'weighted mean'. If the cross-section is divided into a number of elemental areas $A_1, A_2, \ldots, A_i, \ldots, A_n$, at which the stagnation temperatures are $T_{01}, T_{02}, \ldots, T_{0i}, \ldots, T_{0n}$, and the mass flows are $m_1, m_2, \ldots, m_i, \ldots, m_n$, then the weighted mean temperature $T_{0w}$ is defined by

$$T_{0w} = \frac{\Sigma m_i T_{0i}}{\Sigma m_i} = \frac{\Sigma m_i T_{0i}}{m}$$

where the summations are from 1 to $n$. We may assume that $c_p$ is effectively constant over the cross-section. It follows that the product $mc_p T_{0w}$ will be a true measure of the energy passing the section per unit time.

A simple expression for $T_{0w}$ in terms of measured quantities can be derived as follows. The velocity at the centre of each elemental area may be found using a pitot-static tube. Denoting the dynamic head $\rho C^2/2$ by $p_d$, the mass flow for area $A_i$ is then

$$m_i = \rho_i A_i (2p_{di}/\rho_i)^{\frac{1}{2}}$$

If the static pressure is constant over the cross-section, as it will be when there is simple axial flow with no swirl,

$$\rho_i \propto 1/T_i \text{ and } m_i \propto A_i(p_{di}/T_i)^{\frac{1}{2}}$$

and thus

$$T_{0w} = \frac{\Sigma A_i T_{0i}(p_{di}/T_i)^{\frac{1}{2}}}{\Sigma A_i(p_{di}/T_i)^{\frac{1}{2}}}$$

Since $T_i$ has only a second order effect on $T_{0w}$, we can write $T_i = T_{0i}$. Furthermore, it is usual to divide the cross-section into equal areas. The expression then finally reduces to

$$T_{0w} = \frac{\Sigma(p_{di}T_{0i})^{\frac{1}{2}}}{\Sigma(p_{di}/T_{0i})^{\frac{1}{2}}}$$

Thus the weighted mean temperature may be determined directly from measurements of dynamic head and stagnation temperature at the centre of each elemental area. It remains to describe how stagnation temperatures may be measured.

In gas turbine work, temperatures are usually measured by thermocouples. The high accuracy of a pitot-static tube is well known, but considerable difficulty has been experienced in designing thermocouples to operate in a high temperature fast-moving gas stream with a similar order of accuracy. Since the combustion efficiency rarely falls below 98 per cent over much of the operating range, accurate measurements are essential. Chromel–alumel thermocouples have been found to withstand the arduous requirements of combustion chamber testing satisfactorily, and give accurate results up to about 1300 K if special precautions are taken. It must be remembered that the temperature recorded is that of the hot junction of the thermocouple which for various reasons may not be at the temperature of the gas stream in which it is situated, particularly if the velocity of the stream is high.

If it be imagined that the thermocouple junction is moving with the stream of gas, then the temperature of the junction may differ from the static temperature of the gas by an amount depending upon the conduction of heat along the thermocouple wires, the convection between the junction and the gas stream, the radiation from the hot flame to the junction, and the radiation from the junction to the walls of the containing boundary if these are cooler than the junction. There is an additional possible error, because in practice the thermocouple is stationary, and the gas velocity will be reduced by friction in the boundary layer around the thermocouple junction. Kinetic energy is transformed into internal energy, some of which raises the temperature of the junction, while some is carried away by convection. In a high-speed gas stream it is obviously important to know how much of the velocity energy is being measured as temperature. The temperature corresponding to the velocity energy, i.e. the dynamic temperature, is about 40 K for a velocity of 300 m/s.

Since it is the stagnation temperature which is of interest, it is usual to place the thermocouple wires and junction in a metal tube in which the gas stream can be brought to rest adiabatically so that almost the whole of the dynamic temperature is measured, on the same principle as a pitot tube measuring stagnation pressure. Figure 6.5(a) shows one form of stagnation thermocouple which will measure about 98 per cent of the dynamic temperature as against the 60–70 per cent measured by a simple junction placed directly in the gas stream. A large hole facing upstream allows the gas to enter the tube, while a small hole, not more than 5 per cent of the area of the inlet orifice, provides sufficient ventilation without spoiling the pitot effect. This form of thermocouple is excellent for all work where radiation effects are small, such as the measurement of compressor delivery temperature. Where radiation effects are appreciable, as at the outlet of a combustion chamber, it is preferable to use a thermocouple of the type shown in Fig. 6.5(b). A radiation error of the order of 60 K in 1300 K is quite possible with

FIG. 6.5 Stagnation thermocouples

a completely unshielded thermocouple. A short length of polished stainless steel sheet, twisted into a helix and placed in front of the junction, provides an effective radiation shield without impeding the flow of gas into the thermocouple tube. One or more concentric cylindrical shields may also be included. The bending of the wires so that about two or three centimetres run parallel with the direction of the stream, i.e. parallel with an isothermal, reduces the error due to conduction of heat away from the junction along the wires. This is in most cases quite a small error if the wires are of small diameter. If all these precautions are taken, stagnation temperature measurement up to 1300 K is possible to within ±5 K. While these few remarks do not by any means exhaust all the possibilities of thermocouple design, e.g. see Ref. (3), they at least indicate the extreme care necessary when choosing thermocouples for gas turbine temperature measurement.

    The temperature at outlet from the combustion chamber is not often measured on an engine, where there is the added problem that mechanical failure of the thermocouple support system could lead to major damage in the turbine. For this reason, on an engine it is normal to measure the temperature downstream of the turbine. The *jet pipe temperature*, as it is called, is indirectly related to the combustion outlet temperature (i.e. turbine inlet temperature), and the necessary relation can be established by the methods described in Chapter 8.

## Stability limits

For any particular combustion chamber there is both a rich and a weak limit to the air/fuel ratio beyond which the flame is unstable. Usually the limit is taken as the air/fuel ratio at which the flame blows out, although instability often occurs before this limit is reached. Such instability takes the form of rough running, which not only indicates poor combustion, but sets up aerodynamic vibration which reduces the life of the chamber and causes blade vibration problems. The range of air/fuel ratio between the rich and weak limits is reduced with increase of air velocity, and if the air mass flow is increased beyond a certain value it is impossible to initiate combustion at all. A typical stability loop is shown in Fig. 6.6, where the limiting air/

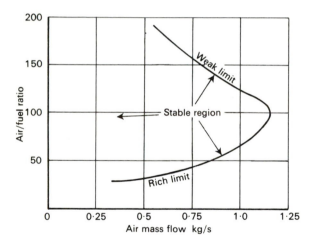

**FIG. 6.6 Stability loop**

fuel ratio is plotted against air mass flow. If a combustion chamber is to be suitable, its operating range defined by the stability loop must obviously cover the required range of air/fuel ratio and mass flow of the gas turbine for which it is intended. Furthermore, allowance must be made for conditions which prevail when the engine is accelerated or decelerated. For example, on acceleration there will be a rapid increase in fuel flow as the 'throttle' is opened while the air flow will not reach its new equilibrium value until the engine has reached its new speed. Momentarily the combustion system will be operating with a very low air/fuel ratio. Most control systems have a built-in device which places an upper limit on the rate of change of fuel flow: not only to avoid blow-out, but also to avoid transient high temperatures in the turbine.

The stability loop is a function of the pressure in the chamber: a decrease in pressure reduces the rate at which the chemical reactions proceed, and consequently it narrows the stability limits. For aircraft engines it is important to check that the limits are sufficiently wide with a chamber pressure equal to the compressor delivery pressure which exists at the highest operating altitude. Engines of high pressure ratio present less of a problem to the combustion chamber designer than those of low pressure ratio. If the stability limits are too narrow, changes must be made to improve the recirculation pattern in the primary zone.

## Combustion intensity

The size of combustion chamber is determined primarily by the rate of heat release required. The nominal heat release rate can be found from $m f Q_{net,p}$ where $m$ is the air mass flow, $f$ the fuel/air ratio and $Q_{net,p}$ the net calorific value of the fuel. Enough has been said for the reader to appreciate that the larger the volume which can be provided the easier it will be to achieve a low pressure drop, high efficiency, good outlet temperature distribution and satisfactory stability characteristics.

The design problem is also eased by an increase in the pressure and temperature of the air entering the chamber, for two reasons. Firstly, an increase will reduce the time necessary for the 'preparation' of the fuel and air mixture (evaporation of droplets, etc.) making more time available for the combustion process itself. Note that since the compressor delivery temperature is a function of the compressor delivery pressure, the pressure (usually expressed in atmospheres) is an adequate measure of both.

Secondly, we have already observed under the heading 'Stability limits' above, that the combustion chamber pressure is important because of its effect on the rate at which the chemical reactions proceed. An indication of the nature of this dependence can be obtained from chemical kinetics, i.e. kinetic theory applied to reacting gases. By calculating the number of molecular collisions per unit time and unit volume which have an energy exceeding a certain activation value $E$, it is possible to obtain the following expression for the rate $r$ at which a simple bimolecular gas reaction proceeds, Ref. (4).

$$r \propto m_j m_k \rho^2 \sigma^2 T^{1/2} M^{-3/2} e^{-E/R_0 T}$$

$\rho$ and $T$ have their usual meaning and the other symbols denote:

| | |
|---|---|
| $m_j, m_k$ | local concentrations of molecules $j$ and $k$ |
| $\sigma$ | mean molecular diameter |
| $M$ | mean molecular weight |
| $E$ | activation energy |
| $R_0$ | universal gas constant |

Substituting for $\rho$ in terms of $p$ and $T$ we can for our purpose simplify the expression to

$$r \propto p^2 f(T)$$

Now $T$ is maintained at a high value by having an approximately stoichiometric mixture in the primary zone: we are concerned here with the independent variable $p$. It is not to be expected that the theoretical exponent 2 will apply to the complex set of reactions occurring when a hydrocarbon fuel is burnt in air, and experiments with homogeneous mixtures in stoichiometric proportions suggest that it should be 1·8. At first sight it appears therefore that the design problem should be eased as the pressure is increased according to the law $p^{1·8}$. In fact there is reason to believe that under design operating conditions the chemical reaction rate is not a limiting factor in an actual combustion chamber where physical mixing processes play such an important role, and that an exponent of unity is more realistic. This is not to say that under extreme conditions—say at high altitude—the performance will not fall off more in accordance with the $p^{1·8}$ law.

A quantity known as the *combustion intensity* has been introduced to take account of the foregoing effects. One definition used is

$$\text{combustion intensity} = \frac{\text{heat release rate}}{\text{comb. vol.} \times \text{pressure}} \text{ kW/m}^3 \text{ atm}$$

Another definition employs $p^{1·8}$, with the units $\text{kW/m}^3 \text{ atm}^{1·8}$. However it is defined, certainly the lower the value of the combustion intensity the easier it is to design a combustion system which will meet all the desired requirements. It is quite inappropriate to compare the performance of different systems on the basis of efficiency, pressure loss, etc., if they are operating with widely differing orders of combustion intensity. In aircraft systems the combustion intensity is in the region $2–5 \times 10^4 \text{ kW/m}^3 \text{ atm}$,† while in industrial gas turbines the figure may be a tenth of this owing to the larger volume of combustion space available and to the fact that when a heat-exchanger is employed a smaller heat release is required.

## 6.5   Some practical problems

To conclude this chapter we will discuss briefly some of the problems which have so far not been mentioned but which are none the less important. These are concerned with (i) flame-tube cooling, (ii) fuel injection, (iii) ignition, (iv) the use of cheaper fuels and (v) pollution.

### Flame-tube cooling

One problem which has assumed greater importance as permissible turbine

---

† Note that 1 $\text{kW/m}^3$ atm = 96·62 Btu/h ft³ atm.

inlet temperatures have increased is that of cooling the flame-tube. The tube receives energy by convection from the hot gases and by radiation from the flame. It loses energy by convection to the cooler air flowing along the outer surface and by radiation to the outer casing, but this loss is not sufficient to maintain the tube wall at a safe temperature. A common practice is to leave narrow annular gaps between overlapping sections of the flame-tube, as shown in Fig. 6.7(a), so that a film of cooling air is swept along the inner surface. Corrugated 'wigglestrip', spot welded to successive lengths of flame-tube, provides a tube of adequate stiffness with annular gaps which do not vary too much with thermal expansion. Another method, illustrated in Fig. 6.7(b), is to use a ring of small holes with an internal splash ring to deflect the jets along the inner surface. A film of cool air both insulates the surface from the hot gases, and removes energy received by radiation.

(a)           (b)

**FIG. 6.7 Film cooling of flame-tube**

Although empirical relations are available from which it is possible to predict convective heat transfer rates when film cooling a plate of known temperature, e.g. see the contribution by Sturgess in Ref. (1), the emissivities of the flame and flame-tube can vary so widely that prediction of the flame-tube temperature from an energy balance is not possible with any accuracy. Even in this limited aspect of combustion chamber design, final development is still a matter of trial and error on the test rig. The emissivity of the flame varies with the type of fuel, tending to increase with the specific gravity. Carbon dioxide and water vapour are the principle radiating components in non-luminous flames, and soot particles in luminous flames. It is worth noting that vaporizer systems ease the problem, because flames from pre-mixed fuel vapour–air mixtures have a lower luminosity than those from sprays of droplets.

Higher turbine inlet temperatures imply the use of lower air/fuel ratios, with consequently less air available for film cooling. Furthermore, the use of a higher cycle temperature is usually accompanied by the use of a higher cycle pressure ratio to obtain the full benefit in terms of cycle efficiency. Thus the temperature of the air leaving the compressor is increased and its cooling potential is reduced. If permissible turbine inlet temperatures increase much beyond 1500 K, some form of porous material may have to be used for the flame-tube: 'effusion cooling', as it is called, is very much more economical in cooling air than film cooling.

## Fuel injection

Although vaporizer systems are receiving increasing attention, most combustion chambers employ high-pressure fuel systems in which the liquid fuel is forced through a small orifice to form a conical spray of fine droplets in the primary zone. The fuel is said to be 'atomized' and the burner is often referred to as an 'atomizer'. It should be realized that even a vaporizer system requires an auxiliary starting burner of this type.

In the simplest form of atomizing burner, fuel is fed to a conical vortex chamber via tangential ports which impart a swirling action to the flow. The vortex chamber does not run full but has a vapour/air core. The combination of axial and tangential components of velocity causes a hollow conical sheet of fuel to issue from the orifice, the ratio of the components determining the cone angle. This conical sheet then breaks up in the air stream into a spray of droplets, and the higher the fuel pressure the closer to the orifice does this break up occur. There will be a certain minimum fuel pressure at which a fully developed spray will issue from the orifice, although for the following reason the effective minimum pressure may well be higher than this.

The spray will consist of droplets having a wide range of diameter, and the degree of atomization is usually expressed in terms of a mean droplet diameter. In common use is the *Sauter mean diameter*, which is the diameter of a drop having the same surface/volume ratio as the mean value for the spray: 50–100 microns is the order of magnitude used in practice. The higher the supply pressure, the smaller the mean diameter. If the droplets are too small they will not penetrate far enough into the air stream, and if too large the evaporation time may be too long. The effective minimum supply pressure is that which will provide the required degree of atomization.

The object is to produce an approximately stoichiometric mixture of air and fuel uniformly distributed across the primary zone, and to achieve this over the whole range of fuel flow from idling to full load conditions. Herein lies the main problem of burner design. If the fuel is metered by varying the pressure in the fuel supply line, the simple type of burner just described (sometimes referred to as the *simplex*) will have widely differing atomizing properties over the range of fuel flow. It must be remembered that the flow through an orifice is proportional to the square root of the pressure difference across it, and thus a fuel flow range of 10:1 implies a supply pressure range of 100:1. If the burner is designed to give adequate atomization at full load, the atomization will be inadequate at low load. This problem has been overcome in a variety of ways.

An early solution, now only of historical interest, was provided by the Lubbock burner, Fig. 6.8. The tangential ports leading to the vortex chamber, in series with and smaller in area than the final orifice, were closed progressively by a spring loaded piston as the pressure decreased. In this manner a reduction in flow was achieved by a decrease in both port area and supply pressure. Thus the required range of fuel flow could be obtained with a smaller range of pressure and consequently with a smaller variation in

**FIG. 6.8 Lubbock burner**

atomization. Difficulties associated with sticking pistons, and the matching of a set of burners to give the same flow at any given supply pressure, proved insurmountable.

Perhaps the most commonly used solution to the problem is that employed in the *duplex* burner, an example of which is shown in Fig. 6.9(a). Two fuel manifolds are required, each supplying independent orifices. The small central orifice is used alone for the lower flows, while the larger annular orifice surrounding it is additionally brought into operation for the higher flows. The sketch also shows a third annulus formed by a shroud through which air passes to prevent carbon deposits building up on the face of the burner. This feature is incorporated in most burners. An alternate form of duplex burner employs a single vortex chamber and final orifice, with the two fuel supply lines feeding two sets of tangential ports in the vortex chamber.

Figure 6.9(b) illustrates a second practical method of obtaining good

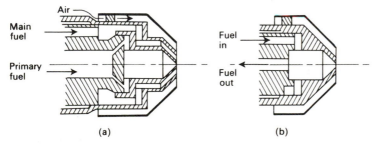

**FIG. 6.9 Duplex and spill burners**

atomization over a wide range of fuel flow: the *spill* burner. It is virtually a simplex burner with a passage from the vortex chamber through which excess fuel can be spilled off. The supply pressure can remain at the high value necessary for good atomization while the flow through the orifice is reduced by reducing the pressure in the spill line. One disadvantage of this system is that when a large quantity of fuel is being recirculated to the pump inlet there may be undesirable heating and consequent deterioration of the fuel.

Burners of the type described here by no means exhaust all the possibilities. For example, for small gas turbines having a single chamber, it has been found possible to use rotary atomizers. Here fuel is fed to a spinning disc or cup and flung into the air stream from the rim. High tip speeds are required for good atomization, but only a low pressure fuel supply is required.

## Ignition

The first step in starting a gas turbine is to accelerate the compressor to a speed which gives an air flow capable of sustaining combustion. In some cases this is achieved by feeding compressed air from an external supply directly to the turbine driving the compressor. It is more common, however, to employ an electric motor or small auxiliary turbine connected to the main shaft by a reduction gear and clutch. The auxiliary turbine may operate from a compressed air supply or, as in the case of military aircraft where independence of ground equipment is desirable, from a cartridge. During the period of acceleration the ignition system is switched on and fuel is fed to the burners. An igniter plug is situated near the primary zone in one or two of the flame-tubes or cans. Once the flame is established, suitably placed interconnecting tubes between the cans permit 'light round', i.e. flame propagation from one flame-tube to another. Light round presents few problems in annular chambers. With an aircraft gas turbine, there is the additional requirement that re-ignition must be possible under windmilling conditions if for any reason the flame is extinguished at altitude.

The ignition performance can be expressed by an ignition loop which is similar to the stability loop of Fig. 6.6 but lying inside it. That is, at any given air mass flow the range of air/fuel ratio within which the mixture can be ignited is smaller than that for which stable combustion is possible once ignition has occurred. The ignition loop is very dependent on combustion chamber pressure, and the lower the pressure the more difficult the problem of ignition. Re-light of an aircraft engine at altitude is thus the most stringent requirement. Although high-tension sparking plugs similar to those used in piston engines are adequate for ground starting, a spark of much greater energy is necessary to ensure ignition under adverse conditions. A *surface-discharge igniter*, yielding a spark having an energy of about three joules at the rate of one per second, is probably the most widely used type for aircraft gas turbines in which fuel is injected as a spray of droplets.

One example of a surface-discharge igniter is shown in Fig. 6.10. It consists of a central and outer electrode separated by a ceramic insulator except near the tip where the separation is by a layer of semiconductor material. When a condenser voltage is applied, current flows through the semiconductor which becomes incandescent and provides an ionized path of low resistance for the energy stored in the capacitor. Once ionization has occurred, the main discharge takes place as an intense flashover. To obtain good performance and long life the location of the igniter is critical: it must protrude through the layer of cooling air on the inside of the flame-tube wall to the outer edge of the fuel spray, but not so far as to be seriously wetted by the fuel.

For vaporizing combustion chambers, some form of *torch igniter* is necessary. This comprises a spark plug and auxiliary spray burner in a common housing. The location is not critical, but it is a bulkier and heavier system than the surface-discharge type. The torch igniter is particularly suitable for industrial gas turbines, and has the advantage that the auxiliary burner can be supplied with distillate fuel from a separate tank when a heavy oil is used as the main fuel.

FIG. 6.10 **Surface-discharge igniter**

## Use of cheap fuels

The gas turbine has found a use in process industries where the power output is desired wholly or in part as compressed air, or steam from a waste-heat boiler. It is particularly suitable when there is a gaseous by-product of the process which can be used as fuel for the gas turbine. One obvious example is the use of a gas turbine in a steel mill for supplying air to the blast furnaces while using blast furnace gas as a fuel. There is little difficulty in designing a combustion system to handle gaseous fuels and it is not this form of cheap fuel with which we are concerned here. A useful discussion of the subject, including the description of a dual fuel burner suitable for both gaseous and liquid fuels, can be found in the paper by Pilkington and Carlisle in Ref. (5).

That the gas turbine has taken so long to compete seriously with other forms of prime mover, except for aircraft applications where small size and low weight are vital, is largely due to the difficulties encountered when trying to use *residual oil.* This cheap fuel is the residue from crude oil following the extraction of profitable light fractions. Some of its undesirable characteristics are:

(a)  high viscosity requiring heating before delivery to the atomizers;
(b)  tendency to polymerize to form tar or sludge when overheated;
(c)  incompatibility with other oils with which it might come into contact, leading to jelly-like substances which can clog the fuel system;
(d)  high carbon content leading to excessive carbon deposits in the combustion chamber;

(e)  presence of vanadium, the vanadium compounds formed during combustion causing corrosion in the turbine;

(f)  presence of alkali metals, such as sodium, which combine with sulphur
in the fuel to form corrosive sulphates;

(g)  relatively large amount of ash, causing build up of deposits on the
nozzle blades with consequent reduction in air mass flow and power
output.

Characteristics (a), (b), (c) and (d) have a nuisance value but the problems to
which they give rise can be overcome without much difficulty. (e), (f) and (g)
on the other hand have proved to be serious.

The rate of corrosion from (e) and (f) increases with turbine inlet temperature, and early industrial gas turbines designed for residual oil operated with
temperatures not much greater than 900 K to avoid the problem. Such a low
cycle temperature inevitably meant a low cycle efficiency. It has now been
found that the alkali metals can be removed by washing the fuel oil with
water and centrifuging the mixture, and that fuel additives such as magnesium
can be used to neutralize the vanadium. Although the latter treatment costs
very little, the former is expensive particularly when carried out by the con-
sumer on a relatively small scale: the fuel is no longer 'cheap' and the gas
turbine becomes barely competitive.

With regard to (g), i.e. build up of deposits, it has been found that applications involving intermittent operation (e.g. locomotive propulsion) are more
suitable than those involving continuous running (e.g. base load electric
power generation). This is because expansions and contractions associated
with thermal cycling cause the deposits to crack off. No satisfactory solution
has been found for continuous running plant, and high maintenance costs
would be an additional factor making the gas turbine uneconomic.

Finally, various experimental coal burning gas turbines have been built
but without success. The problems of ash removal after combustion, or
treatment of the coal itself to remove the impurities leading to ash, have
prevented the gas turbine from competing successfully with coal burning
steam power plant.

## Pollution

From the related facts that a gas turbine uses very high air/fuel ratios and
provides a combustion efficiency of almost 100 per cent over most of the
operating range, it should be clear that the gas turbine is never likely to be a
major contributor to atmospheric pollution. The main trace pollutants as
from any air-breathing engine burning fossil fuel, are (a) unburnt hydrocarbons and carbon monoxide, (b) oxides of nitrogen and (c) oxides of sulphur.

The concentration of unburnt hydrocarbons and carbon monoxide in a
gas turbine exhaust is negligible except under idling conditions when the
combustion chamber operating temperature is at its lowest. With turbojet
engines the order of magnitude is about 20–30 ppm (parts per million) by

weight under normal operating conditions. These pollutants arise due to local chilling of the chemical reactions. The concentration is therefore minimal with an annular combustion chamber or single large tubular chamber because such configurations provide a smaller area of cool flame-tube wall per unit volume than do cannular or multi-tubular designs.

In any air breathing engine a small fraction of the nitrogen is oxidized, and the nitric oxide so formed in the exhaust continues to oxidize slowly in the atmosphere to form nitrogen dioxide. Although non-toxic, it is undesirable because it plays an important part in the formation of 'smog'. The oxidation of nitrogen increases with combustion temperature, and in a turbojet exhaust the concentration of nitric oxide is likely to be negligible under idling conditions rising to approximately 100 ppm by weight at full power.

Oxides of sulphur are detrimental not only to plant and animal life, but also to the engine itself because of corrosive acids to which they give rise. The kerosenes used as fuels for most gas turbines, however, have a very low sulphur content and the sulphur emission is negligible. Furthermore, the latest oil refineries are now able to produce fuel economically with a much lower sulphur content than has been specified by users in the past.

Some aircraft gas turbines have exhibited a smoky exhaust indicating the presence of small particles of unburnt carbon. An amount quite negligible from the point of view of atmospheric pollution can have a pronounced visible effect due to the light scattering properties of the particles. A smoky exhaust can be virtually eliminated during development testing by modifications to the primary zone and fuel burners, and it is unlikely to arise at all when the fuel and air are premixed as in vaporizer systems.

It may be concluded that the good pollution characteristics of the gas turbine are likely to increase its competitiveness for industrial applications in the next decade. Much effort, Ref. (6), is being directed to decreasing the emissions still further for aircraft applications, because of the relatively long periods under idling conditions at take-off and landing, and because of the undesirable effect of NO on the ozone layer at high altitude. The NO problem is particularly severe with the high-pressure ratio engines used for aircraft, because of the high temperature at inlet to the combustion chamber. Unfortunately, design changes which reduce the hydrocarbon and CO emissions at idling (e.g. lower flow velocity and decreased chilling in the mixing zones) tend to increase the NO emissions at full power. One promising approach is to use two-stage combustion, with the pilot stage used alone when the engine is idling.

# 7

# Axial and radial flow turbines

As with the compressor, there are two basic types of turbine—radial flow and axial flow. The vast majority of gas turbines employ the axial flow turbine, so most of the chapter will be devoted to the theory of this type. The simple mean-diameter treatment is described first, and the essential differences between steam and gas turbine design are pointed out. The application of vortex theory is then discussed, followed by a description of the method of applying cascade test data to complete the design procedure and provide an estimate of the isentropic efficiency. Recent developments in the calculation of blade profiles to give specified velocity distributions are also mentioned. Blade stresses are considered briefly because they have a direct impact upon the aerodynamic design. The chapter closes with a section on the cooled turbine, followed by some material on the radial flow turbine.

   The radial turbine can handle low mass flows more efficiently than the axial flow machine and has been widely used in the cryogenic industry as a turbo-expander, and in turbochargers for reciprocating engines. Although for all but the lowest powers the axial flow turbine is normally the more efficient, when mounted back-to-back with a centrifugal compressor the radial turbine offers the benefit of a very short and rigid rotor. This configuration is eminently suitable for gas turbines where compactness is more important than low fuel consumption. Auxiliary power units for aircraft (APUs), generating sets of up to 3 MW, and mobile power plants are typical applications.

## 7.1   Elementary theory of axial flow turbine

Figure 7.1 shows the velocity triangles for one axial flow turbine stage and the nomenclature employed. The gas enters the row of nozzle blades† with a static pressure and temperature $p_1, T_1$ and a velocity $C_1$, is expanded to

---

† These are also known as 'stator blades' and 'nozzle guide vanes'. We shall use suffix $N$ when necessary to denote quantities associated with the nozzle row, but will often use the term 'stator' in the text. Suffix $s$ will be used to denote 'stage'.

$p_2, T_2$ and leaves with an increased velocity $C_2$ at an angle $\alpha_2$.† The rotor blade inlet angle will be chosen to suit the direction $\beta_2$ of the gas velocity $V_2$ relative to the blade at inlet. $\beta_2$ and $V_2$ are found by vectorial subtraction of the blade speed $U$ from the absolute velocity $C_2$. After being deflected, and usually further expanded, in the rotor blade passages, the gas leaves at $p_3, T_3$ with relative velocity $V_3$ at angle $\beta_3$. Vectorial addition of $U$ yields the magnitude and direction of the gas velocity at exit from the stage, $C_3$ and $\alpha_3$. $\alpha_3$ is known as the *swirl angle*.

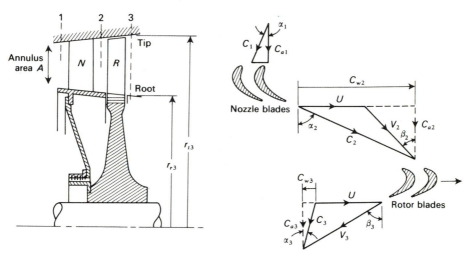

**FIG. 7.1 Axial flow turbine stage**

In a single-stage turbine $C_1$ will be axial, i.e. $\alpha_1 = 0$ and $C_1 = C_{a1}$. If on the other hand the stage is typical of many in a multi-stage turbine, $C_1$ and $\alpha_1$ will probably be equal to $C_3$ and $\alpha_3$ so that the same blade shapes can be used in successive stages: it is then sometimes called a *repeating stage*. Because the blade speed $U$ increases with increasing radius, the shape of the velocity triangles varies from root to tip of the blade. We shall assume in this section that we are talking about conditions at the mean diameter of the annulus, and that this represents an average picture of what happens to the total mass flow $m$ as it passes through the stage. This approach is valid when the ratio of the tip radius to the root radius is low, i.e. for short blades, but for long blades it is essential to account for three-dimensional effects as shown in subsequent sections.

$(C_{w2} + C_{w3})$ represents the change in whirl (or tangential) component of momentum per unit mass flow which produces the useful torque. The change in axial component $(C_{a2} - C_{a3})$ produces an axial thrust on the rotor which may supplement or offset the pressure thrust arising from the pressure drop

† In the early days of gas turbines the blade angles were measured from the tangential direction following steam turbine practice. It is now usual to measure angles from the axial direction as for axial compressor blading.

$(p_2 - p_3)$. In a gas turbine the net thrust on the turbine rotor will be partially balanced by the thrust on the compressor rotor and no large thrust bearings are required. In what follows we shall largely restrict our attention to designs in which the axial flow velocity $C_a$ is constant through the rotor. This will imply an annulus flared as in Fig. 7.1 to accommodate the decrease in density as the gas expands through the stage. With this restriction, when the velocity triangles are superimposed in the usual way we have the velocity diagram for the stage shown in Fig. 7.2.

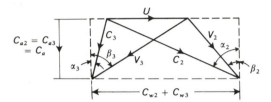

**FIG. 7.2 Velocity diagram**

The geometry of the diagram gives immediately the relations

$$\frac{U}{C_a} = \tan \alpha_2 - \tan \beta_2 = \tan \beta_3 - \tan \alpha_3 \qquad (7.1)$$

Applying the principle of angular momentum to the rotor, the stage work output per unit mass flow is

$$W_s = U(C_{w2} + C_{w3}) = UC_a(\tan \alpha_2 + \tan \alpha_3)$$

Combining with (7.1) we have $W_s$ in terms of the gas angles associated with the rotor blade, namely

$$W_s = UC_a(\tan \beta_2 + \tan \beta_3) \qquad (7.2)$$

Note that the 'work-done factor' required in the case of the axial compressor is unnecessary here. This is because in an accelerating flow the effect of the growth of the boundary layer along the annulus walls is much less than when there is a decelerating flow with an adverse pressure gradient.

From the steady flow energy equation we have $W_s = c_p \Delta T_{0s}$, where $\Delta T_{0s}$ is the stagnation temperature drop in the stage, and hence

$$c_p \Delta T_{0s} = UC_a(\tan \beta_2 + \tan \beta_3) \qquad (7.3)$$

When the stage inlet and outlet velocities are equal, i.e. $C_1 = C_3$, (7.3) also gives the static temperature drop in the stage $\Delta T_s$. In Chapter 2 we used as typical of combustion gases the average values

$$c_p = 1 \cdot 148 \text{ kJ/kg K}; \qquad \gamma = 1 \cdot 333 \text{ or } \gamma/(\gamma - 1) = 4$$

and noted that they were consistent with a gas constant $R$ of 0·287 kJ/kg K. For preliminary design calculations these values are quite adequate and we shall use them throughout this chapter. With velocities in m/s, $\Delta T_{0s}$ in Kelvin units is conveniently given by

$$\Delta T_{0s} = 8.72 \left(\frac{U}{100}\right)\left(\frac{C_a}{100}\right)(\tan \beta_2 + \tan \beta_3) \qquad (7.4)$$

The stagnation pressure ratio of the stage $p_{01}/p_{03}$ can be found from

$$\Delta T_{0s} = \eta_s T_{01}\left[1 - \left(\frac{1}{p_{01}/p_{03}}\right)^{(\gamma-1)/\gamma}\right] \qquad (7.5)$$

where $\eta_s$ is the isentropic stage efficiency based on stagnation (or 'total') temperature. Equation (7.5) is simply equation (2.12) applied to a stage, and $\eta_s$ is often called the *total-to-total stage efficiency*. It is the appropriate efficiency if the stage is followed by others in a multi-stage turbine because the leaving kinetic energy ($C_3^2/2$) is utilized in the next stage. It is certainly also relevant if the stage is part of a turbojet engine, because the leaving kinetic energy is used in the propelling nozzle. Even if it is the last stage of an industrial plant exhausting to atmosphere, the leaving kinetic energy is substantially recovered in a diffuser or volute and, as explained under the heading 'Compressor and turbine efficiencies', section 2.2, we can put $p_{03} = p_a$ and regard $\eta_s$ as the combined efficiency of the last stage and diffuser. [Although we shall not use it, it should be noted that a *total-to-static* isentropic efficiency is sometimes quoted for a turbine as a whole and for a stage, and it would be used where it is desirable to separate the turbine and exhaust diffuser losses. Applied to the stage, we would have

$$\text{total-to-static efficiency} = \frac{T_{01} - T_{03}}{T_{01} - T_3'}$$

where $T_3'$ is the static temperature reached after an isentropic expansion from $p_{01}$ to $p_3$. It assumes that as far as the turbine is concerned all the leaving kinetic energy is wasted, and its value is somewhat less than the total-to-total efficiency which we shall use here.]

There are three dimensionless parameters found to be useful in turbine design. One, which expresses the work capacity of a stage, is called the *blade loading coefficient* or *temperature drop coefficient* $\psi$. We shall adopt NGTE practice and define it as $c_p\Delta T_{0s}/\frac{1}{2}U^2$, although $c_p\Delta T_{0s}/U^2$ is also used. Thus from equation (7.3),

$$\psi = \frac{2c_p\Delta T_{0s}}{U^2} = \frac{2C_a}{U}(\tan \beta_2 + \tan \beta_3) \qquad (7.6)$$

Another useful parameter is the *degree of reaction* or simply the *reaction* $\Lambda$. This expresses the fraction of the stage expansion which occurs in the rotor, and it is usual to define it in terms of static temperature (or enthalpy) drops rather than pressure drops, namely

$$\Lambda = \frac{T_2 - T_3}{T_1 - T_3}$$

For the type of stage we are considering, where $C_{a2} = C_{a3} = C_a$ and $C_3 = C_1$, a simple expression for $\Lambda$ can be derived as follows. From (7.4),

$$c_p(T_1 - T_3) = c_p(T_{01} - T_{03}) = UC_a(\tan \beta_2 + \tan \beta_3)$$

*Relative* to the rotor blades the flow does no work and the steady flow energy equation yields

$$\begin{aligned}
c_p(T_2 - T_3) &= \tfrac{1}{2}(V_3^2 - V_2^2) \\
&= \tfrac{1}{2}C_a^2(\sec^2 \beta_3 - \sec^2 \beta_2) \\
&= \tfrac{1}{2}C_a^2(\tan^2 \beta_3 - \tan^2 \beta_2)
\end{aligned}$$

And thus

$$\Lambda = \frac{C_a}{2U}(\tan \beta_3 - \tan \beta_2) \tag{7.7}$$

The third dimensionless parameter often referred to in gas turbine design appears in both equations (7.6) and (7.7): it is the ratio $C_a/U$ called the *flow coefficient* $\phi$. (It plays the same part as the blade speed ratio $U/C_1$ used by steam turbine designers.) Thus (7.6) and (7.7) can be written as

$$\psi = 2\phi(\tan \beta_2 + \tan \beta_3) \tag{7.8}$$

$$\Lambda = \frac{\phi}{2}(\tan \beta_3 - \tan \beta_2) \tag{7.9}$$

The gas angles can now be expressed in terms of $\psi$, $\Lambda$ and $\phi$ as follows. Adding and subtracting (7.8) and (7.9) in turn we get

$$\tan \beta_3 = \frac{1}{2\phi}(\tfrac{1}{2}\psi + 2\Lambda) \tag{7.10}$$

$$\tan \beta_2 = \frac{1}{2\phi}(\tfrac{1}{2}\psi - 2\Lambda) \tag{7.11}$$

Then using relations (7.1),

$$\tan \alpha_3 = \tan \beta_3 - \frac{1}{\phi} \tag{7.12}$$

$$\tan \alpha_2 = \tan \beta_2 + \frac{1}{\phi} \tag{7.13}$$

Even with the restrictions we have already introduced ($C_{a3} = C_{a2}$ and $C_3 = C_1$), and remembering that stressing considerations will place a limit on the blade speed $U$, there is still an infinite choice facing the designer. For example, although the overall turbine temperature drop will be fixed by cycle calculations, it is open to the designer to choose one or two stages of large $\psi$ or a larger number of smaller $\psi$. To limit still further our discussion at this point, we may observe that any turbine for a gas turbine power plant is essentially a low pressure ratio machine by steam turbine standards (e.g. in the region of 10:1 compared with over 1000:1 even for cycles operating with subcritical steam pressures). Thus there is little case for adopting impulse stages ($\Lambda = 0$) which find a place at the high pressure end of steam turbines. Impulse designs are the most efficient type for that duty, because under such conditions the 'leakage losses' associated with rotor blade tip clearances are excessive in reaction stages. It must be remembered that at the high pressure end of an expansion of large pressure ratio the stage pressure *differences* are considerable even though the stage pressure ratios are modest. Let us then rule out values of $\Lambda$ near zero, and for the moment consider 50 per cent reaction designs. Our general knowledge of the way nature behaves would suggest that the most efficient design is likely to be achieved when the expansion is reasonably evenly divided between the stator and rotor rows. We shall see later that the reaction will vary from root to tip of the blade, but here we are thinking of 50 per cent reaction at the mean diameter.

Putting $\Lambda = 0.5$ in equation (7.9) we have

$$\frac{1}{\phi} = \tan \beta_3 - \tan \beta_2 \tag{7.14}$$

Direct comparison with relations (7.1) then shows that

$$\beta_3 = \alpha_2 \text{ and } \beta_2 = \alpha_3 \tag{7.15}$$

and the velocity diagram becomes symmetrical. Further, if we are considering a repeating stage with $C_3 = C_1$ in direction as well as magnitude, we have $\alpha_1 = \alpha_3 = \beta_2$ also, and the stator and rotor blades then have the same inlet and outlet angles. Finally, from (7.10) and (7.15) we have for $\Lambda = 0.5$,

$$\psi = 4\phi \tan \beta_3 - 2 = 4\phi \tan \alpha_2 - 2 \tag{7.16}$$

and from (7.11) and (7.15) we get

$$\psi = 4\phi \tan \beta_2 + 2 = 4\phi \tan \alpha_3 + 2 \tag{7.17}$$

Equations (7.15), (7.16) and (7.17) give all the gas angles in terms of $\psi$ and $\phi$. Figure 7.3 shows the result of plotting nozzle outlet angle $\alpha_2$ and stage outlet swirl angle $\alpha_3$ on a $\psi$–$\phi$ basis.

Because the blade shapes are determined within close limits by the gas angles, it is possible from results of cascade tests on families of blades to predict the losses in the blade rows and estimate the stage efficiency of the range of 50 per cent reaction designs covered by Fig. 7.3. One such estimate is shown by the efficiency contours superimposed on the $\psi$–$\phi$ plot. The values of $\eta_s$ on the contours represent an average of detailed estimates quoted in Refs. (1) and (2). Many assumptions have to be made about blade profile, blade aspect ratio (height/chord), tip clearance and so on, and no reliance should be placed upon the absolute values of efficiency shown. Nevertheless a knowledge of the general trend is valuable and even essential to the designer. Similar curves for other values of reaction are given in Ref. (2).

We may note that designs having a low $\psi$ and low $\phi$ yield the best stage efficiencies. Referring to the comparative velocity diagrams also given in Fig. 7.3 (drawn for a constant blade speed $U$), we can see that low values of $\psi$ and $\phi$ imply low gas velocities and hence reduced friction losses. But a low $\psi$ means more stages for a given overall turbine output, and a low $\phi$ means a larger turbine annulus area for a given mass flow. For an industrial gas turbine when size and weight are of little consequence and a low s.f.c. is vital, it would be sensible to design with a low $\psi$ and low $\phi$. Certainly in the last stage a low axial velocity and a small swirl angle $\alpha_3$ are desirable to keep down the losses in the exhaust diffuser. For an aircraft propulsion unit, however, it is important to keep the weight and frontal area to a minimum, and this means using higher values of $\psi$ and $\phi$. The most efficient stage design is one which leads to the most efficient power plant for its particular purpose,

**FIG. 7.3  50 per cent reaction designs**

and strictly speaking the optimum $\psi$ and $\phi$ cannot be determined without detailed calculations of the performance of the aircraft as a whole. It would appear from current aircraft practice that the optimum values for $\psi$ range from 3 to 5, with $\phi$ ranging from 0.8 to 1.0. A low swirl angle ($\alpha_3 < 20$ degrees) is desirable because swirl increases the losses in the jet pipe and propelling nozzle; to maintain the required high value of $\psi$ and low value of $\alpha_3$ it might be necessary to use a degree of reaction somewhat less than 50 per cent. The dotted lines in the velocity diagram for $\psi = 4$ indicate what happens when the proportion of the expansion carried out in the rotor is reduced and $V_3$ becomes more equal to $V_2$, while maintaining $U$, $\psi$ and $\phi$ constant.

We will close this section with a worked example showing how a first tentative 'mean-diameter' design may be arrived at. To do this we need some method of accounting for the losses in the blade rows. Two principal parameters are used, based upon temperature drops and pressure drops respectively. These parameters can best be described by sketching the processes in the nozzle and rotor blade passages on the $T$–$s$ diagram as in Fig. 7.4. The

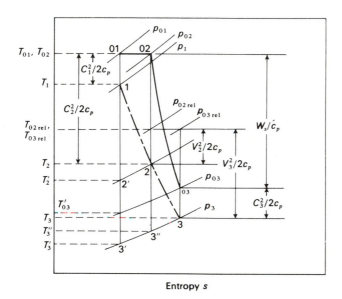

FIG. 7.4 *T-s* **diagram for a reaction stage**

full and chain dotted lines connect stagnation and static states respectively. $T_{02} = T_{01}$ because no work is done in the nozzles; and the short horizontal portion of the full line represents the stagnation pressure drop $(p_{01} - p_{02})$ due to friction in the nozzles. The losses are of course exaggerated in the figure. When obtaining the temperature equivalent of the velocity of the gas leaving the nozzle row, we may say that ideally the gas would be expanded from $T_{01}$ to $T_2'$ but that due to friction the temperature at the nozzle exit is $T_2$, somewhat higher than $T_2'$. The *loss coefficient for the nozzle blades* may be defined either by

$$\lambda_N = \frac{T_2 - T_2'}{C_2^2/2c_p} \text{ or } Y_N = \frac{p_{01} - p_{02}}{p_{01} - p_2} \tag{7.18}$$

Both $\lambda$ and $Y$ express the proportion of the *leaving* energy which is degraded by friction. $Y_N$ can be measured relatively easily in cascade tests (the results being modified to allow for three-dimensional effects on the same lines as described for axial compressors), whereas $\lambda_N$ is the more easily used in design. It will be shown that $\lambda_N$ and $Y_N$ are not very different numerically.

Returning to Fig. 7.4, we see that further expansion in the moving blade passages reduces the pressure to $p_3$. Isentropic expansion in the whole stage would result in a final temperature $T_3'$, and in the rotor blade passages alone $T_3''$. Expansion with friction leads to a final temperature $T_3$. The *rotor blade loss* can be expressed by

$$\lambda_R = \frac{T_3 - T_3''}{V_3^2/2c_p}$$

Note that it is defined as a proportion of the leaving kinetic energy *relative to the row* so that it can be related to cascade test results. As no work is done by the gas relative to the blades, $T_{03rel} = T_{02rel}$. The *loss coefficient in terms of pressure drop for the rotor blades* is defined by

$$Y_R = \frac{p_{02rel} - p_{03rel}}{p_{03rel} - p_3}$$

We may show that $\lambda$ and $Y$ are not very different numerically by the following argument (which applies equally to the stator and rotor rows of blades although given only for the former).

$$Y_N = \frac{p_{01} - p_{02}}{p_{02} - p_2} = \frac{(p_{01}/p_{02}) - 1}{1 - (p_2/p_{02})}$$

Now

$$\frac{p_{01}}{p_{02}} = \frac{p_{01}}{p_2} \frac{p_2}{p_{02}} = \left(\frac{T_{01}}{T_2'}\right)^{\gamma/(\gamma-1)} \left(\frac{T_2}{T_{02}}\right)^{\gamma/(\gamma-1)}$$

$$= \left(\frac{T_2}{T_2'}\right)^{\gamma/(\gamma-1)} \text{ because } T_{01} = T_{02}$$

Hence

$$Y_N = \frac{(T_2/T_2')^{\gamma/(\gamma-1)} - 1}{1 - (T_2/T_{02})^{\gamma/(\gamma-1)}}$$

$$= \frac{\left[1 + \dfrac{T_2 - T_2'}{T_2'}\right]^{\gamma/(\gamma-1)} - 1}{1 - \left[\dfrac{T_2 - T_{02}}{T_{02}} + 1\right]^{\gamma/(\gamma-1)}}$$

Expanding the bracketed expressions binomially and using the first terms only (although not very accurate for the denominator) we have

$$Y_N = \frac{T_2 - T_2'}{T_{02} - T_2} \times \frac{T_{02}}{T_2'} = \lambda_N\left(\frac{T_{02}}{T_2'}\right) \simeq \lambda_N\left(\frac{T_{02}}{T_2}\right) \tag{7.19}$$

From Appendix A, equation (8), we have the general relation

$$\frac{T_{02}}{T_2} = \left(1 + \frac{\gamma - 1}{2} M_2^2\right)$$

Even if the Mach number at the blade exit is unity, as it might be for the nozzle blades of a highly loaded stage, $\lambda = 0.86Y$ and thus $\lambda$ is only 14 per cent less than $Y$.

The type of information which is available for predicting values of $\lambda$ or $Y$ is described briefly in section 7.4. $\lambda_N$ and $\lambda_R$ can be related to the stage isentropic efficiency $\eta_s$ as follows.

$$\eta_s = \frac{T_{01} - T_{03}}{T_{01} - T_{03}'} = \frac{1}{1 + (T_{03} - T_{03}')/(T_{01} - T_{03})}$$

Now a glance at Fig. 7.4 shows that

$$T_{03} - T_{03}' \simeq (T_3 - T_3') = (T_3 - T_3'') + (T_3'' - T_3')$$

But $(T_2'/T_3') = (T_2/T_3'')$ because both equal $(p_2/p_3)^{(\gamma - 1)/\gamma}$. Rearranging and subtracting one from both sides we get

$$\frac{T_3'' - T_3'}{T_3'} = \frac{(T_2 - T_2')}{T_2'} \text{ or } (T_3'' - T_3') \simeq (T_2 - T_2')\frac{T_3}{T_2}$$

Hence

$$\eta_s \simeq \frac{1}{1 + [(T_3 - T_3'') + (T_3/T_2)(T_2 - T_2')]/(T_{01} - T_{03})}$$

$$\simeq \frac{1}{1 + [\lambda_R(V_3^2/2c_p) + (T_3/T_2)\lambda_N(C_2^2/2c_p)]/(T_{01} - T_{03})} \tag{7.20}$$

Alternatively, substituting $V_3 = C_a \sec \beta_3$, $C_2 = C_a \sec \alpha_2$, and

$$c_p(T_{01} - T_{03}) = UC_a(\tan \beta_3 + \tan \beta_2)$$
$$= UC_a[\tan \beta_3 + \tan \alpha_2 - (U/C_a)]$$

equation (7.20) can be written in the form

$$\eta_s \simeq \frac{1}{1 + \frac{1}{2}\frac{C_a}{U}\left[\frac{\lambda_R \sec^2 \beta_3 + (T_3/T_2)\lambda_N \sec^2 \alpha_2}{\tan \beta_3 + \tan \alpha_2 - (U/C_a)}\right]} \tag{7.21}$$

Because $Y \simeq \lambda$, loss coefficients $Y_R$ and $Y_N$ may replace $\lambda_R$ and $\lambda_N$ in equations (7.20) and (7.21) if desired.

For the purpose of the following example we shall assume that $\lambda_N = 0.05$ and $\eta_s = 0.9$. In suggesting $\lambda_N = 0.05$ we are assuming that convergent nozzles are employed and that they are operating with a pressure ratio $(p_{01}/p_2)$ less than the critical pressure ratio $[(\gamma + 1)/2]^{\gamma/(\gamma - 1)}$. Convergent–divergent nozzles as in Fig. 7.5(a) are not used, partly because they tend to be inefficient at pressure ratios other than the design value (i.e. at part load), and partly because high values of $C_2$ usually imply high values of $V_2$. If the Mach number relative to the moving blades at inlet, $M_{V2}$, exceeds about 0·75, additional losses may be incurred by the formation of shock waves in the rotor blade passages (see section A.1 of Appendix A). If $V_2$ is in fact not too high, perhaps because a low value of the flow coefficient $\phi$ is being used as in an industrial gas turbine, there is no reason why convergent nozzles should not be operated at pressure ratios giving an efflux velocity which is slightly supersonic (namely $1 < M_2 < 1·2$): very little additional loss seems to be incurred. The pressure at the throats of the nozzles is then the critical pressure, and there is further semi-controlled expansion to $p_2$ after the throat. As depicted in Fig. 7.5(b), the flow is controlled by the trailing edge on the convex side. On the other side the supersonic stream expands as though turning a corner: it is possible to obtain some idea of the deflection of the stream by treating it as a Prandtl–Meyer expansion and using the method of characteristics (see section A.8 of Appendix A, and Ref. (2) for further details).

(a)                                              (b)

FIG. 7.5 **Convergent-divergent nozzle, and convergent nozzle operating at a pressure ratio greater than the critical value**

## EXAMPLE

Determination of a possible 'mean-diameter' design for the turbine of a small, cheap, turbojet unit, which should be a single-stage turbine if possible. From cycle calculations the following design point specification is proposed for the turbine.

| | |
|---|---|
| Mass flow $m$ | 20 kg/s |
| Isentropic efficiency $\eta_t$ | 0·9 |
| Inlet temperature $T_{01}$ | 1100 K |
| Temperature drop $T_{01} - T_{03}$ | 145 K |
| Pressure ratio $p_{01}/p_{03}$ | 1·873 |
| Inlet pressure $p_{01}$ | 4 bar |

In addition to this information, we are likely to have the rotational speed fixed by the compressor, the design of which is always more critical than the turbine because of the decelerating flow.† Also, experience will suggest an

---

† In practice the compressor and turbine designs interact at this point: a small increase in rotational speed to avoid the use of more than one stage in the turbine would normally be worthwhile even if it meant modifying the compressor design.

upper limit to the blade speed above which stressing difficulties will be severe. Accordingly, we will assume

> Rotational speed $N$      250 rev/s
> Mean blade speed $U$     340 m/s

Finally, we shall assume a nozzle loss coefficient $\lambda_N$ of 0·05 as a reasonable first guess.

We shall start by assuming (a) $C_{a2} = C_{a3}$ and (b) $C_1 = C_3$. As it is to be a single-stage turbine the inlet velocity will be axial, i.e. $\alpha_1 = 0$. From the data, the temperature drop coefficient is

$$\psi = \frac{2c_p \Delta T^0{}_s}{U^2} = \frac{2 \times 1\cdot148 \times 145 \times 10^3}{340^2} = 2\cdot88$$

This is a modest value and there is no difficulty about obtaining the required output from a single stage in a turbojet unit wherein high values of $C_a$ can be used. We will try a flow coefficient $\phi$ of 0·8 and, because swirl increases the losses in the jet pipe, an $\alpha_3$ of zero. The degree of reaction that these conditions imply can be found as follows, remembering that because of assumptions (a) and (b) equations (7.8) to (7.13) are applicable. From equation (7.12)

$$\tan \alpha_3 = 0 = \tan \beta_3 - \frac{1}{\phi}$$

$$\tan \beta_3 = 1\cdot25$$

From equation (7.10)

$$\tan \beta_3 = \frac{1}{2\phi}(\tfrac{1}{2}\psi + 2\Lambda)$$

$$1\cdot25 = \frac{1}{1\cdot6}(1\cdot44 + 2\Lambda)$$

$$\Lambda = 0\cdot28$$

We shall see from section 7.2 that when three-dimensional effects are included the reaction will increase from root to tip of the blades, and a degree of reaction of only 0·28 at the mean diameter might mean too low a value at the root. Negative values must certainly be avoided because this would imply expansion in the nozzles followed by recompression in the rotor and the losses would be large. Perhaps a modest amount of swirl will bring the reaction to a more reasonable value: we will try $\alpha_3 = 10$ degrees.

$$\tan \alpha_3 = 0\cdot1763; \tan \beta_3 = 0\cdot1763 + 1\cdot25 = 1\cdot426$$

$$1\cdot426 = \frac{1}{1\cdot6}(1\cdot44 + 2\Lambda)$$

$$\Lambda = 0\cdot421$$

This is acceptable: the reaction at the root will be checked when the example is continued in section 7.2. The gas angles can now be established. So far we have

$$\alpha_3 = 10°; \beta_3 = \tan^{-1} 1·426 = 54°57'$$

From equations (7.11) and (7.13)

$$\tan \beta_2 = \frac{1}{1·6}(1·44 - 0·842) = 0·3737$$

$$\tan \alpha_2 = 0·3737 + \frac{1}{0·8} = 1·624$$

$$\beta_2 = 20°29'; \alpha_2 = 58°23'$$

The velocity diagram can now be sketched as in Fig. 7.6, and the next task is to calculate the density at stations 1, 2 and 3 so that the blade height $h$ and tip/root radius ratio $(r_t/r_r)$ can be estimated. We shall commence with

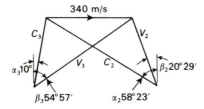

**FIG. 7.6**

station 2 because some modifications will be required if the pressure ratio $p_{01}/p_2$ across the convergent nozzles is much above the critical value, or if the Mach number relative to the rotor blades at inlet $(M_{V2})$ exceeds about 0·75.

From the geometry of the velocity diagram,

$$C_{a2} = U\phi = 340 \times 0·8 = 272 \text{ m/s}$$

$$C_2 = \frac{C_{a2}}{\cos \alpha_2} = \frac{272}{0·5242} = 519 \text{ m/s}$$

The temperature equivalent of the outlet velocity is

$$T_{02} - T_2 = \frac{C_2^2}{2c_p} = \frac{519^2}{2296} = 117·3 \text{ K}$$

Since $T_{02} = T_{01} = 1100$ K, $T_2 = 982·7$ K

$$T_2 - T_2' = \lambda_N \frac{C_2^2}{2c_p} = 0·05 \times 117·3 = 5·9 \text{ K}$$

$$T_2' = 982·7 - 5·9 = 976·8 \text{ K}$$

$p_2$ can be found from the isentropic relation

$$\frac{p_{01}}{p_2} = \left(\frac{T_{01}}{T_2'}\right)^{\gamma/(\gamma-1)} = \left(\frac{1100}{976\cdot8}\right)^4 = 1\cdot607$$

$$p_2 = \frac{4\cdot0}{1\cdot607} = 2\cdot49 \text{ bar}$$

Ignoring the effect of friction on the critical pressure ratio, and putting $\gamma = 1\cdot333$ in equation (A.12) we have

$$\frac{p_{01}}{p_c} = \left(\frac{\gamma+1}{2}\right)^{\gamma/(\gamma-1)} = 1\cdot853$$

The actual pressure ratio is $1\cdot607$, well below the critical value. The nozzles are not choking and the pressure in the plane of the throat is equal to $p_2$.

$$\rho_2 = \frac{p_2}{RT_2} = \frac{100 \times 2\cdot49}{0\cdot287 \times 982\cdot7} = 0\cdot883 \text{ kg/m}^3$$

Annulus area at plane 2 is

$$A_2 = \frac{m}{\rho_2 C_{a2}} = \frac{20}{0\cdot883 \times 272} = 0\cdot0833 \text{ m}^2$$

Throat area of nozzles required is

$$A_{2N} = \frac{m}{\rho_2 C_2} \text{ or } A_2 \cos \alpha_2 = 0\cdot0883 \times 0\cdot524 = 0\cdot0437 \text{ m}^2$$

Note that if the pressure ratio had been slightly above the critical value it would be acceptable if a check (given later) on $M_{V2}$ proved satisfactory. $\rho_2$ and $A_2$ would be unchanged, but the throat area would then be given by $m/\rho_c C_c$, where $\rho_c$ is obtained from $p_c$ and $T_c$, and $C_c$ corresponds to a Mach number of unity so that it can be found from $\sqrt{(\gamma R T_c)}$.

We may now calculate the annulus area required in planes 1 and 3 as follows. Because it is not a repeating stage, we are assuming that $C_1$ is axial and this, together with assumptions (a) and (b) that $C_1 = C_3$ and $C_{a3} = C_{a2}$ yields

$$C_{a1} = C_1 = C_3 = \frac{C_{a3}}{\cos \alpha_3} = \frac{272}{\cos 10°} = 276\cdot4 \text{ m/s}$$

Temperature equivalent of the inlet (and outlet) kinetic energy is

$$\frac{C_1^2}{2c_p} = \frac{276\cdot4^2}{2296} = 33\cdot3 \text{ K}$$

$$T_1 = T_{01} - \frac{C_1^2}{2c_p} = 1100 - 33\cdot3 = 1067 \text{ K}$$

$$\frac{p_1}{p_{01}} = \left(\frac{T_1}{T_{01}}\right)^{\gamma/(\gamma-1)} \text{ or } p_1 = \frac{4\cdot0}{(1100/1067)^4} = 3\cdot54 \text{ bar}$$

$$\rho_1 = \frac{100 \times 3 \cdot 54}{0 \cdot 287 \times 1067} = 1 \cdot 155 \text{ kg/m}^3$$

$$A_1 = \frac{m}{\rho_1 C_{a1}} = \frac{20}{1 \cdot 155 \times 276 \cdot 4} = 0 \cdot 0626 \text{ m}^2$$

Similarly, at outlet from the stage we have

$$T_{03} = T_{01} - \Delta T_{0s} = 1100 - 145 = 955 \text{ K}$$

$$T_3 = T_{03} - \frac{C_3^2}{2c_p} = 955 - 33 \cdot 3 = 922 \text{ K}$$

$p_{03}$ is given in the data by $p_{01}(p_{03}/p_{01})$ and hence

$$p_3 = p_{03}\left(\frac{T_3}{T_{03}}\right)^{\gamma/(\gamma-1)} = \left(\frac{4}{1 \cdot 873}\right)\left(\frac{922}{955}\right)^4 = 1 \cdot 856 \text{ bar}$$

$$\rho_3 = \frac{100 \times 1 \cdot 856}{0 \cdot 287 \times 922} = 0 \cdot 702 \text{ kg/m}^3$$

$$A_3 = \frac{m}{\rho_3 C_{a3}} = \frac{20}{0 \cdot 702 \times 272} = 0 \cdot 1047 \text{ m}^2$$

The blade height and annulus radius ratio at stations 1, 2 and 3 can now be established. At the mean diameter, which we shall now begin to emphasize by the use of suffix $m$,

$$U_m = 2\pi N r_m, \text{ so that } r_m = \frac{340}{2\pi 250} = 0 \cdot 216 \text{ m}$$

Since the annulus area is given by

$$A = 2\pi r_m h = \frac{U_m h}{N}$$

the height and radius ratio of the annulus can be found from

$$h = \frac{AN}{U_m} = \left(\frac{250}{340}\right) A, \text{ and } \frac{r_t}{r_r} = \frac{r_m + (h/2)}{r_m - (h/2)}$$

and the results are as given in the following table.

| Station | 1 | 2 | 3 |
|---|---|---|---|
| $A$, m$^2$ | 0·0626 | 0·0833 | 0·1047 |
| $h$, m | 0·046 | 0·0612 | 0·077 |
| $r_t/r_r$ | 1·24 | 1·33 | 1·43 |

Although we shall leave the discussion of the effects of high and low annulus radius ratio to a later section, we may note here that values in the region of 1·2–1·4 would be regarded as satisfactory. If the rotational speed, which we have assumed to be fixed by the compressor, had led to an ill-proportioned annulus, it would be necessary to rework this preliminary design. For example $r_t/r_r$ could be reduced by increasing the axial velocity, i.e. by using a higher value of the flow coefficient $\phi$. This would also increase the nozzle efflux velocity, but we noted that it was comfortably subsonic and so could be increased if necessary.

The turbine annulus we have arrived at in this example is flared as shown in Fig. 7.7. In sketching this we have assumed a value of blade height/width

$$h = \tfrac{1}{2}(h_2 + h_3)$$

$\simeq 0.25w$          Blade width $w$
$\simeq h/3$

**FIG. 7.7**

ratio of about 3·0 and a space between the stator and rotor blades of about 0·25 of the blade width. The included angle of divergence of the walls then becomes approximately 29 degrees. This might be regarded as rather high, involving a risk of flow separation from the inner wall where the reaction, and therefore the acceleration, is not large. 25 degrees has been suggested as a safe limit, Ref. (3). We shall not pause for adjustment here, because the blade height/width ratio of 3·0 is merely a rough guess to be justified or altered later when the effect of blade stresses on the design has to be considered. Furthermore, the choice of 0·25 for the space/blade width ratio is rather low: a low value is desirable only to reduce the axial length and weight of the turbine. Vibrational stresses are induced in the rotor blades as they pass through the wakes of the nozzle blades, and these stresses increase sharply with decrease in axial space between the blade rows. 0·2 is the lowest value of space/blade width ratio considered to be safe, but a value nearer 0·5 is often used and this would reduce both the vibrational stresses and the annulus flare.

If it was thought desirable to reduce the flare without increasing the axial length of the turbine, then it would be necessary to repeat the calculations allowing the axial velocity to increase through the stage. It would be necessary to check the Mach number at exit from the stage, $M_3$, because if this is too high the friction losses in the jet pipe become unduly large. For the present design we have

$$M_3 = \frac{C_3}{\sqrt{(\gamma R T_3)}} = \frac{276\cdot4}{\sqrt{(1.333 \times 0\cdot287 \times 922 \times 1000)}} = 0\cdot47$$

This could be safely increased to reduce the flare if desired. [It must be remembered that most of the equations we have derived, and in particular (7.8)–(7.13) relating $\psi$, $\phi$ and the gas angles, *are not valid* when $C_{a3} \neq C_{a2}$. It is necessary to revert to first principles or derive new relations. For example, relations (7.1) become

$$\frac{U}{C_{a2}} = \tan \alpha_2 - \tan \beta_2 \text{ and } \frac{U}{C_{a3}} = \tan \beta_3 - \tan \alpha_3$$

and the temperature drop coefficient becomes

$$\psi = \frac{2c_p \Delta T_{0s}}{U^2} = \frac{2}{U}(C_{a2} \tan \alpha_2 + C_{a3} \tan \alpha_3)$$

$$= \frac{2C_{a2}}{U}\left(\tan \beta_2 + \frac{C_{a3}}{C_{a2}} \tan \beta_3\right)$$

Additionally, if the flare is not symmetrical $U$ must be replaced by $U_{m2}$ and $U_{m3}$ as appropriate.]

Finally, for this preliminary design we have taken losses into account via $\lambda_N$ and $\eta_s$ rather than $\lambda_N$ and $\lambda_R$. The value of $\lambda_R$ implied by the design can be found by determining $(T_3 - T_3'')$. Thus

$$\frac{T_2}{T_3''} = \left(\frac{p_2}{p_3}\right)^{(\gamma-1)/\gamma} \text{ or } T_3'' = \frac{982 \cdot 7}{(2 \cdot 49/1 \cdot 856)^{\frac{1}{4}}} = 913 \text{ K}$$

We also require the temperature equivalent of the outlet kinetic energy *relative* to the blading.

$$V_3 = \frac{C_{a3}}{\cos \beta_3} = \frac{272}{\cos 54°57'} = 473 \cdot 5 \text{ m/s}$$

$$\frac{V_3^2}{2c_p} = \frac{473 \cdot 5^2}{2296} = 97 \cdot 8 \text{ K}$$

Then

$$\lambda_R = \frac{T_3 - T_3''}{V_3^2/2c_p} = \frac{922 - 913}{97 \cdot 8} = 0 \cdot 092$$

Had we used the approximate relation between $\eta_s$, $\lambda_N$ and $\lambda_R$, equation (7.20), we would have found $\lambda_R$ to be 0·108 (which is a useful check on the arithmetic). Note that $\lambda_R > \lambda_N$, which it should be by virtue of the tip leakage loss in the rotor blades.

The next steps in the design are

(a) to consider the three-dimensional nature of the flow in so far as it affects the variation of the gas angles with radius;

(b) to consider the blade shapes necessary to achieve the required gas angles, and the effect of centrifugal and gas bending stresses on the design;

(c)   to check the design by estimating $\lambda_N$ and $\lambda_R$ from the results of cascade tests suitably modified to take account of three-dimensional flows.

## 7.2   Vortex theory

Early in the previous section it was pointed out that the shape of the velocity triangles must vary from root to tip of the blade because the blade speed $U$ increases with radius. Another reason is that the whirl component in the flow at outlet from the nozzles causes the static pressure and temperature to vary across the annulus. With a uniform pressure at inlet, or at least with a much smaller variation because the whirl component is smaller, it is clear that the pressure drop across the nozzle will vary giving rise to a corresponding variation in efflux velocity $C_2$. Twisted blading designed to take account of the changing gas angles is called *vortex blading*.

It has been common steam turbine practice, except in low-pressure blading where the blades are very long, to design on conditions at the mean diameter, keep the blade angles constant from root to tip, and assume that no additional loss is incurred by the variation in incidence along the blade caused by the changing gas angles. Comparative tests, Ref. (4), have been conducted on a single-stage gas turbine of radius ratio 1·37, using in turn blades of constant angle and vortex blading. The results showed that any improvement in efficiency obtained with vortex blading was within the margin of experimental error. This contrasts with similar tests on a 6-stage axial compressor, Ref. (5), which showed a distinct improvement from the use of vortex blading. This was, however, not so much an improvement in efficiency (of about 1·5 per cent) as in the delay of the onset of surging which of course does not arise in accelerating flow. It appears, therefore, that steam turbine designers have been correct in not applying vortex theory except when absolutely necessary at the LP end. They have to consider the additional cost of twisted blades for the very large number of rows of blading required, and they know that the Rankine cycle is relatively insensitive to component losses. Conversely, it is not surprising that the gas turbine designer, struggling to achieve the highest possible component efficiency, has consistently used some form of vortex blading which he feels intuitively must give him a better performance however small.

Vortex theory has been outlined in section 5.4 where it was shown that if the elements of fluid are to be in radial equilibrium, an increase in static pressure from root to tip is necessary whenever there is a whirl component of velocity. Figure 7.8 shows why the gas turbine designer cannot talk of impulse or 50 per cent reaction stages. The proportion of the stage pressure or temperature drop which occurs in the rotor must increase from root to tip. Although Fig. 7.8 refers to a single-stage turbine with axial inlet velocity and no swirl at outlet, the whirl component at inlet and outlet of a repeating stage will be small compared with $C_{w2}$: the reaction will therefore still increase from root to tip, if somewhat less markedly.

## Free vortex design

Referring to section 5.4, it was shown that if

(a)   the stagnation enthalpy $h_0$ is constant over the annulus (i.e. $dh_0/dr = 0$),
(b)   the axial velocity is constant over the annulus,
(c)   the whirl velocity is inversely proportional to the radius,

then the condition for radial equilibrium of the fluid elements, namely equation (5.10), is satisfied. A stage designed in accordance with (a), (b) and (c) is called a *free vortex stage*. Applying this to the stage in Fig. 7.8, we can

FIG. 7.8  **Changes in pressure and velocity across the annulus**

see that with uniform inlet conditions to the nozzles then, since no work is done by the gas in the nozzles, $h_0$ must also be constant over the annulus at outlet. Thus condition (a) is fulfilled in the space between the nozzles and rotor blades. Furthermore, if the nozzles are designed to give $C_{a2} = $ constant and $C_{w2}r = $ constant, all three conditions are fulfilled and the condition for radial equilibrium is satisfied in plane 2. Similarly, if the rotor blades are designed so that $C_{a3} = $ constant and $C_{w3}r = $ constant, it is easy to show as follows that condition (a) will be fulfilled, and thus radial equilibrium will be achieved in plane 3 also. Writing $\omega$ for the angular velocity we have

$$W_s = U(C_{w2} + C_{w3}) = \omega(C_{w2}r + C_{w3}r) = \text{constant}$$

But when the work done per unit mass of gas is constant over the annulus, and $h_0$ is constant at inlet, $h_0$ must be constant at outlet also: thus condition (a) is met.

It is apparent that a free vortex design is one in which the work done per unit mass of gas is constant over the annulus, and to obtain the total work output this specific value need only be calculated at one convenient radius and multiplied by the mass flow.

In contrast, we may note that because the density varies from root to tip at exit from the nozzles and the axial velocity is constant, an integration over the annulus will be necessary if the continuity equation is to be used in plane

2. Thus, considering a flow $\delta m$ through an annular element of radius $r$ and width $\delta r$,

$$\delta m = \rho_2 2\pi r \delta r C_{a2}$$

$$m = 2\pi C_{a2} \int_{r_r}^{r_t} \rho_2 r dr \tag{7.22}$$

With the radial variation of density determined from vortex theory, the integration can be performed although the algebra is lengthy. For detailed calculations it would be normal to use a digital computer, permitting ready calculation of the density at a series of radii and numerical integration of equation (7.22) to obtain the mass flow. For preliminary calculations, how-ever, it is sufficiently accurate to take the intensity of mass flow at the mean diameter as being the mean intensity of mass flow. In other words, the total mass flow is equal to the mass flow per unit area calculated using the density at the mean diameter $(\rho_{2m}C_{a2})$ multiplied by the annulus area $(A_2)$. This is one reason why it is convenient to design the turbine on conditions at mean diameter (as was done in the previous example) and use the relations which will now be derived for obtaining the gas angles at other radii.

Using suffix $m$ to denote quantities at mean diameter, the free vortex variation of nozzle angle $\alpha_2$ may be found as follows

$$C_{w2}r = rC_{a2} \tan \alpha_2 = \text{constant}$$
$$C_{a2} = \text{constant}$$

Hence $\alpha_2$ at any radius $r$ is related to $\alpha_{2m}$ at the mean radius $r_m$ by

$$\tan \alpha_2 = \left(\frac{r_m}{r}\right)_2 \tan \alpha_{2m} \tag{7.23}$$

Similarly, when there is swirl at outlet from the stage,

$$\tan \alpha_3 = \left(\frac{r_m}{r}\right)_3 \tan \alpha_{3m} \tag{7.24}$$

The gas angles at inlet to the rotor blade, $\beta_2$, can then be found using equation (7.1), namely

$$\tan \beta_2 = \tan \alpha_2 - \frac{U}{C_{a2}}$$

$$= \left(\frac{r_m}{r}\right)_2 \tan \alpha_{2m} - \left(\frac{r}{r_m}\right)_2 \frac{U_m}{C_{a2}} \tag{7.25}$$

and similarly $\beta_3$ is given by

$$\tan \beta_3 = \left(\frac{r_m}{r}\right)_3 \tan \alpha_{3m} + \left(\frac{r}{r_m}\right)_3 \frac{U_m}{C_{a3}} \tag{7.26}$$

To obtain some idea of the free vortex variation of gas angles with radius, equations (7.23)–(7.26) will be applied to the example of the previous section.

We shall merely calculate the angles at the root and tip, although in practice they would be determined at several stations up the blade to define the twist more precisely.

We shall at the same time clear up two loose ends: we have to check that there is some positive reaction at the root radius, and that the Mach number relative to the rotor blade at inlet, $M_{V2}$, is nowhere higher than say 0·75. From the velocity triangles at root and tip it will be seen that this Mach number is greatest at the root and it is only at this radius that it need be calculated.

## EXAMPLE

The determination of the gas angles at root and tip, and the checking of positive reaction and safe Mach number $M_{V2}$ at the root, for the turbine considered as an example in section 7.1.

From the mean diameter calculation we found that

$$\alpha_{2m} = 58°23', \beta_{2m} = 20°29', \alpha_{3m} = 10°, \beta_{3m} = 54°57'$$

From the calculated values of $h$ and $r_m$ we have $r_r = r_m - (h/2)$ and $r_t = r_m + (h/2)$, and thus

$$\left(\frac{r_m}{r_r}\right)_2 = 1·164, \qquad \left(\frac{r_m}{r_t}\right)_2 = 0·877, \qquad \left(\frac{r_m}{r_r}\right)_3 = 1·217, \qquad \left(\frac{r_m}{r_t}\right)_3 = 0·849$$

Also

$$\frac{U_m}{C_{a2}} = \frac{U_m}{C_{a3}} = \frac{1}{\phi} = 1·25$$

Applying equations (7.23)–(7.26) we get

|       | $\alpha_2$ | $\beta_2$ | $\alpha_3$ | $\beta_3$ |
|-------|-----------|-----------|-----------|-----------|
| Tip   | 54°56'    | 0         | 8°31'     | 58°20'    |
| Root  | 62°9'     | 39°19'    | 12°7'     | 51°8'     |

The variation of gas angles with radius appears as in Fig. 7.9, which also includes the velocity triangles at root and tip drawn to scale. That $M_{V2} = V_2/\sqrt{(\gamma R T_2)}$ is greatest at the root is clear from the velocity triangles: $V_2$ is then a maximum, and $\sqrt{(\gamma R T_2)}$ is a minimum because the temperature drop across the nozzles is greatest at the root. That there is some positive reaction at the root is also clear because $V_{3r} > V_{2r}$. Although there is no need literally to calculate the degree of reaction at the root, we must calculate $(M_{V2})_r$ to ensure that the design implies a safe value. Using data from the example in section 7.1 we have

$$V_{2r} = C_{a2} \sec \beta_{2r} = 272 \sec 39°19' = 352 \text{ m/s}$$
$$C_{2r} = C_{a2} \sec \alpha_{2r} = 272 \sec 62°9' = 583 \text{ m/s}$$

$$T_{2r} = T_{02} - \frac{C_{2r}^2}{2c_p} = 1100 - \frac{583^2}{2294} = 952 \text{ K}$$

$$(M_{V2})_r = \frac{V_{2r}}{\sqrt{(\gamma R T_{2r})}} = \frac{352}{\sqrt{(1 \cdot 333 \times 0 \cdot 287 \times 952 \times 1000)}} = 0 \cdot 58$$

This is a modest value and certainly from this point of view a higher value of the flow coefficient $\phi$ could safely have been used in the design, perhaps instead of introducing swirl at exit from the stage.

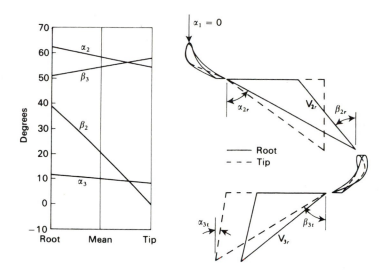

**FIG. 7.9 Variation of gas angles with radius**

## Constant nozzle angle design

As in the case of the axial compressor, it is not essential to design for free vortex flow. Conditions other than constant $C_a$ and $C_w r$ may be used to give some other form of vortex flow, which can still satisfy the requirement for radial equilibrium of the fluid elements. In particular, it may be desirable to make a constant nozzle angle one of the conditions determining the type of vortex, to avoid having to manufacture nozzles of varying outlet angle. This, as will now be shown, requires particular variations of $C_a$ and $C_w$.

The vortex flow equation (5.13) states that

$$C_a \frac{dC_a}{dr} + C_w \frac{dC_w}{dr} + \frac{C_w^2}{r} = \frac{dh_0}{dr}$$

Consider the flow in the space between the nozzles and blades. As before, we assume that the flow is uniform across the annulus at inlet to the nozzles, and so the stagnation enthalpy at outlet must also be uniform, i.e. $dh_0/dr = 0$ in plane 2. Also, if $\alpha_2$ is to be constant we have

$$\frac{C_{a2}}{C_{w2}} = \cot \alpha_2 = \text{constant}$$

$$\frac{dC_{a2}}{dr} = \frac{dC_{w2}}{dr} \cot \alpha_2$$

The vortex flow equation therefore becomes

$$C_{w2} \cot^2 \alpha_2 \frac{dC_{w2}}{dr} + C_{w2} \frac{dC_{w2}}{dr} + \frac{C_{w2}^2}{r} = 0 \qquad (7.27)$$

$$(1 + \cot^2 \alpha_2) \frac{dC_{w2}}{dr} + \frac{C_{w2}}{r} = 0$$

$$\frac{dC_{w2}}{C_{w2}} = -\sin^2 \alpha_2 \frac{dr}{r}$$

Integrating this gives

$$C_{w2} r^{\sin^2 \alpha_2} = \text{constant} \qquad (7.28)$$

And with constant $\alpha_2$, $C_{a2} \propto C_{w2}$ so that the variation of $C_{a2}$ must be the same, namely

$$C_{a2} r^{\sin^2 \alpha_2} = \text{constant} \qquad (7.29)$$

Normally nozzle angles are greater than 60 degrees, and quite a good approximation to the flow satisfying the equilibrium condition is obtained by designing with a constant nozzle angle and constant angular momentum, i.e. $\alpha_2 = $ constant and $C_{w2} r = $ constant. If this approximation is made and the rotor blades are twisted to give constant angular momentum at outlet also, then, as for free vortex flow, the work output per unit mass flow is the same at all radii. On the other hand, if equation (7.28) were used it would be necessary to integrate from root to tip to obtain the work output. We observed early in section 7.2 that there is little difference in efficiency between turbines of low radius ratio designed with twisted and untwisted blading. It follows that the sort of approximation referred to here is certainly unlikely to result in a significant deterioration of performance.

The free vortex and constant nozzle angle types of design do not exhaust the possibilities. For example, one other type of vortex design aims to satisfy the radial equilibrium condition and at the same time meet a condition of *constant mass flow per unit area* at all radii. That is, the axial and whirl velocity distributions are chosen so that the product $\rho_2 C_{a2}$ is constant at all radii. The advocates of this approach correctly point out that the simple vortex theory outlined in section 5.6 assumes no radial component of velocity, and yet even if the turbine is designed with no flare there must be a radial shift of the streamlines as shown in Fig. 7.10. This shift is due to the increase in density from root to tip in plane 2. The assumption that the radial component is zero would undoubtedly be true for a turbine of constant annulus area if the stage were designed for constant mass-flow per unit area. It is argued that the flow is then more likely to behave as intended, so that the gas angles will more closely match the blade angles. Further details can be found in Ref. (2). In view of what has just been said about the dubious benefits of vortex blading for turbines of modest radius ratio, it is very doubtful indeed whether such refinements are more than an academic exercise.

**FIG. 7.10**

## 7.3   Choice of blade profile, pitch and chord

So far in our worked example we have shown how to establish the gas angles at all radii and the blade heights. The next step is to choose stator and rotor blade shapes which will accept the gas incident upon the leading edge, and deflect the gas through the required angle with the minimum loss. An overall blade loss coefficient $Y$ (or $\lambda$) must account for the following sources of friction loss.

(a)  *Profile loss*—associated with boundary layer growth over the blade profile (including separation loss under adverse conditions of extreme angles of incidence or high inlet Mach number).
(b)  *Annulus loss*—associated with boundary layer growth on the inner and outer walls of the annulus.
(c)  *Secondary flow loss*—arising from secondary flows which are always present when a wall boundary layer is turned through an angle by an adjacent curved surface.
(d)  *Tip clearance loss*—near the rotor blade tip the gas does not follow the intended path, fails to contribute its quota of work output, and interacts with the outer wall boundary layer.

The profile loss coefficient $Y_p$ is measured directly in cascade tests similar to those described for compressor blading in section 5.8. Losses (b) and (c) cannot easily be separated, and they are accounted for by a secondary loss coefficient $Y_s$. The tip clearance loss coefficient, which normally arises only for rotor blades, will be denoted by $Y_k$. Thus the total loss coefficient $Y$ comprises the accurately measured two-dimensional loss $Y_p$, plus the three-dimensional loss $(Y_s + Y_k)$ which must be deduced from turbine stage test results. A description of one important compilation of such data will be given in section 7.4; all that is necessary for our present purpose is a knowledge of the sources of loss.

## Conventional blading

Figure 7.1 shows a conventional steam turbine blade profile constructed from circular arcs and straight lines. Gas turbines have until recently used profiles closely resembling this, although specified by aerofoil terminology. One example is shown: the T6 *base profile* which is symmetrical about the centre line. It has a thickness/chord ratio $(t/c)$ of 0·1, a leading edge radius of 12 per cent $t$ and a trailing edge radius of 6 per cent $t$. When scaled up to a $t/c$ of 0·2 and used in conjunction with a parabolic camber line having the point of maximum camber a distance of about 40 per cent $c$ from the leading

Leading edge radius 0·12$t$
Trailing edge radius 0·06$t$
Symmetrical about ¢

**FIG. 7.11 'Conventional' blade profiles: steam turbine section and T6 aerofoil section**

edge, the T6 profile leads to a blade section similar to that shown but with a radiused trailing edge. In particular, the back of the blade after the throat is virtually straight. Other shapes used in British practice have been RAF 27 and C7 base profiles on both circular and parabolic arc camber lines. All such blading may be referred to as *conventional blading*.

It is important to remember that the velocity triangles yield the *gas* angles, not the *blade* angles. Typical cascade results showing the effect of incidence on the profile loss coefficient $Y_p$ for impulse ($\Lambda = 0$ and $\beta_2 \simeq \beta_3$) and reaction type blading are given in Fig. 7.12. Evidently, with reaction blading the angle of incidence can vary from $-15°$ to $+15°$ without increase in $Y_p$. The picture is not very different even when three-dimensional losses are taken into account. This means that a rotor blade could be designed to have an inlet angle $\theta_2$ equal to say $(\beta_{2r} - 5°)$ at the root and $(\beta_{2t} + 10°)$ at the tip to reduce the twist required by a vortex design. It must be remembered, however, that a substantial margin of safe incidence range must be left to cope with part-load operating conditions of pressure ratio, mass flow and rotational speed.

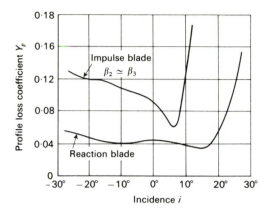

FIG. 7.12 **Effect on incidence upon** $Y_p$

With regard to the outlet angle, it has been common steam turbine practice to take the gas angle as being equal to the blade angle defined by $\cos^{-1}$ (opening/pitch). This takes account of the bending of the flow as it fills up the narrow space in the wake of the trailing edge; there is no 'deviation' in the sense of that obtained with decelerating flow in a compressor cascade. Tests on gas turbine cascades have shown, however, that the $\cos^{-1}$ ($o/s$) rule is an over-correction for blades of small outlet angle operating with low gas velocities, i.e. for some rotor blades. Figure 7.13 shows the relation between

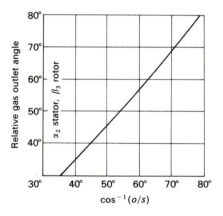

FIG. 7.13 **Relation between gas and blade outlet angles**

the relative gas outlet angle, $\beta_3$ say, and the blade angle defined by $\cos^{-1}$ ($o/s$). The relation does not seem to be affected by incidence within the working range of $\pm 15$ degrees. This curve is applicable to 'straight-backed' conventional blades operating with a relative outlet Mach number below 0·5. With a Mach number of unity the $\cos^{-1}(o/s)$ rule is good for all outlet angles, and at Mach numbers intermediate between 0·5 and 1·0 it can be assumed that $[\cos^{-1}(o/s) - \beta_3]$ varies linearly with Mach number. Reference (3) gives an additional correction for blades with a curved-back trailing edge.

Note that until the *pitch* and *chord* have been established it is not possible to draw a blade section to scale, determine the 'opening', and proceed by trial and error to make adjustments until the required gas outlet angle $\alpha_2$ or $\beta_2$ is obtained. Furthermore, this process must be carried out at a number of radii from root to tip to specify the shape of the blade as a whole. Now the pitch and chord have to be chosen with due regard to (a) the effect of the pitch/chord ratio $(s/c)$ on the blade loss coefficient, (b) the effect of chord upon the aspect ratio $(h/c)$, remembering that $h$ has already been determined, (c) the effect of rotor blade chord on the blade stresses, and (d) the effect of rotor blade pitch upon the stresses at the point of attachment of the blades to the turbine disc. We will consider each effect in turn.

(a) 'Optimum' pitch/chord ratio
In section 7.4 (Fig. 7.24) are presented cascade data on profile loss coefficients $Y_p$, and from such data it is possible to obtain the useful design curves in Fig. 7.14. These curves suggest, as might be expected, that the greater the

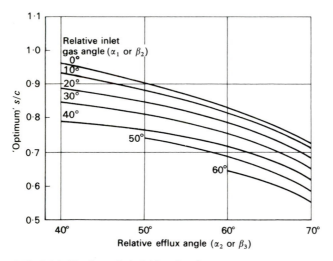

**FIG. 7.14 'Optimum' pitch/chord ratio**

gas deflection required $[(\alpha_1 + \alpha_2)$ for a stator blade and $(\beta_2 + \beta_3)$ for a rotor blade] the smaller must be the 'optimum' $s/c$ ratio to control the gas adequately. The adjective 'optimum' is in inverted commas because it is an optimum with respect to $Y_p$ not to the overall loss $Y$. The true optimum value of $s/c$ could be found only by making a detailed estimate of stage performance (e.g. on the lines described in section 7.4) for several stage designs differing in

$s/c$ but otherwise similar. In fact the $s/c$ value is not very critical.

For the nozzle and rotor blade of our example turbine we have established that

$$\alpha_{1m} = 0, \; \alpha_{2m} = 58°23'; \; \beta_{2m} = 20°29', \; \beta_{3m} = 54°57'$$

From Fig. 7.14 we therefore have at mean diameter

$$(s/c)_N = 0\cdot86 \text{ and } (s/c)_R = 0\cdot83$$

(b)  *Aspect ratio* $(h/c)$

The influence of aspect ratio is open to conjecture, but for our purpose it is sufficient to note that, although not critical, too low a value is likely to lead to secondary flow and tip clearance effects occupying an unduly large proportion of the blade height and so increasing $Y_s$ for the nozzle row and $(Y_s + Y_k)$ for the rotor row. On the other hand, too high a value of $h/c$ will increase the likelihood of vibration trouble: vibration characteristics are difficult to predict and they depend on the damping provided by the method of attaching the blades to the turbine disc.† A value of $h/c$ between 3 and 4 would certainly be very satisfactory, and it would be unwise to use a value below 2.

For our turbine, which is flared, we have the mean heights of the nozzle and rotor blades given by

$$h_N = \tfrac{1}{2}(0\cdot046 + 0\cdot0612) = 0\cdot0536 \text{ m}$$
$$h_R = \tfrac{1}{2}(0\cdot0612 + 0\cdot077) = 0\cdot0691 \text{ m}$$

Adopting an aspect ratio $(h/c)$ of 3 we then have

$$c_N = 0\cdot0175 \text{ m and } c_R = 0\cdot023 \text{ m}$$

Using these values of chord, in conjunction with the chosen $s/c$ values, gives the blade pitches at the mean radius of $0\cdot216$ m as

$$s_N = 0\cdot01\ 506 \text{ m and } s_R = 0\cdot0191 \text{ m}$$

and the numbers of blades, from $2\pi r_m/s$, as

$$n_N = 90 \text{ and } n_R = 71$$

It is usual to avoid numbers with common multiples to reduce the probability of introducing resonant forcing frequencies. The common practice is to use an *even* number for the nozzle blades and a *prime* number for the rotor blades. As it happens the foregoing numbers are satisfactory and there is no need to modify them and re-evaluate the pitch $s$.

(c)  *Rotor blade stresses*

The next step is to check that the stage design is consistent with a permissible level of stress in the rotor blades. The final design must be checked by laying

† Vibration problems with high aspect ratio blading can be significantly reduced by using tip shrouds which prevent vibrations in the cantilever mode. Shrouds are also sometimes used to reduce tip leakage loss.

out the blade cross-sections at several radii between root and tip, and performing an accurate stress analysis on the lines indicated by Sternlicht in
Ref. (6). Although we are not concerned with mechanical design problems
in this book, simple approximate methods adequate for preliminary design
calculations must be mentioned because blade stresses have a direct impact
upon the stage design. There are three main sources of stress: (i) centrifugal
tensile stress (the largest, but not necessarily the most important because it
is a steady stress), (ii) gas bending stress (fluctuating as the rotor blades pass
by the trailing edges of the nozzles) and (iii) centrifugal bending stress when
the centroids of the blade cross-sections at different radii do not lie on a
radial line (any torsional stress arising from this source is small enough to be
neglected).

When the rotational speed is specified, the allowable centrifugal tensile
stress places a limit on the annulus area but does not affect the choice of
blade chord. This somewhat surprising result can easily be shown to be the
case as follows. The maximum value of this stress occurs at the root and is
readily seen to be given by

$$(\sigma_{ct})_{max} = \frac{\rho_b \omega^2}{a_r} \int_r^t ar \, dr$$

where $\rho_b$ is the density of blade material, $\omega$ is the angular velocity, $a$ is the
cross-sectional area of the blade and $a_r$ its value at the root radius. In practice
the integration is performed graphically or numerically, but if the blade were
of uniform cross-section the equation would reduce directly to

$$(\sigma_{ct})_{max} = 2\pi N^2 \rho_b A$$

where $A$ is the annulus area and $N$ is the rotational speed in rev/s. A rotor
blade is usually tapered in chord and thickness from root to tip such that
$a_t/a_r$ is between 1/4 and 1/3. For preliminary design calculations it is sufficiently accurate (and on the safe side) to assume that the taper reduces the
stress to 2/3 of the value for an untapered blade. Thus

$$(\sigma_{ct})_{max} \simeq \tfrac{4}{3}\pi N^2 \rho_b A \qquad (7.30)$$

For the flared turbine of our example we have

$$A = \tfrac{1}{2}(A_2 + A_3) = 0 \cdot 094 \text{ m}^2 \text{ and } N = 250 \text{ rev/s}$$

The density of the Ni–Cr–Co alloys used for gas turbine blading is about
8000 kg/m³, and so equation (7.30) gives

$$(\sigma_{ct})_{max} \simeq 200 \text{ MN/m}^2 \text{ (or 2000 bar)}$$

Judgement as to whether or not this stress is satisfactory must await the
evaluation of the other stresses.

The force arising from the change in angular momentum of the gas in the
tangential direction, which produces the useful torque, also produces a gas
bending moment about the axial direction, namely $M_w$ in Fig. 7.15. There

may be a change of momentum in the axial direction (i.e. when $C_{a3} \neq C_{a2}$), and with reaction blading there will certainly be a pressure force in the axial direction $[(p_2 - p_3)\, 2\pi r/n$ per unit height], so that there will also be a gas bending moment $M_a$ about the tangential direction. Resolving these bending moments into components acting about the principle axes of the blade cross-section, the maximum stresses can be calculated by the method appropriate to asymmetrical sections. A twisted and tapered blade must be divided into strips of height $\delta h$ and the bending moments calculated from the average

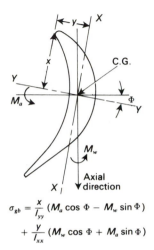

$$\sigma_{gb} = \frac{x}{I_{yy}} (M_a \cos \Phi - M_w \sin \Phi)$$

$$+ \frac{y}{I_{xx}} (M_w \cos \Phi + M_a \sin \Phi)$$

**FIG. 7.15**

force acting on each strip. The gas bending stress $\sigma_{gb}$ will be tensile in the leading and trailing edges and compressive in the back of the blade, and even with tapered twisted blades the maximum value usually occurs at either the leading or trailing edge of the root section. Because $M_w$ is by far the greater bending moment, and the principal axis XX does not deviate widely from the axial direction (angle $\Phi$ is small), a useful approximation for preliminary design purposes is provided by

$$(\sigma_{gb})_{max} \simeq \frac{m(C_{w2m} + C_{w3m})}{n} \times \frac{h}{2} \times \frac{1}{zc^3} \tag{7.31}$$

$n$ is the number of blades, the whirl velocities are evaluated at the mean diameter, and $z$ is the smallest value of the root section modulus ($I_{xx}/y$) of a blade of unit chord. Clearly $\sigma_{gb}$ is directly proportional to the stage work output and blade height, and inversely proportional to the number of blades and section modulus. It is convenient to treat the section modulus as the product $zc^3$ because $z$ is largely a function of blade camber angle ($\simeq$ gas deflection) and thickness/chord ratio.

An unpublished rule for $z$ due to Ainley, useful for approximate calculations, is given in Fig. 7.16. We shall apply this, together with equation (7.31), to our example turbine. Assuming the angle of incidence is zero at the design operating condition, the blade camber angle is virtually equal to the gas deflection, namely at the root

$$\beta_{2r} + \beta_{3r} = 39°19' + 51°8' \simeq 90 \text{ degrees}$$

Then from Fig. 7.16, assuming a blade of $t/c = 0.2$,

$$(z)_{\text{root}} = \frac{(10 \times 0.2)^{1.27}}{570} = 0.00\,423 \text{ mm}^3/\text{mm chord}$$

$$m(C_{w2} + C_{w3}) = mC_a(\tan \alpha_2 + \tan \alpha_3)$$

which at mean diameter yields

$$20 \times 272\,(1.624 + 0.175) = 9800 \text{ kN}$$

FIG. 7.16 **Approximate rule for section moduli**

For the chosen value of $c_R = 0.023$ m, $n_R$ was found to be 71, while $h_R = \frac{1}{2}(h_2 + h_3) = 0.0691$ m. Equation (7.31) can now be evaluated to give

$$(\sigma_{gb})_{\text{max}} \simeq \frac{9800}{71} \times \frac{0.0691}{2} \times \frac{1}{0.00423 \times 0.023^3} \simeq 93 \text{ MN/m}^2$$

By designing the blade with the centroids of the cross-sections slightly off a radial line, as indicated in Fig. 7.17, it is theoretically possible to design for a centrifugal bending stress which will cancel the gas bending stress. It must be remembered, however, that (a) these two stresses would only cancel each other at the design operating condition, (b) $\sigma_{gb}$ is only a quasi-steady stress and (c) the centrifugal bending stress is very sensitive to manufacturing errors in the blade and blade root fixing. $\sigma_{gb}$ is often not regarded as being offset by any centrifugal bending stress and usually the latter is merely calculated using the extreme values of manufacturing tolerances to check that it is small and that at least it does not reinforce $\sigma_{gb}$.

We have now established a steady centrifugal stress of 200 MN/m² and a gas bending stress of 93 MN/m² which is subject to periodic fluctuation with

Gas bending moment

Centrifugal bending moment

**FIG. 7.17**

a frequency dependent on $N$, $n_R$ and $n_N$. Creep strength data for possible blade materials will be available: perhaps in the form of Fig. 7.18(a) which shows the time of application of a steady stress at various temperatures required to produce 0·2 per cent creep strain. Fatigue data (e.g. Gerber diagrams) will also be available from which it is possible to assess the relative capacity of the materials to withstand fluctuating stresses. Such data, together with experience from other turbines in service, will indicate how the fluctuating gas bending stress and the steady centrifugal stress can be combined safely. The designer would hope to have a set of curves of the type shown in Fig. 7.18(b) for several safe working lives. The values of temperature on this plot might refer to turbine inlet stagnation temperature $T_{01}$, allowance having been made for the fact that

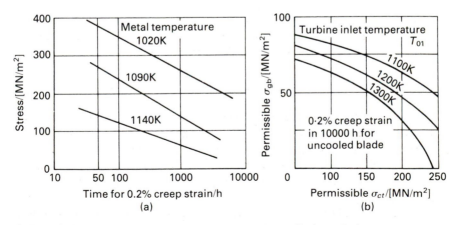

**FIG. 7.18 Creep data and designer's aid for assessing preliminary designs**

(i)   only the leading edge of the rotor blade could theoretically reach stag-
      nation temperature and chordwise conduction in the metal would
      prevent even the local temperature there from reaching $T_{01}$;
(ii)  even in 'uncooled' turbines (those with no cooling passages in the blades),
      some cooling air is bled from the compressor and passed over the turbine
      disc and blade roots: the metal temperature will therefore be appreciably
      less than 1100 K near the root radius for which the stresses have been
      estimated.

Furthermore, the values of permissible $\sigma_{gb}$ and $\sigma_{ct}$ will be conservative, and
include a safety factor to allow for local hot streaks of gas from the combustion
system and for the fact that there will be additional thermal stresses due to
chordwise and spanwise temperature gradients in the blade. In our example
$\sigma_{ct}$ and $\sigma_{gb}$ were found to be 200 and 93 MN/m$^2$ respectively. If a life of
10 000 h was required, the curve relating to our inlet temperature of 1100 K
suggests that the stresses are rather too large. The blade chord could be increased slightly to reduce $\sigma_{gb}$ if the need for reduced stresses is confirmed by
more detailed calculations. As stated earlier, the final design would be subjected to a complete stress analysis, which would include an estimate of the
temperature field in the blade and the consequential thermal stresses.

### (d)   Effect of pitch on the blade root fixing

The blade pitch $s$ at mean diameter has been chosen primarily to be compatible
with required values of $s/c$ and $h/c$, and (via the chord) of permissible $\sigma_{gb}$. A
check must be made to see that the pitch is not so small that the blades
cannot be attached safely to the turbine disc rim. Only in small turbines is
it practicable to machine the blades and disc from a single forging, cast them
integrally, or weld the blades to the rim, and Fig. 7.19 shows the commonly
used fir tree root fixing which permits replacement of blades. The fir trees
are made an easy fit in the rim, being prevented from axial movement only
(e.g. by a lug on one side and peening on the other). When the turbine is
running, the blades are held firmly in the serrations by centripetal force, but
the slight freedom to move can provide a useful source of damping for unwanted vibration. The designer must take into account stress concentrations
at the individual serrations, and manufacturing tolerances are extremely
important; inaccurate machining can result in some of the serrations being
unloaded at the expense of others. Failure may occur by the disc rim yielding
at the base of the stubs left on the disc after broaching (at section $x$); by
shearing or crushing of the serrations; or by tensile stress in the fir tree root
itself. The pitch would be regarded as satisfactory when the root stresses
can be optimized at a safe level. This need not detain us here because calculation of these centrifugal stresses is straightforward once the size of the blade,
and therefore its mass, have been established by the design procedure we
have outlined.

    Finally, the total centrifugal blade loading on the disc and the disc rim
diameter both being known, the disc stresses can be determined to see if the

original assumption of a mean blade speed of 340 m/s is satisfactory. Centrifugal hoop and radial stresses in a disc are proportional to the square of the rim speed. Disc design charts (e.g. the Donath chart) are available to permit the nominal stresses to be estimated rapidly for any disc of arbitrary shape; see Ref. (6). They will be 'nominal' because the real stress pattern will be affected substantially by thermal stresses arising from the large temperature gradient between rim and hub or shaft.

**FIG. 7.19 'Firtree' root**

Before proceeding to make a critical assessment of the stage design used as an example, which we shall do in section 7.4, it is logical to end this section by outlining briefly recent developments in the prediction and construction of more efficient blade profiles. *This will be a digression from the main theme, however, and the reader may prefer to omit it at the first reading.*

## *Theoretical approach to the determination of blade profiles and pitch/chord ratio*

There is little doubt that the approach to turbine design via cascade test results, which has been outlined here, is satisfactory for moderately loaded turbine stages. Recently, however, advanced aircraft propulsion units have required the use of high blade loadings (i.e. high $\psi$ and $\phi$) and cooled blades, which takes the designer into regions of flow involving ever-increasing extrapolation from existing cascade data. Furthermore it is found that under these conditions even minor changes in blade profile, such as the movement of the point of maximum camber towards the leading edge from say 40 to 37 per cent chord, can make a substantial difference to the blade loss coefficient particularly when the turbine is operating away from the design point, i.e. at part load. Rather than repeat the vast number of cascade tests to cover the more arduous range of conditions, the approach now is to run a few such tests to check the adequacy of theoretical predictions and then apply the theory.

The digital computer has made it possible to develop methods of solving the equations of compressible flow through a blade row, even taking radial components into account to cover the case of a flared annulus. (Earlier incompressible flow solutions, adequate for the small pressure changes in compressor blading, are of little use in turbine design.) At first the approach was towards predicting the potential flow for a blade of given profile. That is, the pressure and velocity distributions outside the boundary layer are calculated in a passage bounded by the prescribed concave surface of one blade and the convex surface of the adjacent blade. Conformal transformation theory, methods of distributed sources and sinks, and stream filament theory have all been applied in one form or another and with varying degrees of success. Once the potential flow pattern has been established, boundary layer theory can be applied to predict the profile loss coefficient. The chief difficulties in this part of the analysis are the determination of the point of transition between laminar and turbulent flow on the convex surface, and the determination of a satisfactory mathematical model for the wake downstream of the trailing edge.

Some idea of the complexity of the calculations involved can be formed from Ref. (2), which also gives detailed references to this work. A not unimportant question is which method takes the least computer time consistent with providing solutions of adequate accuracy. While there is still so much controversy as to the best method, it is not practicable to attempt a simple introductory exposition in a book of this kind. One important outcome of the work, however, can be stated in simple terms with reference to the typical pressure and velocity distributions for conventional blading shown in Fig. 7.20.† The vital feature is the magnitude of the opposing pressure gradient on the convex (suction) surface. If too great it will lead to separation of the boundary layer somewhere on the back of the blade, a large wake, and a substantial increase in the profile loss coefficient. When trying to design with high aerodynamic blade loadings, which imply low suction surface pressures, it is desirable to know what limiting pressure or velocity distributions on the suction surface will just give separation at the trailing edge of the blade. One guide is provided by Smith, Ref. (7), who also makes a useful comparison of six different criteria for the prediction of separation. (He sidesteps the difficulty of determining the point of transition from a laminar to turbulent boundary layer, however, by assuming a turbulent layer over the whole of the suction surface.) Using the most conservative criterion for separation,

---

† It is the flow relative to the blade which is referred to in this section: if a rotor blade is under consideration the paramerers would be $(p - p_3)/(p_{0\,3\mathrm{rel}} - p_3)$ and $V/V_3$.

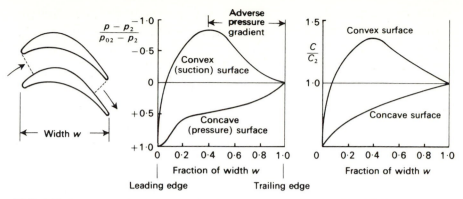

**FIG. 7.20 Pressure and velocity distributions on a conventional turbine blade**

namely that of Stratford, he constructed two extreme families of simplified, limiting, velocity distributions as shown by the two inset sketches in Fig. 7.21. Within each family the distribution is defined by a value of $(C_{max}/C_2)$ and the position of the point of maximum velocity, $A$, along the blade surface. The curves for various Reynolds numbers, only two per family being repro-

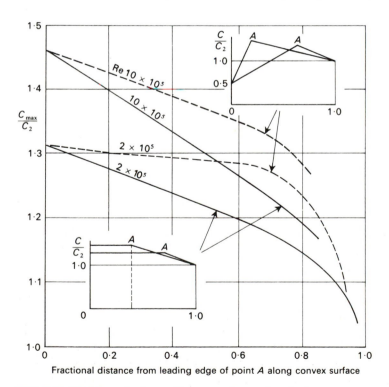

**FIG. 7.21 Limiting velocity profiles on suction surface for boundary layer separation at trailing edge**

duced here, represent the loci of point $A$. Any velocity distribution falling in the region below the relevant curve should imply freedom from separation. (Note that, helpfully, separation is delayed by an increase in Reynolds number.) When this type of information is available the value of being able to predict the velocity distribution is apparent: small changes in a proposed profile or camber line shape can be made until the velocity distribution falls within the safe region.

What we have been referring to so far are approaches that have been made to *the direct problem*: the prediction of the velocity distribution around a given blade in cascade. Attention is now being focused on *the indirect problem*: the theoretical determination of a blade shape which will give a *prescribed blade surface velocity distribution*. What the ideal velocity distribution should be in various circumstances to give the minimum loss is certainly not yet established, but at least enough is known to avoid such obvious weaknesses as boundary layer separation (e.g. near the blade root where the degree of reaction is low) or the formation of unwanted shock waves. Ultimately it is hoped to be able to build into a computer program such restraints on the possible blade shape as those provided by stressing considerations (blade section area and section moduli) and by a minimum trailing edge thickness (dictated by manufacturing necessity or the need to accommodate a cooling passage). The profile loss coefficient of the restricted range of possible profiles can then be evaluated to enable a final choice to be made. The method of solving the indirect problem which seems to have proved the most capable of useful development is that due to Stanitz, although the approach via stream filament theory initiated by Wu is also receiving attention. A helpful summary of the main steps in these solutions is given by Horlock in Ref. (2).

It is to be hoped that enough has been said here to warn the student who wishes to know more of these topics that a first essential is a thorough grounding in aerodynamics and turbulent boundary layer theory. He should follow this with a study of Sections B and C of Ref. (1), on the 'theory of two-dimensional flow through cascades' and 'three-dimensional flow in turbomachines'. To give the reader a feel for the indirect problem, however, we will end this section with a brief description of an approximate solution due to Stanitz which Horlock has put into terms comprehensible to readers of section 5.9.

The case considered is the relatively simple one of compressible flow in a two-dimensional cascade as depicted in Fig. 7.22. We shall find it convenient to refer to the whirl component of the force acting on unit height of blade which was denoted by $F$ in section 5.9. Unlike the treatment in that section, we shall here be concerned with the way $F$ per unit width $(w)$ changes with $x$, and moreover we cannot make the incompressible assumption that $\rho$ is constant. For the indirect problem the following data will be specified:

(a)  upstream conditions $(p_{01}, T_{01}, C_1, \alpha_1)$ and downstream conditions $(C_2, \alpha_2)$;

**FIG. 7.22 The 'indirect' problem for two-dimensional compressible flow in a turbine cascade**

(b)  surface velocity distributions as a function of $x$ between $x = 0$ and $x = 1$, namely $C_s(x)$ and $C_p(x)$ for the suction and pressure surfaces respectively. The directions of $C_s$ and $C_p$ will be unknown because the shape of the surface is unknown, indicating the need for a process of iteration.

Because we are dealing only with the potential flow, the expansion will be isentropic so that $p_0$ and $T_0$ will be constant through the passage. For any local value of $C$, therefore, local values of $p$ and $\rho$ can be found from the isentropic relations

$$T_0 = T + C^2/2c_p; \qquad p_0/p = (T_0/T)^{\gamma/(\gamma-1)}; \qquad \rho_0/\rho = (p_0/p)^{1/\gamma}$$

We note that the blade profile will be completely determined when (a) the pitch/width ratio ($s/w$) is established, and (b) both the camber line angle $\alpha'$ and blade thickness/pitch ratio have been calculated for various values of $x$ between 0 and 1. The procedure is as follows.

*First approximation*
(i)    Determine $\rho_1$ from the given inlet conditions $p_{01}$, $T_{01}$ and $C_1$; and note that for continuity in two-dimensional flow

$$m = \rho_1 s C_1 \cos \alpha_1 = \rho_2 s C_2 \cos \alpha_2$$

where $m$ is the mass flow per unit height of passage.
(ii)   Using the isentropic relations, calculate the pressures $p_p$ and $p_s$ for a series of values of $x$ between 0 and 1 from the given surface velocity distributions of $C_p$ and $C_s$.
(iii)  Integrate numerically $p_p(x)$ and $p_s(x)$ from $x = 0$ to $x = 1$, and so determine the mean force per unit height of blade in the whirl direction from $F_m = \Delta p_m \times w$.
(iv)   Determine $s/w$ by equating $F_m$ to the overall change of momentum in the whirl direction, i.e.

$$\Delta p_m \times w = m(C_{w1} + C_{w2}) = \rho_1 s C_1 \cos \alpha_1 (C_1 \sin \alpha_1 + C_2 \sin \alpha_2)$$

$$\frac{s}{w} = \frac{\Delta p_m}{\rho_1 C_1 \cos \alpha_1 (C_1 \sin \alpha_1 + C_2 \sin \alpha_2)}$$

(v)   For various values of $x$ between 0 and 1, integrate the pressure distributions from the leading edge to $x$, and so obtain the pressure differences
$\Delta p$ which act over the series of areas $(xw) \times 1$.

(vi)  Determine the mean whirl velocity $C_w$ at the values of $x$ used in (v)
from

$$\Delta p \times xw = \rho_1 s C_1 \cos \alpha_1 (C_1 \sin \alpha_1 \pm C_w)$$

(The negative sign refers to values near the leading edge where $C_w$ is
in the same direction as $C_{w1}$.)

(vii) Finding the corresponding mean values of $C$ from $(C_p + C_s)/2$, the values
of $\alpha'$ at the various values of $x$ are given by

$$\sin \alpha' = C_w/C$$

(viii) For convenience the blade thickness is measured in the whirl direction.
Thus $t_w$ at the various values of $x$ can be found as a fraction of the pitch
$s$ from the continuity equation:

$$\rho_1 C_1 \cos \alpha_1 = \rho \left[ 1 - \left( \frac{t_w}{s} \right) \right] C \cos \alpha'$$

where the mean density $\rho$ at any $x$ is found from the corresponding
mean values of $p$ and $C$ by using the isentropic relations.

In this first approximation the directions of the surface velocities $C_p$ and
$C_s$ have been assumed the same and equal to that of the mean velocity $C$,
namely $\alpha'$. Furthermore the properties have been assumed constant across
the passage at the mean value. Having established an approximate blade
profile it is possible to refine these assumptions and obtain a better approximation to the true profile. The value of $s/w$ did not depend on the assumptions
and remains unchanged, but the calculation of $\alpha'$ and $t_w/s$ as functions of $x$
must be repeated.

*Second approximation*
(i)   The flow directions $\alpha_p$ and $\alpha_s$ of the surface velocities $C_p$ and $C_s$ are
determined by the geometry of the concave and convex surfaces resulting
from the first approximation, and initially the properties are considered
to vary linearly across the passage from one surface to the other.

(ii)  The position of the streamline which divides the flow equally between
the surfaces, is determined by assuming that the axial velocity $C \cos \alpha$
varies linearly across the passage from $C_p \cos \alpha_p$ to $C_s \cos \alpha_s$. [See Fig.
7.22(c).]

(iii) The velocity on the central streamline, $C_m$, is determined from the fact that potential flow is irrotational. The criterion of irrotationality is expressed by

$$\frac{\partial}{\partial y}(C \cos \alpha) - \frac{\partial}{\partial x}(C \sin \alpha) = 0$$

where $y$ is the co-ordinate in the whirl direction. The assumption in (ii) implies that the first term is constant. With the additional assumption that $\partial C/\partial x$ varies linearly with $y$ across the passage, the equation can be integrated numerically to give $C$ as a function of $y$ and in particular the value of $C$ on the central streamline $(C_m)$.

(iv) The value of the whirl velocity on the central streamline, $C_{wm}$, is again determined by equating the change of momentum, from the leading edge to $x$, to the corresponding pressure force; and the camber line angle is then obtained from $\tan \alpha' = C_{wm}/C_m$.

(v) The thickness in terms of $t_w/s$ is determined from the continuity equation as before.

When integrating across the passage for steps (iv) and (v), parabolic variations of the relevant properties are used instead of linear variations. The parabolas pass through the values at the two surfaces and at the central streamline.

A third approximation on the same lines as the second can be made if necessary, and the final shape is then determined by rounding off the leading and trailing edges, and subtracting an estimated displacement thickness of the boundary layer which may be appreciable on the back of the blade where the flow is decelerating.

## 7.4 Estimation of stage performance

The last step in the process of arriving at the preliminary design of a turbine stage is to check that the design is likely to result in values of nozzle loss coefficient and stage efficiency which were assumed at the outset. If not, the design calculations may be repeated with more probable values of loss coefficient and efficiency. When satisfactory agreement has been reached, the final design may be laid out on the drawing board and accurate stressing calculations can be performed.

Before proceeding to describe a method of estimating the design point performance of a stage, however, the main factors limiting the choice of design, which we have noted during the course of the worked example, will be summarized. The reason we considered a turbine for a turbojet engine was simply that we would thereby be working near those limits to keep size and weight to a minimum. The designer of an industrial gas turbine has a somewhat easier task: he will be using lower temperatures and stresses to obtain a longer working life, and this means lower mean blade speeds, more stages, and much less stringent aerodynamic limitations. A power turbine, not mechanically coupled to the gas generator, is another case where much less difficulty will be encountered in arriving at a satisfactory solution. The choice of gear ratio between the power turbine and driven component is normally at the disposal of the turbine designer, and thus the rotational speed can be varied to suit the turbine, instead of the compressor as we have assumed here.

## Limiting factors in turbine design

(a) *Centrifugal stresses* in the blades are proportional to the square of the rotational speed $N$ and the annulus area: when $N$ is fixed they place an upper limit on the *annulus area*.

(b) *Gas bending stresses* are (1) inversely proportional to the number of blades and blade section moduli, while being (2) directly proportional to the blade height and specific work output.

   (1) The *number of blades* cannot be increased beyond a point set by blade fixing considerations, but the *section moduli* are roughly proportional to the cube of the *blade chord* which might be increased to reduce $\sigma_{gb}$. There is an aerodynamic limit on the *pitch/chord ratio*, however, which if too small will incur a high loss coefficient (friction losses increase because a reduction in $s/c$ increases the blade surface area swept by the gas).

   (2) There remains the *blade height*: but reducing this while maintaining the same annulus area (and therefore the same axial velocity for the given mass flow), implies an increase in the *mean diameter of the annulus*. For a fixed $N$, the mean diameter cannot be increased without increasing the *centrifugal disc stresses*. There will also be an aerodynamic limit set by the need to keep the *blade aspect ratio* ($h/c$) and *annulus radius ratio* ($r_t/r_r$) at values which do not imply disproportionate losses due to *secondary flows, tip clearance* and *friction on the annulus walls* (say not less than 2 and 1·2 respectively). The blade height might be reduced by reducing the annulus area (with the added benefit of reducing the centrifugal blade stresses) but, for a given mass flow, only by increasing the *axial velocity*. An aerodynamic limit on $C_a$ will be set by the need to keep the maximum *relative Mach number at the blade inlet* (namely at the root radius), and the *Mach number at outlet from the stage*, below the levels which mean high friction losses in the blading and jet pipe respectively.

(c) *Optimizing the design*, so that it just falls within the limits set by all these conflicting mechanical and aerodynamic requirements, will lead to an efficient turbine of minimum weight. If it proves to be impossible to meet one or more of the limiting conditions, the required work output must be split between *two stages*. The second design attempt would be commenced on the assumption that the efficiency is likely to be a maximum when the work, and hence the temperature drop, is equally divided between the stages.

(*d*)   The velocity triangles, upon which the rotor blade section depends, are partially determined by the desire to work with an average *degree of reaction of* 50 *per cent* to obtain low blade loss coefficients and *zero swirl* for minimum loss in the jet pipe. To avoid the need for two stages in a marginal case, particularly if it means adding a bearing on the downstream side, it would certainly be preferable to design with a lower degree of reaction and some swirl. An aerodynamic limit on the minimum value of the reaction at mean diameter is set by the need to ensure some *positive reaction at the blade root radius*.

For what follows in the next section, it will be helpful to have a summary of the results of the design calculations for the turbine of our worked example: such a summary is given in Fig. 7.23 and over the page.

Mean diameter stage parameters

$$\psi = 2c_p\Delta T_{0s}/U^2 = 2{\cdot}88, \quad \phi = C_a/U = 0{\cdot}8, \quad \Lambda = 0{\cdot}421$$
$$U = 340 \text{ m/s}, \quad C_{a1} = C_1 = C_3 = 276{\cdot}4 \text{ m/s}, \quad C_{a2} = C_{a3} = C_a$$
$$= 272 \text{ m/s}$$
$$C_2 = 519 \text{ m/s}, \quad V_3 = 473{\cdot}5 \text{ m/s}$$

| Gas angles | $\alpha_1$ | $\alpha_2$ | $\alpha_3$ | $\beta_2$ | $\beta_3$ |
|---|---|---|---|---|---|
| root | 0° | 62°9′ | 8°31′ | 39°19′ | 51°8′ |
| mean | 0° | 58°23′ | 10° | 20°29′ | 54°57′ |
| tip | 0° | 54°56′ | 12°7′ | 0° | 58°20′ |

| Plane | 1 | 2 | 3 | |
|---|---|---|---|---|
| $p$ | 3·54 | 2·49 | 1·856 | bar |
| $T$ | 1067 | 982·7 | 922 | K |
| $\rho$ | 1·155 | 0·883 | 0·702 | kg/m³ |
| $A$ | 0·0626 | 0·0833 | 0·1047 | m² |
| $r_m$ | ← | 0·216 | → | m |
| $r_t/r_r$ | 1·24 | 1·33 | 1·43 | |
| $h$ | 0·046 | 0·0612 | 0·077 | m |

| Blade row | Nozzle | Rotor |
|---|---|---|
| $s/c$ | 0·86 | 0·83 |
| $h$ (mean) | 0·0536 | 0·0691 m |
| $h/c$ | 3·0 | 3·0 |
| $c$ | 0·0175 | 0·023 m |
| $s$ | 0·01 506 | 0·0191 m |
| $n$ | 90 | 71 |

**FIG. 7.23  Summary of data from preliminary design calculations of single-stage turbine**

Rotor blade designed for $i = 0°$ with conventional profile having $t/c = 0.2$. At the root section; camber $\simeq \beta_{2r} + \beta_{3r} \simeq 90°$ and hence $z_r \simeq 0.00423$ mm$^3$/mm chord giving $(\sigma_{gb})_{max} \simeq 93$ MN/m$^2$. $(\sigma_{ct})_{max} \simeq f(N,A) \simeq 200$ MN/m$^2$

## Estimation of design point performance

The method to be outlined here is that due to Ainley and Mathieson, Ref. (8), which estimates the performance on flow conditions at the mean diameter of the annulus. Reference (8) describes how to calculate the performance of a turbine over a range of operating conditions, but we shall be concerned here only to find the efficiency at the design point. A start is made using the two correlations for profile loss coefficient $Y_p$ obtained from cascade data, which are shown in Fig. 7.24. These refer to nozzle-type blades ($\beta_2 = 0$) and impulse-type blades ($\beta_2 = \beta_3$) of conventional profile (e.g. T6) having a thickness/ chord ratio ($t/c$) of 0.2 and a trailing edge thickness/pitch ratio ($t_e/s$) of 0.02. Rotor blade notation is used in Fig. 7.24 and in what follows, to emphasize

**FIG. 7.24  Profile loss coefficient for conventional blading with $t/c = 0.20$**

that we are thinking of the flow *relative* to any blade row. When the nozzle row is being considered $\beta_2$ becomes $\alpha_1$ and $\beta_3$ becomes $\alpha_2$: there is no need to duplicate equations which apply equally to both rows. The values of $Y_p$ in Fig. 7.24 refer to blades operating at zero incidence, i.e. when the gas inlet angle $\beta_2$ is also the blade inlet angle.

*Step* 1
Estimate $(Y_p)_N$ and $(Y_p)_R$ from the gas angles of the proposed design by using Fig. 7.24 in conjunction with the interpolation formula

$$Y_p = \left\{ Y_{p(\beta_2 = 0)} + \left(\frac{\beta_2}{\beta_3}\right)^2 \left[ Y_{p(\beta_2 = \beta_3)} - Y_{p(\beta_2 = 0)} \right] \right\} \left(\frac{t/c}{0.2}\right)^{\beta_2/\beta_3} \quad (7.32)$$

This equation represents a correction for a change in inlet angle at a constant outlet angle, so that $Y_{p(\beta_2 = 0)}$ and $Y_{p(\beta_2 = \beta_3)}$ are the values for a nozzle and impulse-type blade having the same *outlet* gas angle $\beta_3$ as the actual blade. Equation (7.32) also includes a correction for $t/c$ if it differs from 0·2, a reduction in $t/c$ leading to reduced profile loss for all blades other than nozzle-type blades ($\beta_2 = 0$). The degree of acceleration of the flow in the blading decreases with the degree of reaction as $\beta_2/\beta_3 \to 1$, and the influence of blade thickness becomes more marked as the acceleration is diminished. The correction is considered reliable only for $0·15 < t/c < 0·25$.

For the nozzle blades of our example, $\alpha_1 = 0$ and so $(Y_p)_N$ can be read straight from Fig. 7.24. $\alpha_2 = 58°23'$ and $(s/c)_N = 0·86$, and hence

$$(Y_p)_N = 0·024.$$

For the rotor blades, $\beta_2 = 20°29'$, $\beta_3 = 54°57'$, $(s/c)_R = 0·83$ and $(t/c) = 0·2$, so that

$$(Y_p)_R = \left\{ 0·023 + \left(\frac{20·5}{55}\right)^2 \left[ 0·087 - 0·023 \right] \right\} = 0·032$$

*Step* 2
If it had been decided to design the blades to operate with some incidence at the design point, a correction to $Y_p$ would be required. As this correction is really only important when estimating performance at part load, we shall refer the reader to Ref. (8) for the details. Briefly, it involves using correlations of cascade data to find the stalling incidence $i_s$ for the given blade (i.e. incidence at which $Y_p$ is twice the loss for $i = 0$); and then using a curve of $Y_p/Y_{p(i = 0)}$ versus $i/i_s$ to find $Y_p$ for the given $i$ and value of $Y_{p(i = 0)}$ calculated in Step 1.

*Step* 3
Secondary and tip clearance loss data for $Y_s$ and $Y_k$ have been correlated using the concepts of lift and drag coefficient which were introduced in section 5.9 for axial compressors. Without repeating the whole argument leading to equation (5.34), it should be possible to see by glancing at section 5.9 that for a turbine cascade (with rotor blade notation)

$$C_L = 2(s/c)(\tan \beta_2 + \tan \beta_3) \cos \beta_m$$

where

$$\beta_m = \tan^{-1} \left[ (\tan \beta_3 - \tan \beta_2)/2 \right]$$

Now as stated at the beginning of section 7.3, it is convenient to treat $Y_s$ and $Y_k$ simultaneously. The proposed correlation is

$$Y_s + Y_k = \left[\lambda + B\left(\frac{k}{h}\right)\right]\left[\frac{C_L}{s/c}\right]^2\left[\frac{\cos^2 \beta_3}{\cos^3 \beta_m}\right] \tag{7.33}$$

Reference back to equations (5.35) and (5.38) will make the appearance of the last two bracketed terms seem logical: they occur in the expression for the combined secondary and tip clearance loss coefficient for a compressor blade row. Considering now the first bracketed term, the tip clearance component is proportional to $k/h$ where $k$ is the clearance and $h$ the blade height. The constant $B$ is 0·5 for a radial tip clearance, and 0·25 for a shrouded blade with side clearance; see Fig. 7.25. The secondary loss component $\lambda$ is

$B = 0·5$        $B = 0·25$

**FIG. 7.25**

more complex. We have suggested that secondary flow and annulus wall friction might be affected by the aspect ratio $(h/c)$ and/or annulus radius ratio $(r_t/r_r)$. As we shall see, $r_t/r_r$ is thought by Ainley and Mathieson to be the more relevant parameter. (They argue that $h/c$ is only important in so far as there is a change in $h$, not in $c$.) Also, like the profile loss, $Y_s$ is considerably affected by the amount of acceleration of the flow in the blade passage. In general terms, the larger the acceleration the thinner and more stable are the boundary layers, the smaller is the chance of boundary layer separation, and the smaller is the effect of a curved neighbouring surface in setting up the secondary flows. The degree of acceleration is conveniently indicated by the ratio of the area *normal* to the flow at outlet to that at inlet, i.e. $A_3 \cos \beta_3/A_2 \cos \beta_2$ where $A$ is the annulus area. It is found that the quantity $\lambda$ in equation (7.33) is given approximately by

$$\lambda = f\left\{\left(\frac{A_3 \cos \beta_3}{A_2 \cos \beta_2}\right)^2 \Big/ \left(1 + \frac{r_r}{r_t}\right)\right\} \tag{7.34}$$

where the function f is given by the curve in Fig. 7.26.

Let us now evaluate $(Y_s + Y_k)$ for the blades in our example.
*Nozzle blades*: We shall assume that the nozzles are shrouded, with seals supported by a diaphragm at the shaft radius so that the leakage loss is very small. Then $B$ in equation (7.33) can be assumed zero. $\lambda$ is found as follows.

      $A_2 = 0·0833$ m$^2$, $A_1 = 0·0626$ m$^2$
      $\cos \alpha_2 = \cos 58°23' = 0·524$, $\cos \alpha_1 = \cos 0 = 1·0$
      mean $r_t/r_r$ between planes 1 and 2 $= 1·29$

$$\left(\frac{A_2 \cos \alpha_2}{A_1 \cos \alpha_1}\right)^2 \Big/ \left(1+\frac{r_r}{r_t}\right) = \left(\frac{0\cdot0833 \times 0\cdot524}{0\cdot0626 \times 1\cdot0}\right)^2 \Big/ \left(1+\frac{1}{1\cdot29}\right) = 0\cdot274$$

From Fig. 7.26, $\lambda = 0\cdot012$.

$$\frac{C_L}{s/c} = 2(\tan \alpha_1 + \tan \alpha_2) \cos \alpha_m$$

$$\alpha_m = \tan^{-1}\left[(\tan \alpha_2 - \tan \alpha_1)/2\right]$$
$$= \tan^{-1}\left[(\tan 58°23' - \tan 0°)/2\right] = 39°5'$$
$$\frac{C_L}{s/c} = 2(\tan 58°23' + \tan 0°) \cos 39°5' = 2\cdot52$$

$$\frac{\cos^2 \alpha_2}{\cos^3 \alpha_m} = \frac{0\cdot524^2}{0\cdot776^3} = 0\cdot589$$

$$\left[Y_s + Y_k\right]_N = 0\cdot012 \times 2\cdot52^2 \times 0\cdot589 = 0\cdot0448$$

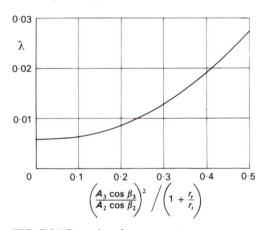

FIG. 7.26 Secondary loss parameter

*Rotor blades*: We will assume unshrouded rotor blades with radial tip clearance equal to 2 per cent of the mean blade height, so that

$$B(k/h) = 0\cdot5 \times 0\cdot02 = 0\cdot01$$

For these blades, using the data from Fig. 7.23,

$$\left(\frac{A_3 \cos \beta_3}{A_2 \cos \beta_2}\right)^2 \Big/ \left(1+\frac{r_r}{r_t}\right) = \left(\frac{0\cdot1047 \times \cos 54°57'}{0\cdot0833 \times \cos 20°29'}\right)^2 \Big/ \left(1+\frac{1}{1\cdot38}\right) = 0\cdot334$$

From Fig. 7.26, $\lambda = 0\cdot015$

$$\beta_m = \tan^{-1}\left[(\tan 54°57' - \tan 20°29')/2\right] = 27°45'$$

$$\frac{C_L}{s/c} = 2(\tan 54°57' + \tan 20°29') \cos 27°45' = 3\cdot18$$

$$\frac{\cos^2 \beta_3}{\cos^3 \beta_m} = \frac{0.574^2}{0.805^3} = 0.475$$

$$[Y_s + Y_k]_R = (0.015 + 0.01)\, 3.18^2 \times 0.475 = 0.120$$

*Step* 4

The total loss coefficients become

$$Y_N = (Y_p)_N + [Y_s + Y_k]_N = 0.024 + 0.0448 = 0.0688$$
$$Y_R = (Y_p)_R + [Y_s + Y_k]_R = 0.032 + 0.120 = 0.152$$

If the trailing edge thickness/pitch ratio ($t_e/s$) differs from 0.02, it is at this point that a correction is made for the effect on the losses. 0.02 was the value for the blading to which Fig. 7.24 relates, but trailing edge thickness affects all the losses, not merely the profile loss. The correction curve in Fig. 7.27 has

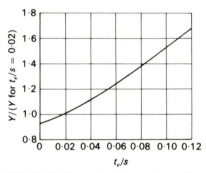

**FIG. 7.27 Correction factor for trailing edge thickness**

been deduced from turbine test results. There is no reason to suppose that the normal value of 0.02 would be unsuitable for the nozzle and rotor blades of the turbine of our example and no correction is required. The modest turbine inlet temperature indicates a low-cost, long-life, application by aircraft standards and certainly there will be no need to thicken the trailing edge to accommodate cooling passages. Early development tests may indicate vibration troubles which might be overcome by increasing the thickness, and Fig. 7.27 enables the penalty to be paid in loss of performance to be estimated.

*Step* 5

The stage efficiency can now be calculated using equations (7.19) and (7.20). We first calculate the equivalent loss coefficients defined in terms of temperature. For the nozzles,

$$\lambda_N = \frac{Y_N}{(T_{02}/T_2')} = \frac{0.0688}{(1100/976.8)} = 0.0611$$

For the rotor,

$$\lambda_R = \frac{Y_R}{(T_{03\mathrm{rel}}/T_3'')}$$

We have previously calculated $T_3''$ to be 913 K, but have not found the value of $T_{03rel}$. We did however find that $(V_3^2/2c_p) = 97.8$ K and $T_3 = 922$ K, so that

$$T_{03rel} = T_3 + (V_3^2/2c_p) = 1020 \text{ K}$$

$$\lambda_R = \frac{0.152}{1020/913} = 0.136$$

Equation (7.20) now becomes

$$\eta_s = \cfrac{1}{1 + \left[0.136 \times 97.8 + \left(\dfrac{922}{982.7}\right)0.0611 \times 117.3\right] \Big/ 145} = 0.88$$

Thus the design yields $\lambda_N = 0.061$ and $\eta_s = 0.88$, in comparison with the values of 0.05 and 0.9 assumed at the outset. This can be regarded as satisfactory agreement, but minor changes would be looked for to improve the efficiency: perhaps a slight increase in degree of reaction with the reduction in work due to this compensated by designing with some progressive increase in $C_a$ through the stage. The latter would have the added advantage of reducing the flare of the annulus.

*Step 6*

In conclusion, it must be emphasized that the cascade data and other loss correlations are strictly applicable only to designs where the Mach numbers are such that no shock losses are incurred in the blade passages. It has recently been suggested, Ref. (13), that the additional loss incurred by designing with a blade outlet relative Mach number greater than unity can be accounted for by adjusting the profile loss coefficient $Y_p$ of the blade row concerned. The correction is given by

$$Y_p = \left[Y_p \text{ from eqn. (7.32)}\right] \times \left[1 + 60(M-1)^2\right]$$

where $M$ is $M_{V3}$ for the rotor blades and $M_{C2}$ for the nozzles. There is another restriction on the applicability of the data not yet mentioned: the Reynolds number of the flow should be in the region of $1 \times 10^5$ to $3 \times 10^5$, with $Re$ defined in terms of blade chord, and density and *relative* velocity at *outlet* of a blade row. If the mean Reynolds number for a turbine, taken as the arithmetic mean of $Re$ for the first nozzle row and the last rotor row (to cover multistage turbines), differs much from $2 \times 10^5$, an approximate correction can be made to the overall isentropic efficiency by using the expression

$$(1 - \eta_t) = \left(\frac{Re}{2 \times 10^5}\right)^{-0.2} (1 - \eta_t)_{Re = 2 \times 10^5} \tag{7.35}$$

To calculate $Re$ for the nozzle and rotor rows of our example, we need the viscosity of the gas at temperatures $T_2 = 982.7$ K and $T_3 = 922$ K. Using data for air, which will be sufficiently accurate for this purpose,

$$\mu_2 = 4 \cdot 11 \times 10^{-5} \text{ kg/m s and } \mu_3 = 3 \cdot 95 \times 10^{-5} \text{ kg/m s}$$

$$(Re)_N = \frac{\rho_2 C_2 c_N}{\mu_2} = \frac{0 \cdot 883 \times 519 \times 0 \cdot 0175}{4 \cdot 11 \times 10^{-5}} = 1 \cdot 95 \times 10^5$$

$$(Re)_R = \frac{\rho_3 V_3 c_R}{\mu_3} = \frac{0 \cdot 702 \times 473 \cdot 5 \times 0 \cdot 023}{3 \cdot 95 \times 10^{-5}} = 1 \cdot 93 \times 10^5$$

Thus no Reynolds number correction is required.

The Ainley–Mathieson method outlined here has been found to predict efficiencies to within ±3 per cent of the measured values for aircraft turbines, but to be not so accurate for small turbines which tend to have blades of rather low aspect ratio. Dunham and Came, in Ref. (13), suggest that the method becomes applicable to a wider range of turbines if the secondary and tip clearance loss correlation, equation (7.33), is modified as follows.

(a)   $\lambda$, instead of being given by the function expressed in Fig. 7.26, is replaced by

$$0 \cdot 0334 \left( \frac{c}{h} \right) \left( \frac{\cos \beta_3}{\cos \beta_2} \right)$$

(with $\beta_3 = \alpha_2$ and $\beta_2 = \alpha_1$ for nozzles).
(b)   $B(k/h)$ is replaced by

$$B \left( \frac{c}{h} \right) \left( \frac{k}{c} \right)^{0 \cdot 78}$$

with $B$ equal to 0·47 for radial tip clearances and 0·37 for side clearances on shrouded blades.

When this modification is applied to the turbine of our example there is, as might be expected, very little difference in the predicted efficiency, namely 0·89 as compared with 0·88. Larger differences of up to 5 per cent are obtained with low aspect ratios of about unity. Further refinements to the method are described in Ref. (14).

## 7.5   Overall turbine performance

In the previous section we have described a method of estimating the stage efficiency $\eta_s$. If the whole turbine comprises a large number of similar stages, it would be a reasonable approximation to treat the stage efficiency as being equal to the polytropic efficiency $\eta_{\infty t}$ and obtain the overall isentropic efficiency from equation (2.18). This was the approach suggested in section (5.7) for the preliminary design of multi-stage axial compressors. Turbines have few stages, however, and it is preferable to work through the turbine stage by stage, with the outlet conditions from one stage becoming the inlet conditions of the next, until the outlet temperature is established. The overall efficiency $\eta_t$ is then obtained from the ratio of the actual to isentropic overall temperature drop.

As stated under the heading 'Estimation of design point performance', it is possible to calculate the performance of a turbine over a range of operating conditions. Whether calculated, or measured on a test rig, the performance is normally expressed by plotting $\eta_t$ and $m\sqrt{T_{03}}/p_{03}$ against pressure ratio $p_{03}/p_{04}$ for various values of $N/\sqrt{T_{03}}$ as in Fig. 7.28. We are here reverting to cycle notation, using suffixes 3 and 4 to denote turbine inlet and outlet conditions respectively. The efficiency plot shows that $\eta_t$ is sensibly constant over a wide range of rotational speed and pressure ratio. This is because the accelerating nature of the flow permits turbine blading to operate over a wide range of incidence without much increase in the loss coefficient.

The maximum value of $m\sqrt{T_{03}}/p_{03}$ is reached at a pressure ratio which produces choking conditions at some point in the turbine. Choking may occur in the nozzle throats or, say, in the annulus at outlet from the turbine depending on the design. The former is the more normal situation and then the constant speed lines merge into a single horizontal line as indicated on the mass flow plot of Fig. 7.28. (If choking occurs in the rotor blade passages

FIG. 7.28 Turbine characteristics

or outlet annulus the maximum mass flow will vary slightly with $N/\sqrt{T_{03}}$.) Even in the unchoked region of operation the separation of the $N/\sqrt{T_{03}}$ lines is not great, and the larger the number of stages the more nearly can the mass flow characteristics be represented by a single curve independent of $N/\sqrt{T_{03}}$. Such an approximation is very convenient when predicting the part-load performance of a complete gas turbine unit as will be apparent from Chapter 8. Furthermore, the approximation then yields little error because when the turbine is linked to the other components the whole operating range shown in Fig. 7.28 is not used. Normally both the pressure ratio and mass flow increase simultaneously as the rotational speed is increased, as indicated by the dotted curve.

## 7.6   The cooled turbine

It has always been the practice to pass a quantity of cooling air over the turbine disc and blade roots. When speaking of the *cooled turbine*, however, we mean the application of a substantial quantity of coolant to the nozzle and rotor blades themselves. Chapters 2 and 3 should have left the reader in no doubt as to the benefits in reduced s.f.c. and increased specific power output (or increased specific thrust in the case of aircraft propulsion units) which follow from an increase in permissible turbine inlet temperature. The benefits are still substantial even when the additional losses introduced by the cooling system are taken into account.

Figure 7.29 illustrates the methods of blade cooling that have received serious attention and research effort. Apart from the use of spray cooling for

**FIG. 7.29   Methods of blade cooling**

thrust boosting in turbojet engines, the liquid systems have not proved to be practicable. There are difficulties associated with channelling the liquid to and from the blades—whether as primary coolant for forced convection or free convection open thermosyphon systems, or as secondary coolant for closed thermosyphon systems. It is impossible to eliminate corrosion or the formation of deposits in open systems, and very difficult to provide adequate secondary surface cooling area at the base of the blades for closed systems. The only method used successfully in production engines has been internal, forced convection, air cooling. With 1·5–2 per cent of the air mass flow used for cooling per blade row, the blade temperature can be reduced by between 200 and 300°C. Using current alloys, this permits turbine inlet temperatures of up to 1650 K to be used. The blades are either cast, using cores to form the cooling passages, or forged with holes of any desired shape produced by electrochemical drilling. The next step forward is likely to be achieved by transpiration cooling, where the cooling air is forced through a porous blade wall. This method is by far the most economical in cooling air, because not only does it remove heat from the wall more uniformly, but the effusing layer of air insulates the outer surface from the hot gas stream and so reduces the rate of heat transfer to the blade. Successful application awaits further development of suitable porous materials and techniques of blade manufacture.

We are here speaking mainly of rotor blade cooling because this presents the most difficult problem. Nevertheless it should not be forgotten that, with high gas temperatures, oxidation becomes as significant a limiting factor as creep, and it is therefore equally important to cool even relatively unstressed components such as nozzle blades and annulus walls. A typical distribution of cooling air required for a turbine stage designed to operate at 1500 K might be as follows. The values are expressed as fractions of the entry gas mass flow.

| | |
|---|---|
| annulus walls | 0·016 |
| nozzle blades | 0·025 |
| rotor blades | 0·019 |
| rotor disc | 0·005 |
| | 0·065 |

Nozzle blades may be manufactured in a similar manner to rotor blades, or constructed of bent sheet metal with inserts to form the cooling passages, as in Fig. 7.30. The nozzle cooling air is often introduced in such a way as to

**FIG. 7.30 Nozzle cooling**

provide jet impingement cooling of the inside surface of the very hot leading edge, the spent air leaving through slots or holes in the blade surface (to provide some film cooling) or in the trailing edge.

There are two distinct aspects of cooled turbine design. Firstly, there is the problem of choosing an aerodynamic design which requires the least amount of cooling air for a given cooling performance. One cooling perfor-mance parameter in common use is the *blade relative temperature* defined by

$$\text{blade relative temperature} = \frac{T_b - T_{cr}}{T_g - T_{cr}}$$

where $T_b$ = mean blade temperature

$T_{cr}$ = coolant temperature at inlet (i.e. at the root radius $r_r$)

$T_g$ = mean effective gas temperature relative to the blade ($\simeq$ static temperature $+ 0.85 \times$ dynamic temperature)

Relative to an uncooled turbine, the optimum design might well involve the use of a higher blade loading coefficient $\psi$ (to keep the number of stages to a minimum), a higher pitch/chord ratio (to reduce the number of blades in a row), and a higher flow coefficient $\phi$ (which implies a blade of smaller camber and hence smaller surface area). The importance of these, and other para-meters such as gas flow Reynolds number, are discussed in detail in Ref. (9).

The second aspect is the effect on the cycle efficiency of losses incurred by the cooling process: a pertinent question is whether it is advantageous overall to sacrifice some aerodynamic efficiency to reduce such losses. The sources of loss are as follows.

(a)   There is a direct loss of turbine work due to the reduction in turbine mass flow.

(b)   The expansion is no longer adiabatic; and furthermore there will be a negative reheat effect in multi-stage turbines.

(c)   There is a pressure loss, and a reduction in enthalpy, due to the mixing of spent cooling air with the main gas stream at the blade tips. (This has been found to be partially offset by a reduction in the normal tip leakage loss.)

(d)   Some 'pumping' work is done by the blades on the cooling air as it passes radially outwards through the cooling passages.

(e)   When considering cooled turbines for cycles with heat-exchange, account must be taken of the reduced temperature of the gas leaving the turbine which makes the heat-exchanger less effective.

Losses (a) and (e) can be incorporated directly into any cycle calculation, while the effect of (b), (c) and (d) can be taken into account by using a reduced value of turbine efficiency. One assessment of the latter, Ref. (10), suggests that the turbine efficiency is likely to be reduced by from 1 to 3 per cent of the uncooled efficiency, the lower value referring to near-impulse designs and the higher to 50 per cent reaction designs. The estimate for reaction designs is substantially confirmed by the tests on an experimental cooled turbine reported in Ref. (11). Cycle calculations have shown that even when all these losses are accounted for, there is a substantial advantage to be gained from using a cooled turbine.†

Before either of these two aspects of cooled turbine design can be investigated it is necessary to be able to estimate the cooling air flow required to achieve a specified blade relative temperature for any given aerodynamic stage design. We will end this section with an outline of an approximate one-dimensional treatment, and further refinements can be found in Ref. (9). Figure 7.31 shows the notation employed and the simplifying assumptions made.

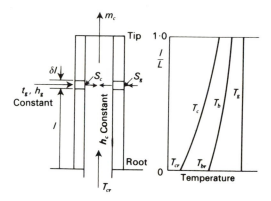

FIG. 7.31 Forced convection air cooling

Consider the heat flow to and from an elemental length of blade $\delta l$ a distance $l$ from the root. As the cooling air passes up the blade it increases in temperature and becomes less effective as a coolant, so that the blade temperature increases from root to tip. There must therefore be some conduction of heat along the blade to and from the element $\delta l$ due to this spanwise temperature gradient. Because turbine blade alloys have a low thermal conductivity, the conduction term will be small and we shall neglect it here. The heat balance for the elemental length $\delta l$ is then simply

$$h_g S_g (T_g - T_b) = h_c S_c (T_b - T_c) \qquad (7.36)$$

where $h_g$ and $h_c$ are the gas-side and coolant-side heat transfer coefficients, and $S_g$ and $S_c$ are the wetted perimeters of the blade profile and combined coolant passages respectively. For the internal air flow $m_c$ we also have

$$m_c c_{pc} \frac{dT_c}{dl} = h_c S_c (T_b - T_c) \qquad (7.37)$$

† It is worth remembering that there may be special applications where a cooled turbine might be employed not to raise the cycle temperature but to enable cheaper material to be used at ordinary temperatures: the first researches in blade cooling were carried out in Germany during the Second World War with this aim in mind.

We may first find the variation of $T_b$ with $l$ by eliminating $T_c$ between the two equations. From (7.36) we have

$$T_c = T_b - \frac{h_g S_g}{h_c S_c}(T_g - T_b) \tag{7.38}$$

and hence

$$\frac{dT_c}{dl} = \left(1 + \frac{h_g S_g}{h_c S_c}\right)\frac{dT_b}{dl}$$

Substituting in (7.37), remembering that $dT_b/dl = -d(T_g - T_b)/dl$, we get

$$\left(1 + \frac{h_g S_g}{h_c S_c}\right)\frac{d(T_g - T_b)}{dl} + \frac{h_g S_g}{m_c c_{pc}}(T_g - T_b) = 0$$

The solution of this differential equation, with $T_b = T_{br}$ at $l = 0$, is

$$T_g - T_b = (T_g - T_{br})e^{-kl/L} \tag{7.39}$$

where

$$k = \frac{h_g S_g L}{m_c c_{pc}[1 + (h_g S_g/h_c S_c)]}$$

To obtain the variation of $T_c$ with $l$, we may write (7.38) in the form

$$T_g - T_c = (T_g - T_b)\left[1 + \frac{h_g S_g}{h_c S_c}\right]$$

and substitute (7.39) for $(T_g - T_b)$ to give

$$T_g - T_c = (T_g - T_{br})\left[1 + \frac{h_g S_g}{h_c S_c}\right]e^{-kl/L} \tag{7.40}$$

when $l = 0$, $T_c = T_{cr}$ and hence

$$T_g - T_{cr} = (T_g - T_{br})\left[1 + \frac{h_g S_g}{h_c S_c}\right] \tag{7.41}$$

Combining (7.40) and (7.41) we have the variation of $T_c$ given by

$$T_g - T_c = (T_g - T_{cr})e^{-kl/L} \tag{7.42}$$

Finally, subtracting (7.39) from (7.41),

$$T_b - T_{cr} = (T_g - T_{br})\left[1 + \frac{h_g S_g}{h_c S_c} - e^{-kl/L}\right]$$

and dividing this by (7.41) we have the blade relative temperature given by

$$\frac{T_b - T_{cr}}{T_g - T_{cr}} = 1 - \frac{e^{-kl/L}}{[1 + (h_g S_g/h_c S_c)]} \tag{7.43}$$

We may note that $h_c$ will be a function of coolant flow Reynolds number and hence of $m_c$, and that $m_c$ also appears in the parameter $k$. Thus equation (7.43) is not explicit in $m_c$, and it is convenient to calculate values of blade relative temperature for various values of $m_c$ rather than vice versa.

The next step is the evaluation of the heat transfer coefficients: we will consider $h_c$ first. For straight cooling passages of uniform cross-section, pipe flow formulae may be used. For this application one recommended correlation is

$$Nu = 0.034 \, (L/D)^{-0.1} \, (Pr)^{0.4} \, (Re)^{0.8} \, (T_c/T_b)^{0.55} \qquad (7.44)$$

with fluid properties calculated at the mean bulk temperature. The $L/D$ term accounts for the entrance length effect. The $T_c/T_b$ term is necessary when the difference in temperature between fluid and wall is large, to allow for the effect of variation in fluid properties with temperature. The characteristic dimension $D$ can be taken as the equivalent diameter ($4 \times$ area/periphery) if the cooling passages are of non-circular cross-section. For air ($Pr \simeq 0.71$) and practicable values of $L/D$ between 30 and 100, the equation reduces to the simpler form

$$Nu = 0.020 \, (Re)^{0.8} \, (T_c/T_b)^{0.55} \qquad (7.45)$$

The correlation is for turbulent flow, with the accuracy decreasing at Reynolds numbers below 8000. Note that the mean value of $T_c$ at which the fluid properties should be evaluated, and the mean value of $T_b$ for the $T_c/T_b$ term, are unknown at this stage in the calculations. Guessed values must be used, to be checked later by evaluating $T_c$ and $T_b$ at $l/L = 0.5$ from equations (7.42) and (7.43).

Data for mean blade heat transfer coefficients $h_g$ are available both from cascade tests and turbine tests on a wide variety of blades. The latter yield higher values than the former, presumably because of the greater intensity of the turbulence in a turbine. The full line in Fig. 7.32, taken from Ref. (9), is a useful design curve for the mean value of Nusselt number round the blade surface in terms of the most significant blade shape parameter which is the ratio of inlet/outlet angle $\beta_2/\beta_3$ (or $\alpha_1/\alpha_2$ for nozzle blades). $Nu_g$ decreases as the degree of acceleration of the flow increases, because the point of transition from a laminar to a turbulent layer on the convex surface is delayed by accelerating flow. The curve applies to conventional turbine blade profiles, and gives nominal values denoted by $Nu_g^*$ for operating conditions of $Re_g = 2 \times 10^5$ and $T_g/T_b \rightarrow 1$. $Nu_g$ can then be found from

$$Nu_g = Nu_g^* \left( \frac{Re_g}{2 \times 10^5} \right)^x \left( \frac{T_g}{T_b} \right)^y$$

where the exponent $x$ is given by the supplementary curve, and $y$ is given by

$$y = 0.14 \left( \frac{Re_g}{2 \times 10^5} \right)^{-0.4}$$

The characteristic dimension in $Nu_g$ and $Re_g$ is the blade chord, and the fluid properties should be evaluated at the temperature $T_g$. The velocity in $Re_g$ is the gas velocity relative to the blade at outlet ($V_3$ or $C_2$ as the case may be). The quantities required will be known for any given stage design, with the exception of the mean value of $T_b$ for which the guessed value must be used.

FIG. 7.32 Heat transfer data for convectional blade profiles

All the information has now been obtained to permit the calculation of $(T_b - T_{cr})/(T_g - T_{cr})$, for various values of coolant flow $m_c$, from equation (7.43). Typical curves of spanwise variation, for values of $m_c$ of 1 and 2 per cent of the gas mass flow per blade, are shown in Fig. 7.33. Note that the distribution matches requirements because the blade stresses decrease from root to tip. The quantity of heat extracted from the blade row can be found by calculating $T_{ct}$, i.e. $T_c$ at $l = L$, from equation (7.42), and then evaluating $m_c c_{pc}(T_{ct} - T_{cr})$.

FIG. 7.33 Typical spanwise temperature distributions

To conclude, we may note that the final design calculation for a cooled blade will involve an estimation of the two-dimensional temperature distribution over the blade cross-section at several values of $l/L$. Finite difference methods are used to solve the differential equations, and conduction within the blade is taken into account. Figure 7.34 shows a typical temperature

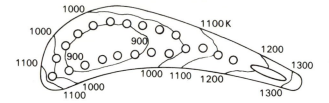

**FIG. 7.34 Typical temperature distribution (from Ref. [12])**

distribution at the mid-span of a blade designed to operate with $T_g = 1500$ K and $T_{cr} = 320$ K. It emphasizes one of the main problems of blade cooling, i.e. that of obtaining adequate cooling at the trailing edge. Finally, an estimation will be made of the thermal stresses incurred with due allowance for redistribution of stress by creep: with cooled blades the thermal stresses can dominate the gas bending stresses and be comparable with the centrifugal tensile stresses. References to the literature dealing with these more advanced aspects can be found in the paper by Barnes and Dunham in Ref. (12).

Finally, mention must be made of an alternative approach to the high-temperature turbine—the use of ceramic materials which obviates the need for elaborate cooling passages. Much effort has been expended on the development of silicon nitride and silicon carbide materials for small turbine rotors (both axial and radial) in which it would be difficult to incorporate cooling passages. Adequate reliability and life are difficult to achieve, but demonstrator engines have been run for short periods. The use of ceramic turbines in production engines, however, remains an elusive goal.

## 7.7 The radial flow turbine

Figure 7.35 illustrates a rotor having the back-to-back configuration mentioned in the introduction to this chapter, and Ref. (15) discusses the development of a successful family of radial industrial gas turbines. In a radial flow turbine, gas flow with a high tangential velocity is directed inwards and leaves the rotor with as small a whirl velocity as practicable near the axis of rotation. The result is that the turbine looks very similar to the centrifugal compressor, but with a ring of nozzle vanes replacing the diffuser vanes as in Fig. 7.36. Also, as shown there would normally be a diffuser at the outlet to reduce the exhaust velocity to a negligible value.

The velocity triangles are drawn for the normal design condition in which

**FIG. 7.35 Back-to-back rotor [by courtesy of Kongsberg Ltd]**

**FIG. 7.36 Radial inflow turbine**

the relative velocity at the rotor tip is radial (i.e. the incidence is zero) and the absolute velocity at exit is axial. Because $C_{w3}$ is zero, the specific work output $W$ becomes simply

$$W = c_p(T_{01} - T_{03}) = C_{w2}U_2 = U_2^2 \tag{7.46}$$

In the ideal isentropic turbine with perfect diffuser the specific work output would be

$$W' = c_p(T_{01} - T_4') = C_o^2/2$$

where the velocity equivalent of the isentropic enthalpy drop, $C_o$, is sometimes called the 'spouting velocity' by analogy with hydraulic turbine practice. For this ideal case it follows that $U_2^2 = C_o^2/2$ or $U_2/C_o = 0.707$. In practice, it is

FIG. 7.37 *T–s* diagram for a radial flow turbine

found that a good overall efficiency is obtained if this velocity ratio lies between 0·68 and 0·71, Ref. (16). In terms of the turbine pressure ratio, $C_o$ is given by

$$\frac{C_o^2}{2} = c_p T_{01} \left[ 1 - \left( \frac{1}{p_{01}/p_a} \right)^{(\gamma - 1)/\gamma} \right]$$

Figure 7.37 depicts the processes in the turbine and exhaust diffuser on the *T–s* diagram. The overall isentropic efficiency of the turbine and diffuser may be expressed by

$$\eta_o = \frac{T_{01} - T_{03}}{T_{01} - T_4'} \tag{7.47}$$

because $(T_{01} - T_4')$ is the temperature equivalent of the maximum work that could be produced by an isentropic expansion from the inlet state $(p_{01}, T_{01})$ to $p_a$. Considering the turbine alone, however, the efficiency is more suitably expressed by

$$\eta_t = \frac{T_{01} - T_{03}}{T_{01} - T_3'} \tag{7.48}$$

which is the 'total-to-static' isentropic efficiency referred to in section 7.1.

Following axial flow turbine practice, the nozzle loss coefficient may be defined in the usual way by

$$\lambda_N = \frac{T_2 - T_2'}{C_2^2/2c_p} \tag{7.49}$$

Similarly, the rotor loss coefficient is given by

$$\lambda_R = \frac{T_3 - T_3''}{V_3^2/2c_p} \tag{7.50}$$

Plane 2 is located at the periphery of the rotor, so that the nozzle loss includes not only the loss in the inlet volute but also any friction loss in the vaneless space between nozzle vane trailing edge and rotor tip. Bearing in mind that for the small constant pressure processes denoted by 2'–2 and 3'–3'' we can write $c_p \delta T = T \delta s$, $\lambda_N$ can be alternatively expressed by

$$\lambda_N = \frac{T_3'' - T_3'}{C_2^2/2c_p} \cdot \frac{T_2'}{T_3'} \tag{7.51}$$

A useful expression for $\eta_t$ in terms of the nozzle and rotor loss coefficients can be found as follows. The denominator in equation (7.48) may be expanded to yield

$$T_{01} - T_3' = (T_{01} - T_{03}) + (T_{03} - T_3) + (T_3 - T_3'') + (T_3'' - T_3')$$

$$= (T_{01} - T_{03}) + \frac{C_3^2}{2c_p} + \lambda_R \frac{V_3^2}{2c_p} + \lambda_N \frac{C_2^2}{2c_p} \frac{T_3'}{T_2'}$$

Consequently $\eta_t$ becomes

$$\eta_t = \left[ 1 + \frac{1}{2c_p(T_{01} - T_{03})} \left\{ C_3^2 + \lambda_R V_3^2 + \lambda_N C_2^2 \frac{T_3'}{T_2'} \right\} \right]^{-1}$$

From the velocity triangles,

$$C_2 = U_2 \operatorname{cosec} \alpha_2, \ V_3 = U_3 \operatorname{cosec} \beta_3, \ C_3 = U_3 \cot \beta_3$$

$\alpha_2$ is the gas angle at inlet to the impeller and hence the effective outlet angle of the nozzle vanes, while $\beta_3$ is the outlet angle of the impeller vanes. Furthermore, $U_3 = U_2 r_3/r_2$ and $c_p(T_{01} - T_{03}) = U_2^2$, so that the expression for efficiency finally becomes

$$\eta_t = \left[ 1 + \tfrac{1}{2} \left\{ \left( \frac{r_3}{r_2} \right)^2 (\cot^2 \beta_3 + \lambda_R \operatorname{cosec}^2 \beta_3) + \lambda_N \frac{T_3'}{T_2'} \operatorname{cosec}^2 \alpha_2 \right\} \right]^{-1}$$

$$\tag{7.52}$$

Here $\eta_t$ is expressed in terms of the nozzle and impeller outlet angles, radius ratio of impeller, loss coefficients, and the temperature ratio $T_3'/T_2'$. The temperature ratio $T_3'/T_2'$ can in turn be expressed in terms of the major design variables, although it is usually ignored because it is sufficiently near unity for it to have little effect on $\eta_t$. Thus we can write

$$\frac{T_3'}{T_2'} \approx \frac{T_3''}{T_2} = 1 - \frac{T_2 - T_3''}{T_2} = 1 - \frac{1}{T_2} [(T_2 - T_3) + (T_3 - T_3'')]$$

Now $(T_2 - T_3)$ may be found by expanding equation (7.46) and making use of the velocity triangles in Fig. 7.36. Thus, since $T_{01} = T_{02}$,

$$U_2^2 = c_p(T_2 - T_3) + \tfrac{1}{2}(C_2^2 - C_3^2)$$

$$= c_p(T_2 - T_3) + \tfrac{1}{2}(V_2^2 + U_2^2) - \tfrac{1}{2}(V_3^2 - U_3^2)$$

$$(T_2 - T_3) = \frac{1}{2c_p}[(V_3^2 - V_2^2) + (U_2^2 - U_3^2)]$$

(Note that $T_{0\text{rel}} = T + V^2/2c_p$ is not the same at inlet and outlet of the rotor as it is in the axial flow machine, because $U_3 \neq U_2$. This is the main difference between Fig. 7.37 and Fig. 7.4.) It follows that

$$\frac{T_3'}{T_2'} \approx \frac{T_3''}{T_2} = 1 - \frac{1}{2c_p T_2}[(V_3^2 - V_2^2) + (U_2^2 - U_3^2) + \lambda_R V_3^2]$$

$$= 1 - \frac{U_2^2}{2c_p T_2}\left[1 + \left(\frac{r_3}{r_2}\right)^2 \{(1 + \lambda_R)\operatorname{cosec}^2 \beta_3 - 1\} - \cot^2 \alpha_2\right]$$

$$(7.53)$$

And, finally, $T_2$ may be expressed as

$$T_2 = T_{01} - \frac{U_2^2}{2c_p}\operatorname{cosec}^2 \alpha_2$$

$\lambda_N$ is usually obtained from separate tests on the inlet volute and nozzle vane assembly, enabling $\lambda_R$ to be deduced from overall efficiency measurements with the aid of equation (7.52) as in the following example.

## EXAMPLE

A radial flow turbine of the following geometry is tested under design conditions with a pressure ratio $p_{01}/p_3$ of 2·0 and an inlet temperature of 1000 K. The work output, after allowance has been made for mechanical losses, is 45·9 kW when the rotational speed is 1000 rev/s and the gas mass flow is 0·322 kg/s.

| | |
|---|---|
| Rotor inlet tip diameter | 12·7 cm |
| Rotor exit tip diameter | 7·85 cm |
| Hub/tip ratio at exit | 0·30 |
| Nozzle efflux angle $\alpha_2$ | 70° |
| Rotor vane outlet angle $\beta_3$ | 40° |

Separate tests on the volute/nozzle combination showed the nozzle loss coefficient $\lambda_N$ to be 0·070. The turbine isentropic efficiency $\eta_t$ and the rotor loss coefficient $\lambda_R$ are to be evaluated from the results.

$$T_{01} - T_3' = T_0 \left[ 1 - \left( \frac{1}{p_{01}/p_3} \right)^{(\gamma-1)/\gamma} \right] = 1000 \left[ 1 - \left( \frac{1}{2 \cdot 0} \right)^{\frac{1}{4}} \right] = 159 \cdot 1 \text{ K}$$

Since $W = m c_p (T_{01} - T_{03})$,

$$T_{01} - T_{03} = \frac{45 \cdot 9}{0 \cdot 322 \times 1 \cdot 148} = 124 \cdot 2 \text{ K}$$

$$\eta_t = \frac{124 \cdot 2}{159 \cdot 1} = 0 \cdot 781$$

Putting $T_3'/T_2' = 1 \cdot 0$ in equation (7.52), with $r_3/r_2$ given by

$$\frac{r_3}{r_2} = \frac{0 \cdot 3 \times 7 \cdot 85 + 7 \cdot 85}{2 \times 12 \cdot 7} = 0 \cdot 402$$

we get

$$0 \cdot 781 \approx [1 + \tfrac{1}{2} \{ 0 \cdot 402^2 (\cot^2 40 + \lambda_R \operatorname{cosec}^2 40) + 0 \cdot 07 \operatorname{cosec}^2 70 \}]^{-1}$$

$$1 \cdot 280 = 1 + 0 \cdot 1148 + 0 \cdot 1956 \lambda_R + 0 \cdot 0396$$

$$\lambda_R = 0 \cdot 64$$

It can be seen from the foregoing that the term containing $T_3'/T_2'$ is small. From the data, $U_2 = \pi \times 1000 \times 12 \cdot 7 = 390$ m/s. Using this in equation (7.53) we get $T_3'/T_2'$ equal to $(0 \cdot 921 - 0 \cdot 028) \lambda_R$. Substituting this result in equation (7.52) and recalculating, we have $\lambda_R = 0 \cdot 66$. The approximate value of $0 \cdot 64$ is within the margin of experimental uncertainty.

One of the more comprehensive research programmes on radial flow turbines was that carried out by Ricardo and Co. and reported in Ref. (17). Turbine efficiencies of up to 90 per cent were obtained under optimum running conditions with a 12·5 cm diameter rotor having 12 blades. The optimum number of nozzle vanes was found to be 17. Values of $\lambda_N$ varied from 0·1 to 0·05, decreasing steadily with increase of nozzle angle from 60 to 80 degrees. $\lambda_R$ varied more widely, from 0·5 to 1·7, increasing rapidly as the nozzle angle was increased above 70 degrees. In spite of this variation in $\lambda_R$ the overall turbine efficiency was relatively insensitive to nozzle angle: it fell by only 2 per cent as $\alpha_2$ was increased from 70 to 80 degrees. Also studied in the programme were the effects of varying the axial width of the vanes at the rotor tip, radial width of the vaneless space, and clearance between vanes and casing. The optimum width of vane at the rotor tip was about 10 per cent of the rotor diameter, and the performance seemed insensitive to the radial width of the vaneless space. There was a fall in efficiency of 1 per cent for every increase in clearance of 1 per cent of the rotor vane width (averaged between rotor inlet and outlet). This implies that the radial flow turbine is rather less sensitive to clearance loss than its axial flow counterpart. Because clearances cannot be reduced in proportion to blade height as the size is reduced, this lower sensitivity is probably the basic reason why radial flow has the advantage for very small turbines.

Finally, various major alterations to the geometry of the rotor were made, including scalloping the disc between the blades to reduce disc stresses, weight and inertia: the reader must turn to Ref. (17) for the consequences of such changes. All we can do here is to give a rough guide to the major dimensions determining the basic shape of the rotor, viz. the hub/tip diameter ratio of the vanes at the rotor exit and the ratio of the vane tip diameter at the exit to the rotor disc diameter. The former should not be much less than 0·3 to avoid excessive blade blockage at the hub, and the latter should be limited to a maximum of about 0·7 to avoid excessive curvature of the rotor passages.

Methods of dealing with the losses other than by the simple use of $\lambda_N$ and $\lambda_R$ have been devised, and Ref. (18) provides a useful comparison of the various coefficients that have been used. In particular, if the relative velocity is not radial at inlet to the rotor there is an additional loss, variously described as an 'incidence loss' or 'shock loss', for which a separate loss coefficient is desirable. The use of such a coefficient becomes essential when trying to predict off-design performance because then the flow relative to the rotor vanes may depart substantially from the radial direction. There is no shock in the gas dynamic sense, but there is a shock in the ordinary sense of the word when the fluid impinging on the rotor vanes at some incidence $\beta_2$ (Fig. 7.38) is suddenly constrained to move in the radial direction. There is a resulting drop of

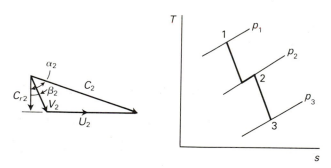

**FIG. 7.38  Effect of incidence loss**

stagnation pressure and increase in entropy which moves line 2–3 to the right on the *T–s* diagram. Bridle and Boulter [Ref. (19)] suggest that in the incidence range $\beta_2 = \pm 65$ degrees, this stagnation pressure loss can be accounted for by the expression

$$\frac{\Delta p_0}{(p_{02} - p_2)\cos^2 \beta_2} = (\tan \beta_2 + 0 \cdot 1)^2$$

Making use of the test results in Ref. (17), Bridle and Boulter have also deduced equations for the other components of the rotor loss, i.e. friction loss in the rotor passages and clearance loss, in an attempt to make possible the prediction of off-design performance. Benson in Ref. (20) describes one method of tackling this problem, and includes in an Appendix the Fortran programs necessary for the calculation.

## NOMENCLATURE

For velocity triangle notation $(U, C, V, \alpha, \beta)$ see Fig. 7.1 and Fig. 7.36
For blade geometry notation $(s, c, o, t, \theta, i)$ see Fig. 7.11

| | |
|---|---|
| $C_L$ | blade lift coefficient |
| $h$ | blade height, heat transfer coefficient |
| $n$ | number of blades |
| $S$ | perimeter |
| $z$ | section modulus of blade |
| $h/c$ | aspect ratio |
| $k/h$ | tip clearance/blade height ratio |
| $s/c$ | pitch/chord ratio |
| $t/c$ | thickness/chord ratio |
| $t_e/s$ | trailing edge thickness/pitch ratio |
| $Y_N$ | nozzle blade loss coefficient $[(p_{01} - p_{02})/(p_{02} - p_2)]$ |
| $Y_R$ | rotor blade loss coefficient $[(p_{02\mathrm{rel}} - p_{03\mathrm{rel}})/(p_{03\mathrm{rel}} - p_3)]$ |
| $Y_k$ | tip clearance loss coefficient |
| $Y_p$ | profile loss coefficient |
| $Y_s$ | secondary loss coefficient |
| $\lambda_N$ | nozzle blade loss coefficient $[(T_2 - T_2')/(C_2^2/2c_p)]$ |
| $\lambda_R$ | rotor blade loss coefficient $[(T_3 - T_3'')/(V_3^2/2c_p)]$ |
| $\sigma_{ct}$ | centrifugal tensile stress |
| $\sigma_{gb}$ | gas bending stress |
| $\Lambda$ | degree of reaction $[(T_2 - T_3)/(T_1 - T_3)]$ |
| $\phi$ | flow coefficient $(C_a/U)$ |
| $\psi$ | stage temperature drop coefficient $(2c_p\Delta T_{0s}/U^2)$ |

*Suffixes*

| | |
|---|---|
| $a, w$ | axial, whirl, component |
| $b$ | blade |
| $c$ | coolant |
| $m, r, t$ | at mean, root, tip, radius |
| $N, R$ | nozzle, rotor, blades |
| $p, s$ | pressure, suction, surface of blade |

# 8

# Prediction of performance of simple gas turbines

From cycle calculations such as those of Chapter 2, it is possible to determine the pressure ratio which for any given maximum cycle temperature will give the greatest overall efficiency, and the mass flow required to give the desired power. When such preliminary calculations have been made, the most suitable design data for any particular application may be chosen. It is then possible to design the individual components of a gas turbine so that the complete unit will give the required performance when running at the *design point*, that is, when it is running at the particular speed, pressure ratio and mass flow for which the components were designed. The problem then remains to find the variation of performance of the gas turbine over the complete operating range of speed and power output, which is normally referred to as *off-design performance*.

The performance characteristics of the individual components may be estimated on the basis of previous experience or obtained from actual tests. When the components are linked together in an engine, the range of possible operating conditions for each component is considerably reduced. The problem is to find corresponding operating points on the characteristics of each component when the engine is running at a steady speed, or in *equilibrium* as it is frequently termed. The equilibrium running points for a series of speeds may be plotted on the compressor characteristic and joined up to form an *equilibrium running line* (or *zone*, depending upon the type of gas turbine and load), the whole forming an *equilibrium running diagram*. When once the operating conditions have been determined, it is a relatively simple matter to obtain performance curves of power output or thrust, and specific fuel consumption.

The equilibrium running diagram also shows the proximity of the operating line or zone to the compressor surge line. If it intersects the surge line the gas turbine will not be capable of being brought up to full speed without some remedial action being taken. It is this phenomenon which was referred to when speaking of 'stability of operation' in the Introduction. Finally, it shows whether the engine is operating in a region of adequate compressor efficiency; ideally the operating line or zone should lie near the locus of points of maximum compressor efficiency shown in Fig. 4.9(a).

The variation of specific fuel consumption with reduction in power, sometimes referred to as *part-load performance*, is of major importance in applications where considerable running at low power settings is required. This would be the case for any vehicular gas turbine, and the poor specific fuel consumption at part load is probably the biggest disadvantage of the gas turbine for vehicular use. The fuel consumption of aircraft gas turbines at reduced power is of critical importance when extensive delays occur either at the point of departure or the destination. In one case the idling fuel consumption when taxying is important, and in the other the fuel flow at low flight speeds and medium altitudes is critical.

When determining the off-design performance it is important to be able to predict not only the effect on specific fuel consumption of operation at part load, but also the effect of ambient conditions on maximum output. The effects of high and low ambient temperatures and pressures must all be considered. Land based gas turbines may operate between ambient temperatures of $-60\,°C$ in the Arctic and $50\,°C$ in the tropics, and at altitudes from sea level to about 3000 metres, while aircraft gas turbines have to operate over much wider ranges of inlet temperature and pressure. The variation of maximum power with ambient conditions is clearly of prime importance to the customer, and the manufacturer must be prepared to guarantee the performance available at any specified condition. If, for example, we consider the use of gas turbines for peak load generation of electricity, the peak loads will occur in the coldest days of winter in Europe, but are much more likely to occur in the summer in the United States because of the heavy demand for air conditioning and refrigeration systems. The effect of ambient conditions on the performance of aircraft gas turbines has a critical effect on the runway length required and the payload that can be carried, and hence on both safety and economics.

The basic methods for determining equilibrium off-design performance of simple gas turbines will be described here, while more complex arrangements and transient operation are dealt with in Chapter 9. The types of gas turbine discussed in this chapter will be (a) the single-shaft unit delivering shaft power, (b) the free turbine engine, where the gas generator turbine drives the compressor and the power turbine drives the load and (c) the simple jet engine where the useful output is a high velocity jet produced in the propelling nozzle. Schematics of these layouts are shown in Fig. 8.1, and it can readily be observed that the gas generator performs exactly the same function for both the free turbine engine and the jet engine. The flow characteristics of a free turbine and a nozzle are similar, and they impose the same restrictions on the operation of the gas generator, with the result that the free turbine engine and the jet engine are thermodynamically similar and differ only in the manner in which the output is utilized. Several successful jet engines have been converted to shaft power use by substituting a free power turbine for the propelling nozzle, and this approach has been widely used for peak-load electricity generation and propulsion of naval vessels.

All off-design calculations depend on satisfying the essential conditions of compatibility of mass flow, work and rotational speed between the various components. It is logical to deal with the single-shaft engine first, and then proceed to the free turbine engine where there is the added complication of flow compatibility between the gas generator turbine and the power turbine. Lastly, we will deal with the jet engine where there is the further complication of forward speed and altitude effects.

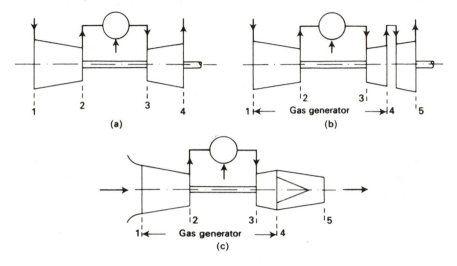

FIG. 8.1 **Simple gas turbine units**

## 8.1   Component characteristics

The variation of mass flow, pressure ratio and efficiency with rotational speed of the compressor and turbine is obtained from the compressor and turbine characteristics, examples of which are given in Chapters 4, 5 and 7. It is convenient to represent the compressor characteristic as shown in Fig. 8.2, with the variation of efficiency along each constant speed line plotted against both mass flow and pressure ratio. With high performance axial compressors the constant speed lines become vertical on a mass flow basis when the inlet is choked, and in this region it is essential to plot the efficiency as a function of pressure ratio. The turbine characteristic can be used in the form given in Fig. 7.28. It is often found in practice, however, that turbines do not exhibit any significant variation in non-dimensional flow with non-dimensional speed, and in most cases the turbine operating region is severely restricted by another component downstream of it. In explaining the method used for off-design performance calculations it will initially be assumed that the mass flow function can be represented by a single curve as in Fig. 8.3. The modification necessary to account for a family of constant speed curves will be discussed in section 8.3.

**FIG. 8.2 Compressor characteristics**

For accurate calculations it is necessary to consider the variation of pressure losses in the inlet ducting, the combustion chamber and the exhaust ducting. These are essentially secondary effects, however, and the off-design calculations will be introduced on the basis of negligible inlet and exhaust losses and a combustion chamber pressure loss which is a fixed percentage of the compressor delivery pressure. Such approximations are quite adequate for many purposes. For detailed calculations it would be normal to use a digital computer, when methods of allowing for variable pressure losses are easily introduced. Further discussion of this matter will be deferred until section 8.7.

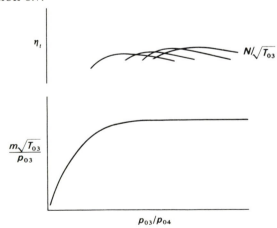

Note: parameters will be $\dfrac{m\sqrt{T_{04}}}{p_{04}}$, $\dfrac{p_{04}}{p_a}$, $\dfrac{N_p}{\sqrt{T_{04}}}$ and $\eta_{tp}$ for a power turbine

**FIG. 8.3 Turbine characteristics**

## 8.2  Off-design operation of the single-shaft gas turbine

Referring to the single-shaft gas turbine shown in Fig. 8.1(a) it can readily be seen that when inlet and exhaust pressure losses are ignored, the pressure ratio across the turbine is determined by the compressor pressure ratio and the pressure loss in the combustion chamber. The mass flow through the turbine will be equal to the mass flow delivered by the compressor, less any bleeds, but supplemented by the fuel flow; it has been pointed out earlier that the bleeds are often approximately equal to the fuel flow. In general terms, the procedure for obtaining an equilibrium running point is as follows.

(a)  Select a constant speed line on the compressor characteristic and choose any point on this line; the values of $m\sqrt{T_{01}}/p_{01}$, $p_{02}/p_{01}$, $\eta_c$ and $N\sqrt{T_{01}}$ are then determined.

(b)  The corresponding point on the turbine characteristic is obtained from consideration of compatibility of rotational speed and flow.

(c)  Having matched the compressor and turbine characteristics, it is necessary to ascertain whether the work output corresponding to the selected operating point is compatible with that required by the driven load; this requires knowledge of the variation of power with speed, which depends on the manner in which the power is absorbed.

The compressor and turbine are directly coupled together, so that compatibility of rotational speed requires

$$\frac{N}{\sqrt{T_{03}}} = \frac{N}{\sqrt{T_{01}}} \times \sqrt{\frac{T_{01}}{T_{03}}} \tag{8.1}$$

Compatibility of flow between the compressor and turbine can be expressed in terms of the non-dimensional flows by the identity

$$\frac{m_3\sqrt{T_{03}}}{p_{03}} = \frac{m_1\sqrt{T_{01}}}{p_{01}} \times \frac{p_{01}}{p_{02}} \times \frac{p_{02}}{p_{03}} \times \sqrt{\frac{T_{03}}{T_{01}}} \times \frac{m_3}{m_1}$$

The pressure ratio $p_{03}/p_{02}$ can be obtained directly from the combustion pressure loss, i.e. $p_{03}/p_{02} = 1-(\Delta p_b/p_{02})$. It will normally be assumed that $m_1 = m_3 = m$, but variation in mass flow at different points in the engine can easily be included if required. Rewriting the previous equation in terms of $m$, we get

$$\frac{m\sqrt{T_{03}}}{p_{03}} = \frac{m\sqrt{T_{01}}}{p_{01}} \times \frac{p_{01}}{p_{02}} \times \frac{p_{02}}{p_{03}} \times \sqrt{\frac{T_{03}}{T_{01}}} \tag{8.2}$$

Now $m\sqrt{T_{01}}/p_{01}$ and $p_{02}/p_{01}$ are fixed by the chosen operating point on the compressor characteristic, $p_{03}/p_{02}$ is assumed to be constant and $m\sqrt{T_{03}}/p_{03}$ is a function of the turbine pressure ratio $p_{03}/p_{04}$. Neglecting inlet and exhaust pressure losses $p_a = p_{01} = p_{04}$, so that the turbine pressure ratio can be calculated from $p_{03}/p_{04} = (p_{03}/p_{02})(p_{02}/p_{01})$. Thus all the terms of equation (8.2) with the exception of $\sqrt{(T_{03}/T_{01})}$ can be obtained from the compressor and turbine characteristics. The turbine inlet temperature $T_{03}$ can therefore be obtained from equation (8.2) when the ambient temperature, which is equal to $T_{01}$, is specified.

Having determined the turbine inlet temperature, the turbine non-dimensional speed $N/\sqrt{T_{03}}$ is obtained from equation (8.1). The turbine efficiency can then be obtained from the turbine characteristic using the known values of $N/\sqrt{T_{03}}$ and $p_{03}/p_{04}$, and the turbine temperature drop can be calculated from

$$\Delta T_{034} = \eta_t T_{03} \left[ 1 - \left( \frac{1}{p_{03}/p_{04}} \right)^{(\gamma-1)/\gamma} \right] \tag{8.3}$$

The compressor temperature rise for the point selected on the compressor characteristic can be similarly calculated as

$$\Delta T_{012} = \frac{T_{01}}{\eta_c} \left[ \left( \frac{p_{02}}{p_{01}} \right)^{(\gamma-1)/\gamma} - 1 \right] \tag{8.4}$$

The net power output corresponding to the selected operating point is then found from

$$\text{net power output} = mc_{pg}\Delta T_{034} - \frac{1}{\eta_m} mc_{pa}\Delta T_{012}\dagger \tag{8.5}$$

where $\eta_m$ is the mechanical efficiency of the compressor–turbine combination, and $m$ is given by $(m\sqrt{T_{01}}/p_{01})(p_a/\sqrt{T_a})$ for prescribed ambient conditions.

Finally, it is necessary to consider the characteristics of the load to determine whether the compressor operating point selected represents a valid solution. If, for example, the engine were run on a test bed coupled to a hydraulic or electrical dynamometer, the load could be set independently of the speed and it would be possible to operate at any point on the compressor characteristic within the temperature limit imposed by safety considerations. With a propeller load, however, the power absorbed varies as the cube of the rotational speed of the propeller. When the transmission efficiency and gear ratio are known, the load characteristic in terms of the net power actually required from the turbine and the turbine speed can be plotted as in Fig. 8.4. The problem then becomes one of finding the *single* point on each constant speed line of the compressor characteristic which will give the required net power output at that speed; this can only be done by trial and error, taking several operating points on the compressor characteristic and establishing the power output corresponding to each one. If the calculated net power output for any point on the compressor characteristic is not equal to the power required at the selected speed, the engine will not be in equilibrium and will either accelerate or decelerate depending on whether there is a surplus or deficiency of power. Repeating this procedure for a series of

† Distinguishing suffixes will be added to $c_p$ only in equations where both $c_{pa}$ and $c_{pg}$ appear simultaneously. In other cases it will be clear from the context which mean $c_p$ should be used. $\gamma$ will be treated similarly.

constant speed lines, a series of points are obtained, which can be joined up to form the equilibrium running line as shown in Fig. 8.5.

The most common type of load used with a single-shaft gas turbine is the electric generator which runs at a constant rotational speed with the load varied electrically. Thus the equilibrium running line for a generator set

FIG. 8.4 **Load characteristics**

would correspond to a particular line of constant non-dimensional speed, as shown in Fig. 8.5, and each point on this line would represent a different value of turbine inlet temperature and power output. At each speed it is possible to find, by trial and error, the compressor operating point corresponding to zero net output and the no-load running line for a generator set is also shown in Fig. 8.5.

The equilibrium running lines depicted show that a propeller load implies operation in a zone of high compressor efficiency over a wide range of out-

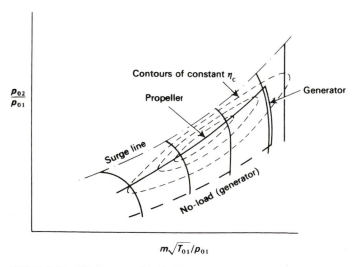

FIG. 8.5 **Equilibrium running lines**

put, whereas the generator load results in a rapid drop in compressor efficiency as load is reduced. The location of the equilibrium running line relative to the surge line indicates whether the engine can be brought up to full power without any complications. In the case of the propeller, the equilibrium running line lies close to the surge line, and may even intersect it, in which case the engine could not be accelerated to full power. This can be overcome by incorporating a blow-off valve towards the rear of the compressor. Section 8.6 deals with the matter in more detail. The running line for the generator at no-load can be seen to be well away from surge, and the generator could be accelerated to full speed before applying the load without any surge problem being encountered.

The calculations described above determine the values of all the parameters required for a complete performance calculation for any point within the operating range. $T_{03}$ is known, and $T_{02}$ is found from $(\Delta T_{012} + T_{01})$. Thus the combustion temperature rise is known and it is possible to obtain the fuel/air ratio $f$ from the curves of Fig. 2.15 and an assumed value of combustion efficiency. The fuel flow is then given by $\dot{m}f$. From the fuel flow and power output at each operating point the variation in specific fuel consumption (or thermal efficiency) with load can be determined. The results refer to operation at the assumed value of $T_{01}$ $(= T_a)$ and $p_{01}$ $(= p_a)$, but the process could be repeated over the range of values of ambient temperature and pressure likely to be encountered.

The matching calculations for a single-shaft gas turbine are illustrated in the following example.

## EXAMPLE

The following data refer to a single-shaft gas turbine operating at its design speed.

| Compressor characteristic | | | Turbine characteristic | |
|---|---|---|---|---|
| $p_{02}/p_{01}$ | $m\sqrt{T_{01}}/p_{01}$ | $\eta_c$ | $m\sqrt{T_{03}}/p_{03}$ | $\eta_t$ |
| 5·0 | 329·0 | 0·84 | 139·0 | 0·87 |
| 4·5 | 339·0 | 0·79 | (both constant over | |
| 4·0 | 342·0 | 0·75 | range of pressure ratio | |
| | | | considered) | |

Assuming ambient conditions of 1·013 bar and 288 K, a mechanical efficiency of 98 per cent, and neglecting all pressure losses, calculate the turbine inlet temperature required for a power output of 3800 kW. The 'non-dimensional' flows are expressed in terms of kg/s, K and bar.

The method of solution is to establish, for each point given on the compressor characteristic, the turbine inlet temperature from equation (8.2), the compressor temperature rise from equation (8.4) and the turbine temperature drop from equation (8.3). Once these have been established the power output can be obtained from equation (8.5); it is then necessary to plot turbine inlet temperature against power output to find the required temperature for an output of 3800 kW.

Taking the compressor operating point at a pressure ratio of 5·0,

$$\sqrt{\frac{T_{03}}{T_{01}}} = \frac{(m\sqrt{T_{03}/p_{03}})(p_{03}/p_{01})}{m\sqrt{T_{01}/p_{01}}}$$

With pressure losses neglected $p_{03} = p_{02}$, and hence

$$\sqrt{\frac{T_{03}}{T_{01}}} = \frac{139\cdot0 \times 5\cdot0}{329\cdot0} = 2\cdot11$$

giving $T_{03} = 1285$ K
The compressor temperature rise is given by

$$\Delta T_{012} = \frac{288}{0\cdot84}\left[5\cdot0^{1/3\cdot5} - 1\right] = 200\cdot5 \text{ K}$$

The turbine temperature drop is given by

$$\Delta T_{034} = 0\cdot87 \times 1285\left[1 - \frac{1}{5\cdot0^{1/4}}\right] = 370\cdot0 \text{ K}$$

The air mass flow is obtained from the non-dimensional flow entering the compressor,

$$m = 329 \times \frac{1\cdot013}{\sqrt{288}} = 19\cdot64 \text{ kg/s}$$

The power output can now be obtained, giving

$$\text{power output} = 19\cdot64 \times 1\cdot148 \times 370\cdot0 - \left(\frac{19\cdot64 \times 1\cdot005 \times 200\cdot5}{0\cdot98}\right)$$
$$= 8340 - 4035$$
$$= 4305 \text{ kW}$$

Thus for a power output of 4305 kW the turbine inlet temperature is 1285 K.

Repeating the calculation for the three points given on the compressor characteristic yields the following results

| $p_{02}/p_{01}$ | $T_{03}$ | $\Delta T_{012}$ | $\Delta T_{034}$ | $m$ | Power output |
|---|---|---|---|---|---|
| | (K) | (K) | (K) | (kg/s) | (kW) |
| 5·0 | 1285 | 200·5 | 370·0 | 19·64 | 4305 |
| 4·5 | 982 | 196·1 | 267·0 | 20·25 | 2130 |
| 4·0 | 761 | 186·7 | 194·0 | 20·4 | 635 |

Plotting the value of $T_{03}$ against power output it is found that for an output of 3800 kW the required turbine inlet temperature is 1215 K.

## 8.3   Equilibrium running of a gas generator

It was pointed out at the start of this chapter that the gas generator performs the same function for both the free turbine engine and the jet engine, namely the generation of a continuous flow of gas at high pressure and temperature which can be expanded to a lower pressure to produce either shaft work or a high velocity propulsive jet. Before considering either type of engine, it is appropriate to consider the behaviour of the gas generator alone.

Considerations of compatibility of speed and flow are the same as for the single-shaft engine described in the previous section, and equations (8.1) and (8.2) are applicable. This time, however, the pressure ratio across the turbine is not known, and it must be determined by equating the turbine work to the compressor work. The required turbine temperature drop, in conjunction with the turbine inlet temperature and efficiency, determines the turbine pressure ratio. Thus, instead of equation (8.5) the work requirement is expressed by

$$\eta_m c_{pg} \Delta T_{034} = c_{pa} \Delta T_{012}$$

Re-writing in terms of non-dimensional groups

$$\frac{\Delta T_{034}}{T_{03}} = \frac{\Delta T_{012}}{T_{01}} \times \frac{T_{01}}{T_{03}} \times \frac{c_{pa}}{c_{pg}\eta_m} \qquad\qquad (8.6)$$

Equations (8.1), (8.2) and (8.6) are all linked by the temperature ratio $T_{03}/T_{01}$ and it is necessary to determine, by trial and error, the turbine inlet temperature required for operation at any arbitrary point on the compressor characteristic. The procedure is as follows:

(a)   Having selected a point on the compressor characteristic, the values of $N/\sqrt{T_{01}}$, $p_{02}/p_{01}$, $m\sqrt{T_{01}}/p_{01}$, and $\eta_c$ are determined, and $\Delta T_{012}/T_{01}$ can be calculated from equation (8.4).

(b)   If a value of $p_{03}/p_{04}$ is guessed, the value of $m\sqrt{T_{03}}/p_{03}$ can be obtained from the turbine characteristic, enabling the temperature ratio $T_{03}/T_{01}$ to be obtained from the flow compatibility equation (8.2).

(c)   This value of $T_{03}/T_{01}$ can now be used to calculate $N/\sqrt{T_{03}}$ from equation (8.1).

(d)   With $N/\sqrt{T_{03}}$ and $p_{03}/p_{04}$ known, the turbine efficiency can be obtained from the turbine characteristic.

(e)   The non-dimensional temperature drop $\Delta T_{034}/T_{03}$ can be calculated from equation (8.3) and used in conjuction with equation (8.6) to calculate another value of $T_{03}/T_{01}$.

(f)   This second value of $T_{03}/T_{01}$ will not, in general, agree with the first value obtained from equation (8.2), indicating that the guessed value of the pressure ratio $p_{03}/p_{04}$ is not valid for an equilibrium running point.

(g)   A new value of $p_{03}/p_{04}$ must now be assumed and the above calculations repeated until the same value of $T_{03}/T_{01}$ is obtained from both equation (8.2) and (8.6).

(h)  Agreement signifies that the turbine operating point is compatible with the originally chosen compressor operating point when the fuel flow is such as to produce the iterated value of $T_{03}/T_{01}$.

The procedure is summarized in the information flow chart of Fig. 8.6(a).

It would be possible to carry out this calculation for a large number of points on the compressor characteristic, and express the results by joining

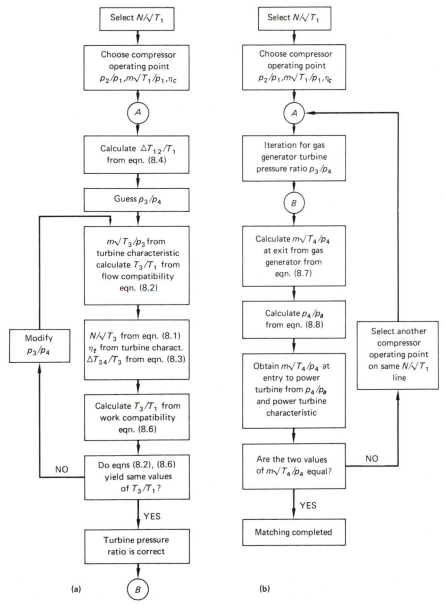

FIG. 8.6(a)  Iteration for gas generator

FIG. 8.6(b)  Overall iteration procedure for free power turbine unit up points of constant $T_{03}/T_{01}$ on the compressor characteristic as indicated by the dotted lines in Fig. 8.7. In practice this is unnecessary. The further requirement for flow compatibility with the component downstream of the gas generator, whether it be a power turbine or propelling nozzle, seriously restricts the operating zone on the compressor characteristic. The foregoing procedure thus comprises only one part of the calculation for the whole unit, and when the whole unit is considered only relatively few points may be needed. The free turbine engine will be considered first.

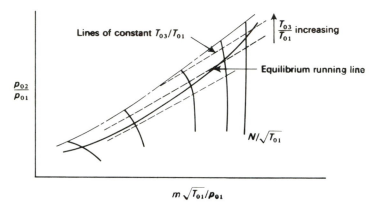

**FIG. 8.7 Equilibrium running line for free turbine**

Before doing so, it is necessary to emphasize that the matching procedure just outlined has been developed on the assumption that the turbine non-dimensional flow is independent of the non-dimensional speed and is a function only of pressure ratio. If the turbine characteristic does exhibit a variation of $m\sqrt{T_{03}}/p_{03}$ with $N/\sqrt{T_{03}}$, as in Fig. 7.28, the procedure must be modified. There is no change for any turbine operating point falling on the choking point of the mass flow curve, but for others the process would be as follows. Immediately after step ($a$) it is necessary to guess a value of $T_{03}/T_{01}$, which permits calculation of $N/\sqrt{T_{03}}$ from equation (8.1) and $m/\sqrt{T_{03}}/p_{03}$ from equation (8.2). $p_{03}/p_{04}$ and $\eta_t$ can then be obtained from the turbine characteristic, enabling $\Delta T_{034}/T_{03}$ to be calculated from equation (8.3). The work compatibility equation (8.6) is then used to provide a value of $T_{03}/T_{01}$ for comparison with the original guess. To avoid obscuring the main principles involved, the use of multi-line turbine flow characteristics will not be discussed further.† Enough has been said to show that they can be accommodated without undue difficulty.

---

† The characteristic shown in Fig. 7.28 is typical of that for a turbine in which choking occurs in the stator passages. In some designs choking may occur initially at exit from the rotor passages; this will result in a small variation of the choking value of $m\sqrt{T_{03}}/p_{03}$ with $N/\sqrt{T_{03}}$. The modification to the calculation would then apply also to the choking region.

## 8.4 Off-design operation of free turbine engine

### *Matching of gas generator with free turbine*

The gas generator is matched to the power turbine by the fact that the mass flow leaving the gas generator must equal that at entry to the power turbine, coupled with the fact that the pressure ratio available to the power turbine is fixed by the compressor and gas generator turbine pressure ratios. The power turbine characteristic will have the same form as Fig. 8.3, but the parameters will be $m\sqrt{T_{04}}/p_{04}$, $p_{04}/p_a$, $N_p/\sqrt{T_{04}}$ and $\eta_{tp}$.

The preceding section described how the gas generator operating conditions can be determined for any point on the compressor characteristic. The value of $m\sqrt{T_{04}}/p_{04}$ at exit from the gas generator can then be calculated from

$$\frac{m\sqrt{T_{04}}}{p_{04}} = \frac{m\sqrt{T_{03}}}{p_{03}} \times \frac{p_{03}}{p_{04}} \times \sqrt{\frac{T_{04}}{T_{03}}} \tag{8.7}$$

where

$$\sqrt{\frac{T_{04}}{T_{03}}} = \sqrt{\left(1 - \frac{\Delta T_{034}}{T_{03}}\right)} \text{ and } \frac{\Delta T_{034}}{T_{03}} = \eta_t\left[1 - \left(\frac{1}{p_{03}/p_{04}}\right)^{(\gamma-1)/\gamma}\right]$$

The corresponding pressure ratio across the power turbine can also be established from

$$\frac{p_{04}}{p_a} = \frac{p_{02}}{p_{01}} \times \frac{p_{03}}{p_{02}} \times \frac{p_{04}}{p_{03}} \tag{8.8}$$

(It must be remembered that in the case of a stationary gas turbine with inlet and exit duct losses ignored $p_{01} = p_a$, and the power turbine outlet pressure is also equal to $p_a$.)

Having found the pressure ratio across the power turbine, the value of $m\sqrt{T_{04}}/p_{04}$ can be found from the power turbine characteristic for comparison with the value obtained from equation (8.7). If agreement is not reached it is necessary to choose another point on the same constant speed line of the compressor characteristic and repeat the procedure until the requirement of flow compatibility between the two turbines is satisfied. The overall procedure for the free turbine engine, including the iteration for the gas generator, is summarized in the information flow chart of Fig. 8.6(b).

For each constant $N/\sqrt{T_{01}}$ line on the compressor characteristic there will be only one point which will satisfy both the work requirement of the gas generator and flow compatibility with the power turbine. If the foregoing calculations are carried out for each constant speed line, the points obtained can be joined up to form the equilibrium running line as shown in Fig. 8.7. *The running line for the free turbine engine is independent of the load and is determined by the swallowing capacity of the power turbine.* This is in contrast to the behaviour of the single-shaft unit, where the running line depends on the characteristic of the load as indicated in Fig. 8.5.

The next step is to calculate the power output and specific fuel consumption for the equilibrium running points. Before discussing this, however, a useful approximation which simplifies the foregoing procedure will be mentioned. It arises from the behaviour of two turbines in series. We introduce it here because it facilitates a better physical understanding of some of the phenomena discussed in subsequent sections.

## Matching of two turbines in series

The iterative procedure required for the matching of a gas generator and a free turbine can be considerably simplified if the behaviour of two turbines in series is considered; this approach is also valuable for the analysis of more complex gas turbines considered in Chapter 9. It was shown in the previous sub-section that by using equation (8.7) the value of $m\sqrt{T_{04}}/p_{04}$ at exit from the gas generator turbine can be obtained for any gas generator operating point, and in particular that it is a function of $m\sqrt{T_{03}}/p_{03}$, $p_{03}/p_{04}$ and $\eta_t$. The value of $\eta_t$ could be read from the turbine characteristic because $N/\sqrt{T_{03}}$ had been determined for the operating point in the course of the gas generator calculation. Now in practice the variation of $\eta_t$ at any given pressure ratio is not large (see Fig. 8.3), particularly over the restricted range of operation of the gas generator turbine. Furthermore, such a variation has little effect on $m\sqrt{T_{04}}/p_{04}$ because the resulting change in $\sqrt{(T_{03}/T_{04})}$ is very small. It is often sufficiently accurate to take a mean value of $\eta_t$ at any given pressure ratio, so that $m\sqrt{T_{04}}/p_{04}$ becomes a function only of $m\sqrt{T_{03}}/p_{03}$ and $p_{03}/p_{04}$. If this is done, a single curve representing the turbine outlet flow characteristic can readily be obtained by applying equation (8.7) to points on the single curve of the inlet flow characteristic. The result is shown by the dotted curve in Fig. 8.8.

The effect of operating two turbines in series is also shown in Fig. 8.8 where it can be seen that the requirement for flow compatibility between the two turbines places a major restriction on the operation of the gas generator

**FIG. 8.8 Operation of turbines in series**

turbine. In particular, as long as the *power turbine is choked* the gas generator turbine will operate at a *fixed non-dimensional point*, i.e. at the pressure ratio marked (*a*). With the power turbine unchoked, the gas generator will be restrained to operate at a fixed pressure ratio for each power turbine pressure ratio, e.g. (*b*) and (*c*). Thus the maximum pressure ratio across the gas generator turbine is controlled by choking of the power turbine, and at all times the pressure ratio is controlled by the swallowing capacity of the power turbine.

A further consequence of the fixed relation between the turbine pressure ratios is that it is possible to plot the gas generator pressure ratio $p_{03}/p_{04}$ against compressor pressure ratio $p_{02}/p_{01}$ by using the identity

$$\frac{p_{03}}{p_{04}} = \frac{p_{03}}{p_{02}} \times \frac{p_{02}}{p_{01}} \times \frac{p_a}{p_{04}} \qquad (8.9)$$

$p_{03}/p_{02}$ is determined by the assumed combustion pressure loss, and $p_{04}/p_a$ is obtained from Fig. 8.8 for any value of $p_{03}/p_{04}$. Such a curve is shown in Fig. 8.9. From it, for any value of compressor pressure ratio, the pressure

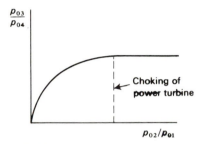

FIG. 8.9

ratio of the gas generator turbine can be determined; this in turn fixes the values of $m\sqrt{T_{03}}/p_{03}$ and $\Delta T_{034}/T_{03}$ which are required for use with equations (8.2) and (8.6). Thus it is no longer necessary to carry out the iteration for the pressure ratio of the gas generator turbine and, for each constant speed line considered, only a *single* iteration is required to find the correct equilibrium running point.

## Variation of power output and s.f.c. with output speed

The net power output of the free turbine engine is simply the output from the power turbine, namely

$$\text{power output} = mc_p\Delta T_{045} \qquad (8.10)$$

where

$$\Delta T_{045} = \eta_{tp}T_{04}\left[1 - \left(\frac{1}{p_{04}/p_a}\right)^{(\gamma-1)/\gamma}\right]$$

For each equilibrium running point established (one for each compressor speed line) $p_{04}/p_a$ will be known, and $T_{04}$ can be readily calculated from

$$T_{04} = T_{03} - \Delta T_{034} \tag{8.11}$$

The mass flow is obtained from $m\sqrt{T_{01}}/p_{01}$ for assumed values of $p_a$ and $T_a$. The outstanding unknown is the power turbine efficiency $\eta_{tp}$. It can be obtained from the power turbine characteristic, but it depends not only on the pressure ratio $p_{04}/p_a$ but also upon $N_p/\sqrt{T_{04}}$, i.e. upon the power turbine speed $N_p$. Now free turbine engines are used to drive a variety of loads such as pumps, propellers and electric generators, each with a different power versus speed characteristic. For this reason it is usual to calculate the power output over a range of power turbine speed for each equilibrium running point (i.e. for each compressor speed). The results could be plotted as in Fig. 8.10. Any

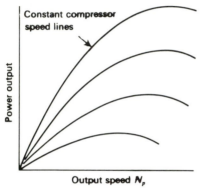

**FIG. 8.10**

curve corresponding to a given compressor speed will be fairly flat in the useful upper half of the output speed range where $\eta_{tp}$ does not vary much with $N_p/\sqrt{T_{04}}$ (see Fig. 8.3).

The fuel consumption can also be calculated for each equilibrium running point, in the same manner as described in section 8.2 for the single-shaft unit. Since it depends only on the gas generator parameters there will be only one value for each compressor speed. When combined with the power output data to give the s.f.c., however, it is clear that the s.f.c., like the power output, will be a function of both compressor speed and power turbine speed. It is convenient to express the off-design performance by plotting s.f.c. against power output for several power turbine speeds as shown in Fig. 8.11. This type of presentation permits ·the customer to evaluate the performance of the unit when coupled to his type of load by superimposing the load characteristic upon it. The dotted curve in Fig. 8.11 indicates a particular variation of power and speed imposed by a load, and the points of intersection with the $N_p$ curves give the s.f.c. versus power output curve for the free turbine engine driving that particular load. Figure 8.11 relates to operation at one ambient condition, and normally the performance calculations would be repeated for extreme values of $p_a$ and $T_a$.

It should be appreciated that although for convenience of calculation the compressor speed has been chosen as the independent variable, in practice the fuel flow is the independent variable. A chosen value of fuel flow (and hence $T_{03}$) determines the compressor speed and ultimately the power output. The s.f.c. curves of Fig. 8.11 exhibit an increase in s.f.c. as the power is

s.f.c.

Constant $N_p$ lines

$N_p$

Power v. $N_p$ load characteristic

Power output

**FIG. 8.11**

reduced because the reduction in fuel flow leads to a reduction in compressor speed and gas generator turbine inlet temperature. And it will be remembered from Chapter 2 that the efficiency of a real cycle falls as the turbine inlet temperature is reduced. This poor part-load economy is a major disadvantage of the simple gas turbine. Consideration of the substantial improvement arising from the use of more complex cycles will be deferred until Chapter 9. Referring again to Fig. 8.11, the change in s.f.c. with $N_p$ at any given power output is not very marked, because the gas generator parameters change only slightly under this operating condition.

It is useful to consider the variation of the key variables (power, turbine inlet temperature and fuel flow) with gas generator speed as shown in Fig. 8.12. It can be seen that all increase rapidly as the gas generator approaches its maximum permissible value. The change in turbine inlet temperature is especially critical because of the effect on creep life of the first-stage turbine blades and, in general, operation at the maximum speed would only be for limited periods. In electricity generation, for example, this maximum speed would be used for emergency peak duty, and base-load operation would utilize a reduced gas generator speed with correspondingly increased blade life.

## Single-shaft versus twin-shaft engines

The choice of whether to use a single-shaft or twin-shaft (free turbine) power

FIG. 8.12 **Variation of key parameters with gas generator speed**

plant is largely determined by the characteristics of the driven load. If the load speed is constant, as in the case of an electric generator, a single-shaft unit is often specified; an engine specifically designed for electric power generation would make use of a single-shaft configuration. An alternative, however, is the use of an aircraft derivative with a free power turbine in the place of the propelling nozzle. With this arrangement it is possible to design a power turbine of substantially larger diameter than the gas generator, using an elongated duct between the gas generator and the power turbine; this then permits the power turbine to operate at the required electric generator speed without the need for a reduction gear box. Turboprops may use either configuration, as shown in Figs. 1.10 and 1.11.

The running lines for single-shaft and twin-shaft units were shown in Figs 8.5 and 8.7. It should be noted that in the case of the single-shaft engine driving a generator, reduction in output power results in a slight increase in compressor mass flow; although there is some reduction in compressor pressure ratio, there is little change in compressor temperature rise because the efficiency is also reduced. This means that the compressor power remains essentially fixed. With a twin-shaft engine, however, as Fig. 8.7 shows, reducing net power output involves a reduction in compressor speed and hence in air flow, pressure ratio and temperature rise. The compressor power needed is therefore appreciably lower than for the single-shaft engine. It should also be evident from a comparison of Figs 8.5 and 8.7 that the compressor operates over a smaller range of efficiency in a twin-shaft engine. For these reasons the part-load fuel consumption of a twin-shaft engine is superior when driving a constant speed load.

The two types also have different characteristics regarding the supply of waste heat to a cogeneration plant, primarily due to the differences in exhaust flow as load is reduced; the essentially constant air flow and compressor power in a single-shaft unit results in a larger decrease of exhaust temperature for a given reduction in power, which might necessitate the burning of supplementary fuel in the waste heat boiler under operating conditions where it would be unnecessary with a twin-shaft. In both cases, the exhaust temperature may be increased by the use of variable inlet guide vanes. Cogeneration systems have been successfully built using both single shaft and twin shaft units, the latter often being aircraft derivative engines.

Most gas turbines do not spend long periods at low power and the s.f.c. differences may not be critical in practice. The *torque characteristics*, however, are very different and the variation of torque with output speed at a given power may well determine the engine's suitability for certain applications; e.g. a high starting torque is particularly important for traction purposes.

The compressor of a *single-shaft engine* is constrained to turn at some multiple of the load speed, fixed by the transmission gear ratio, so that a reduction in load speed implies a reduction in compressor speed. This results in a reduction in mass flow and hence of output torque as shown by curve (*a*) of Fig. 8.13. This type of turbine is clearly unsuitable for traction purposes. The normal flat torque curve of an internal combustion engine is shown dotted for comparison.

The *free power turbine unit*, however, has a torque characteristic even more favourable than the internal combustion engine. The variation of power output with load speed, at any given compressor speed determined by the fuel flow, is shown in Fig. 8.10. It can be seen that the output power remains relatively constant over a wide load speed range for a fixed compressor speed. This is due to the fact that the compressor can supply an essentially constant flow at a given compressor speed, irrespective of the free turbine speed. Thus at fixed gas generator operating conditions, reduction in output speed results in an increase in torque as shown by curve (*b*) in Fig. 8.13. It is quite possible

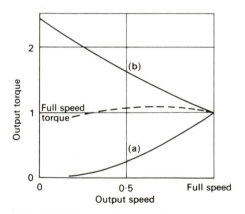

FIG. 8.13 Torque characteristics

to obtain a stall torque of two to three times the torque delivered at full speed.

The actual range of speed over which the torque conversion is efficient depends on the efficiency characteristic of the power turbine. The typical turbine efficiency characteristic shown in Fig. 8.3 suggests that the fall in efficiency will not be greater than about 5 or 6 per cent over a speed range from half to full speed. Thus quite a large increase in torque can be obtained efficiently when the output speed is reduced to 50 per cent of its maximum value. The efficiency of the torque conversion at low speeds, for example when accelerating a vehicle from rest, will be very low. In some applications a simple two-speed gearbox might be sufficient to overcome this defect, but it is probable that gas turbines for heavy road vehicles will incorporate some form of automatic transmission with five or six speeds.

Further differences between single- and twin-shaft units are brought out in section 9.5 where the transient behaviour of gas turbines is discussed.

## EXAMPLE

A gas turbine with a free power turbine is designed for the following conditions:

| | |
|---|---|
| air mass flow | 30 kg/s |
| compressor pressure ratio | 6·0 |
| compressor isentropic efficiency | 0·84 |
| turbine inlet temperature | 1200 K |
| turbine isentropic efficiency (both turbines) | 0·87 |
| combustion pressure loss | 0·20 bar |
| mechanical efficiency | 0·99 |
|  (applies to gas generator and load) | |
| ambient conditions | 1·01 bar, 288 K |

Calculate the power developed and the turbine non-dimensional flows $m\sqrt{T_{03}}/p_{03}$ and $m\sqrt{T_{04}}/p_{04}$ at the design point.

If the engine is run at the same mechanical speed at an ambient temperature of 268 K calculate the values of turbine inlet temperature, pressure ratio and power output, assuming the following:

(a) combustion pressure loss remains constant at 0·20 bar;

(b) both turbines choking, with values of $m\sqrt{T_{03}}/p_{03}$ and $m\sqrt{T_{04}}/p_{04}$ as calculated above and no change in turbine efficiency;

(c) at 268 K and the same $N$, the $N/\sqrt{T_{01}}$ line on the compressor characteristic is a vertical line with a non-dimensional flow 5 per cent greater than the design value;

(d) variation of compressor efficiency with pressure ratio at the relevant value of $N/\sqrt{T_{01}}$ is

| $p_{02}/p_{01}$ | 6·0 | 6·2 | 6·4 | 6·6 |
|---|---|---|---|---|
| $\eta_c$ | 0·837 | 0·843 | 0·845 | 0·840 |

The design point calculation is straightforward and only the salient results are presented:

Gas generator turbine

| | |
|---|---|
| pressure ratio | 2·373 |
| inlet pressure | 5·86 bar |
| temperature drop | 203 K |

Power turbine

| | |
|---|---|
| pressure ratio | 2·442 |
| inlet pressure | 2·47 bar |
| temperature drop | 173·5 K |
| inlet temperature | 997 K |

The power output is then $(30 \times 1·148 \times 173·5 \times 0·99)$ kW or 5910 kW.
The design point values of $m\sqrt{T_{03}}/p_{03}$ and $m\sqrt{T_{04}}/p_{04}$ are

$$\frac{30 \times \sqrt{1200}}{5·86} = 177·4 \text{ and } \frac{30 \times \sqrt{997}}{2·47} = 383·5 \text{ respectively}$$

At 268 K the value of $m\sqrt{T_{01}}/p_{01}$ is

$$1·05 \times \left[\frac{30 \times \sqrt{288}}{1·01}\right] = 529·5$$

If the power turbine remains choked, the gas generator turbine will be constrained to operate at a fixed non-dimensional point, and thus the value of $\Delta T_{034}/T_{03} = 203/1200 = 0·169$, as for the design condition.
  For work compatibility

$$\frac{\Delta T_{012}}{T_{01}} = \frac{\Delta T_{034}}{T_{03}} \times \frac{T_{03}}{T_{01}}\left[\frac{c_{pg}\eta_m}{c_{pa}}\right] = \frac{0·169 \times 1·148 \times 0·99}{1·005} \times \frac{T_{03}}{T_{01}}$$

and hence

$$\frac{T_{03}}{T_{01}} = 5·23 \frac{\Delta T_{012}}{T_{01}} \tag{A}$$

For flow compatibility

$$\frac{m\sqrt{T_{03}}}{p_{03}} = \frac{m\sqrt{T_{01}}}{p_{01}} \times \frac{p_{01}}{p_{03}} \times \sqrt{\frac{T_{03}}{T_{01}}}$$

$$177·4 = 529·5 \frac{p_{01}}{p_{03}} \times \sqrt{\frac{T_{03}}{T_{01}}}$$

$$\sqrt{\frac{T_{03}}{T_{01}}} = 0·335 \frac{p_{03}}{p_{01}} \tag{B}$$

  The problem is thus to find the compressor operating point that will give the same value of $T_{03}/T_{01}$ for both equations (A) and (B). With the variation in efficiency prescribed the value of $\Delta T_{012}/T_{01}$ can readily be calculated from equation (8.4), and with the constant value of combustion pressure loss the value of $p_{03}$ can also be calculated. The solution is best carried out in tabular form as shown.

| $\dfrac{p_{02}}{p_{01}}$ | $\left[\left(\dfrac{p_{02}}{p_{01}}\right)^{(\gamma-1)/\gamma}-1\right]$ | $\eta_c$ | $\dfrac{\Delta T_{012}}{T_{01}}$ | $\left(\dfrac{T_{03}}{T_{01}}\right)_A$ | $p_{02}$ | $p_{03}$ | $\dfrac{p_{03}}{p_{01}}$ | $\sqrt{\dfrac{T_{03}}{T_{01}}}$ | $\left(\dfrac{T_{03}}{T_{01}}\right)_B$ |
|---|---|---|---|---|---|---|---|---|---|
| 6·0 | 0·669 | | 0·837 | 0·799 | 4·18 | 6·06 | 5·86 | 5·80 | 1·943 | 3·78 |
| 6·2 | 0·684 | | 0·843 | 0·812 | 4·25 | 6·26 | 6·06 | 6·00 | 2·010 | 4·05 |
| 6·4 | 0·700 | | 0·845 | 0·828 | 4·33 | 6·46 | 6·26 | 6·20 | 2·078 | 4·32 |
| 6·6 | 0·715 | | 0·840 | 0·851 | 4·45 | 6·66 | 6·46 | 6·40 | 2·144 | 4·60 |

Solving graphically the required pressure ratio is found to be 6·41, with the value of $T_{03}/T_{01} = 4\cdot34$; the corresponding turbine inlet temperature is $4\cdot34 \times 268 = 1163$ K.

Having established the compressor pressure ratio and turbine inlet temperature, it is a straightforward matter to calculate the power developed. Remembering that the gas generator turbine will still operate at the same non-dimensional point ($\Delta T_{034}/T_{03} = 0\cdot169$, $p_{03}/p_{04} = 2\cdot373$) the power turbine entry conditions and temperature drop can readily be calculated; the resulting temperature drop is 179·6 K. The mass flow is obtained from the non-dimensional flow and the ambient conditions:

$$\frac{m\sqrt{T_{01}}}{p_{01}} = 529\cdot5, \text{ hence } m = \frac{529\cdot5 \times 1\cdot01}{\sqrt{268}} = 32\cdot7 \text{ kg/s}$$

Hence power output $= 32\cdot7 \times 1\cdot148 \times 179\cdot6 \times 0\cdot99 = 6680$ kW.

It can be seen that operation at the design mechanical speed on a cold day results in a decrease of maximum cycle temperature from 1200 to 1163 K, even though the value of $T_{03}/T_{01}$ has increased from 4·17 to 4·34 due to the increase in $N/\sqrt{T_{01}}$. The power has increased from 5910 to 6680 kW and this can be seen to be due to the simultaneous increase in air mass flow and overall pressure ratio. The beneficial effect of low ambient temperature on gas turbine operation is evident; conversely, high ambient temperatures result in significant penalties. The effect of increased ambient temperature on turbojet operation is discussed under the heading 'Variation of thrust with rotational speed, forward speed and altitude'.

## 8.5   Off-design operation of the jet engine

### Propelling nozzle characteristics

The propelling nozzle area for a jet engine is determined from the design point calculations as described in Chapter 3, and once the nozzle size has been fixed it has a major influence on the off-design operation. The characteristic for a nozzle, in terms of 'non-dimensional' flow $m\sqrt{T_{04}}/p_{04}$ and pressure ratio $p_{04}/p_{05}$, can readily be calculated as follows.

The mass flow parameter is given by

$$\frac{m\sqrt{T_{04}}}{p_{04}} = C_5 A_5 \rho_5 \frac{\sqrt{T_{04}}}{p_{04}}$$

$$= \frac{C_5}{\sqrt{T_{04}}} \times \frac{A_5}{R} \times \frac{p_5}{p_{04}} \times \frac{T_{04}}{T_5} \tag{8.12}$$

where $A_5$ is the effective nozzle area. Making use of equation (3.12), $C_5/\sqrt{T_{04}}$ can be found from

$$\frac{C_5^2}{T_{04}} = 2c_p \eta_j \left[ 1 - \left( \frac{1}{p_{04}/p_5} \right)^{(\gamma-1)/\gamma} \right] \tag{8.13}$$

and $T_5/T_{04}$ from

$$\frac{T_5}{T_{04}} = 1 - \frac{T_{04} - T_5}{T_{04}} = 1 - \eta_j \left[ 1 - \left( \frac{1}{p_{04}/p_5} \right)^{(\gamma-1)/\gamma} \right] \tag{8.14}$$

It follows that for a nozzle of given area and efficiency, $m\sqrt{T_{04}}/p_{04}$ can be calculated as a function of the pressure ratio $p_{04}/p_5$. But equations (8.13) and (8.14) are valid for pressure ratios only up to the critical value, given by equation (3.14), namely

$$\frac{p_{04}}{p_c} = 1 \left/ \left[ 1 - \frac{1}{\eta_j} \left( \frac{\gamma-1}{\gamma+1} \right) \right]^{\gamma/(\gamma-1)} \right. \tag{8.15}$$

and up to this point $p_5 = p_a$. For pressure ratios $p_{04}/p_a$ greater than the critical, $m\sqrt{T_{04}}/p_{04}$ remains constant at the maximum (choking) value, i.e. it is independent of $p_{04}/p_a$ (and incidentally $p_5 = p_c > p_a$). Thus $m\sqrt{T_{04}}/p_{04}$ can be plotted against the *overall* nozzle pressure ratio $p_{04}/p_a$ as in Fig. 8.14, and the similarity between the nozzle flow characteristic and that of a turbine is evident.

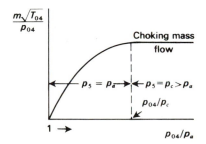

FIG. 8.14  **Nozzle characteristics**

Up to the choking condition, $T_5/T_{04}$ is given by equation (8.14), whereas when the nozzle is choking it is given by equation (3.13), namely

$$\frac{T_c}{T_{04}} = \frac{2}{\gamma+1} \tag{8.16}$$

Likewise, with the nozzle unchoked $C_5/\sqrt{T_{04}}$ is given by equation (8.13), whereas when it is choked $C_5$ is the sonic velocity and the Mach number $M_5$ is unity. Now, recalling that $C = M\sqrt{(\gamma RT)}$ and $T_0 = T[1+(\gamma-1)M^2/2]$, we have the general relation

$$\frac{C}{\sqrt{T_0}} = \frac{M\sqrt{(\gamma R)}}{\sqrt{\left(1+\frac{\gamma-1}{2}M^2\right)}} \tag{8.17}$$

Thus when the nozzle is choked (i.e. $M_5 = 1$) we have

$$\frac{C_5^2}{T_{04}} = \frac{C_5^2}{T_{05}} = \frac{2\gamma R}{\gamma+1} \tag{8.18}$$

We shall make use of equations (8.15) to (8.18) later when discussing the determination of the thrust.

## Matching of gas generator with nozzle

The similarity between the flow characteristic of a nozzle and a turbine means that the nozzle will exert the same restriction on the operation of the gas generator as a free power turbine. Thus if the operation of a jet engine under *static* conditions is considered, there can be no difference between its behaviour and that of a free power turbine unit. The equilibrium running line can be determined according to the procedure specified in the flow chart of Fig. 8.6(b), with the nozzle characteristic replacing the power turbine characteristic. It follows that Fig. 8.7 can be taken as representative of a typical equilibrium running diagram for a jet engine under static conditions.

The jet engine, of course, is intended for flight at high speeds and it is necessary to consider the effect of forward speed on the equilibrium running line. It is most convenient to express the forward speed in terms of Mach number for the matching calculations, and this can readily be converted to velocity for calculation of the momentum drag and thrust as shown later in the next sub-section.

Forward speed produces a ram pressure ratio which is a function of both flight Mach number and intake efficiency. This ram effect will give rise to an increase in compressor delivery pressure, which will in turn lead to a higher pressure before the propelling nozzle, thereby increasing the nozzle pressure ratio. Once the nozzle is choked, however, the nozzle non-dimensional flow will reach its maximum value and will then be independent of the nozzle pressure ratio and therefore of forward speed. The significance of this is that the turbine operating point will then also be unchanged, because of the requirement for compatibility of flow between the turbine and the nozzle. It follows that as long as the *nozzle is choking* the equilibrium running line will be uniquely determined by the fixed turbine operating point and will be *independent of flight speed.*†

At current levels of cycle pressure ratio, virtually all jet engines operate with the nozzle choked during take-off, climb and cruise, and the nozzle only becomes unchoked when thrust is significantly reduced. Thus the nozzle is liable to be unchoked only when preparing to land or when taxying. Nevertheless, it is important to consider the effect of forward speed on the running line under these conditions because it is at low rotational engine speeds that the running line is in close proximity to the surge line.

Now the nozzle pressure ratio $p_{04}/p_a$ is linked to the ram pressure ratio by the identity

$$\frac{p_{04}}{p_a} = \frac{p_{04}}{p_{03}} \times \frac{p_{03}}{p_{02}} \times \frac{p_{02}}{p_{01}} \times \frac{p_{01}}{p_a} \tag{8.19}$$

This differs from the corresponding equation for the free turbine unit, equation (8.8), only by the inclusion of the ram pressure ratio $p_{01}/p_a$. The ram pressure ratio in terms of intake efficiency $\eta_i$ and flight Mach number $M_a$ is given by equation (3.10(b)), namely

$$\frac{p_{01}}{p_a} = \left[ 1 + \eta_i \left( \frac{\gamma-1}{2} \right) M_a^2 \right]^{\gamma/(\gamma-1)} \tag{8.20}$$

It follows from equations (8.19) and (8.20) that for a given intake efficiency, $p_{04}/p_a$ is a function of the gas generator parameters and flight Mach number. The procedure of Fig. 8.6(b) can be followed, with equation (8.19) substituted for equation (8.8), but for each compressor speed line the calculation is repeated for several values of $M_a$ covering the desired range of flight speed. The result is a fan of equilibrium running lines of constant $M_a$. As shown in Fig. 8.15, they merge into the single running line obtained with the higher compressor speeds for which the nozzle is choked.

It should be noted that increasing Mach number pushes the equilibrium running line away from the surge line at low compressor speeds. Fundamentally, this is because the ram pressure rise allows the compressor to utilize a lower pressure ratio for pushing the required flow through the nozzle.

## Variation of thrust with rotational speed, forward speed and altitude

We have seen in section 3.1 that when the whole of the expansion from $p_{04}$ to $p_a$ occurs in the propelling nozzle, the net thrust of the jet unit is simply the

---

† It should be noted that the argument based on Fig. 8.8 is equally applicable here. Thus when the nozzle is choked the turbine operating point, and hence $m\sqrt{T_{03}}/p_{03}$, $p_{03}/p_{04}$ and $\Delta T_{034}/T_{03}$ are fixed and any increase in the overall expansion ratio $p_{03}/p_a$ due to increasing ram effect will therefore result in an increase in overall nozzle pressure ratio $p_{04}/p_a$ with the turbine pressure ratio unaffected.

$\dfrac{p_{02}}{p_{01}}$

**FIG. 8.15 Jet engine running lines**

overall rate of change of momentum of the working fluid, namely $F = m(C_5 - C_a)$ where $C_a$ is the aircraft speed. On the other hand, when part of the expansion occurs outside the propelling nozzle, i.e. when the nozzle is choking and $p_5$ is greater than $p_a$, there will be an additional pressure thrust. In this case the net thrust is given by the more general expression

$$F = m(C_5 - C_a) + (p_5 - p_a)A_5 \tag{8.21}$$

To obtain the curves which show the thrust delivered by the jet engine over the complete operating range of inlet pressure and temperature, flight speed and rotational speed, information obtained from the points used to establish the equilibrium running diagram can be used. For each running point on this diagram the values of all the thermodynamic variables such as

$$\frac{p_{01}}{p_a}, \quad \frac{m\sqrt{T_{01}}}{p_{01}}, \quad \frac{p_{02}}{p_{01}}, \quad \frac{T_{03}}{T_{01}}, \quad \frac{T_{04}}{T_{03}}, \quad \frac{p_{03}}{p_{04}} \text{ and } \frac{p_{04}}{p_a}$$

are determined for specified values of $N/\sqrt{T_{01}}$ and $M_a$ (and $N/\sqrt{T_{01}}$ alone in the choked nozzle range). The thrust can ultimately be expressed in terms of these non-dimensional quantities. Thus equation (8.21) can immediately be written in the form

$$\frac{F}{p_a} = \frac{m\sqrt{T_{01}}}{p_{01}}\frac{p_{01}}{p_a}\left[\frac{C_5}{\sqrt{T_{04}}}\sqrt{\left(\frac{T_{04}}{T_{03}}\times\frac{T_{03}}{T_{01}}\right)} - \frac{C_a}{\sqrt{T_{01}}}\right] + \left(\frac{p_5}{p_a}-1\right)A_5$$

$$\tag{8.22}$$

(A dimensional check will show that the true non-dimensional thrust is $F/(p_aD^2)$, but for an engine with fixed geometry the characteristic dimension can be omitted.) Making use of equation (8.17),

$$\frac{C_a}{\sqrt{T_{01}}} = \frac{C_a}{\sqrt{T_{0a}}} = \frac{M_a\sqrt{(\gamma R)}}{\sqrt{\left(1 + \dfrac{\gamma - 1}{2}M_a^2\right)}}$$

*When the nozzle is unchoked,* $C_5/\sqrt{T_{04}}$ can be obtained from equation (8.13) with $p_{04}/p_5$ put equal to $p_{04}/p_a$, and the pressure thrust is zero because $p_5/p_a = 1$.

*When the nozzle is choked,* $C_5/\sqrt{T_{04}}$ is given by equation (8.18), and $p_5/p_a$ can be calculated from

$$\frac{p_5}{p_a} = \frac{p_5}{p_{04}} \times \frac{p_{04}}{p_a} = \frac{p_c}{p_{04}} \times \frac{p_{04}}{p_a}$$

$p_c/p_{04}$ is the reciprocal of the critical pressure ratio given by equation (8.15), i.e. it is a function only of $\gamma$ for the gas and the nozzle efficiency $\eta_j$.

$p_{04}/p_a$ is one of the known parameters having been calculated using equation (8.19).

A typical variation of thrust with engine speed and flight speed is illustrated in Fig. 8.16, which shows separate curves of thrust for each Mach number considered. The abscissa can be left as $N/\sqrt{T_{01}}$ as shown, or be put in terms of $N/\sqrt{T_a}$ by making use of the relations

$$\frac{N}{\sqrt{T_a}} = \frac{N}{\sqrt{T_0}} \times \sqrt{\frac{T_{01}}{T_a}} \text{ and } \frac{T_{01}}{T_a} = \left(1 + \frac{\gamma-1}{2} M_a^2\right)$$

It should be noted that although a unique running line is obtained on the compressor characteristic when the propelling nozzle is choked, the thrust for a given value of $N/\sqrt{T_{01}}$ does depend on the flight Mach number. There is a direct dependence due to the increase in momentum drag ($mC_a$) with increasing flight speed, and an indirect dependence due to the increase in compressor inlet pressure arising from the ram compression. At low rotational speeds the effect of momentum drag predominates and an increase in $M_a$ causes the thrust to decrease, whereas at high values of $N/\sqrt{T_{01}}$ the beneficial effect of ram pressure rise predominates.

Although it is convenient to express the performance in terms of non-dimensional engine speed, it is the actual *mechanical* speed upon which a limit is set by the turbine stresses and which must be controlled by a governor. The strong dependence of thrust upon engine speed indicates that accurate control is essential.† If the speed is controlled at a value below the limit, take-off thrust will be substantially reduced. The situation is more serious if the speed exceeds the correct limit: not only do the centrifugal stresses increase with the square of the speed but there is also a rapid increase in turbine inlet temperature. The latter can be seen from the way the running line crosses the $T_{03}/T_{01}$ lines on Fig. 8.15. Typically, an increase in rotor speed of only 2 per cent above the limit may result in an increase in $T_{03}$ of 50 K. With blade life determined by creep, the time for which high speeds are permitted must be strictly controlled. The maximum permissible speed is normally restricted to periods of less than 5 min, giving the *take-off rating*. The *climb rating* is obtained with a small reduction in fuel flow and hence in

† It should be realized that the same arguments are applicable in the case of a free turbine engine delivering shaft power.

rotor speed, and can usually be maintained for a period of 30 min. The *cruise rating* requires a further reduction in fuel flow and rotor speed, resulting in stress and temperature conditions permitting unrestricted operation. Ratings for the Rolls-Royce Viper 20 at SLS conditions are shown in the following table; this is a simple turbojet with a low pressure ratio. The very rapid drop in thrust with reduction of speed is evident.

| Rolls-Royce Viper 20, sea level static, I.S.A. conditions | | | |
|---|---|---|---|
| | Engine speed ($\% N_{max}$) | Thrust (kN) | s.f.c. (kg/kN h) |
| Take off (5 min) | 100 | 13·35 | 100·4 |
| Climb (30 min) | 98 | 12·30 | 98·2 |
| Cruise | 95 | 10·90 | 95·1 |

The effect of ambient temperature on the take-off rating is important both to the manufacturer and the user. With the engine running at its maximum mechanical speed an *increase* in ambient temperature will cause a decrease in $N/\sqrt{T_a}$ and hence $N/\sqrt{T_{01}}$. This will cause the operating point on the compressor characteristic to move along the equilibrium running line to lower values of both $m\sqrt{T_{01}}/p_{01}$ and $p_{02}/p_{01}$, and it can readily be seen to be equivalent to a decrease in mechanical speed. The actual mass flow entering the engine will be further reduced due to the increase in ambient temperature, because $m$ is given by $(m\sqrt{T_{01}}/p_{01})/(\sqrt{T_{01}}/p_{01})$. The well known result of all these effects is that an *increase in ambient temperature* results in an appreciable *loss of thrust* as indicated in Fig. 8.16. But the picture is not yet complete. $T_{03}/T_{01}$ also decreases with increase in ambient temperature when $N$ is held constant. The actual temperature $T_{03}$ is given by $(T_{03}/T_{01})T_{01}$, and the decrease in $(T_{03}/T_{01})$ will be more than offset by the increase in $T_{01}$. In general, for a fixed mechanical speed $T_{03}$ *will increase with increase in ambient temperature*, and the allowable turbine inlet temperature may be exceeded on

**FIG. 8.16 Thrust curves**

a hot day. To keep within the limit it is then necessary to reduce the mechanical speed, giving an even greater reduction in $N/\sqrt{T_{01}}$ and hence thrust.

Turning our attention next to variation in ambient pressure, Fig. 8.16 indicates that the thrust will change in direct proportion to the ambient pressure; no change in the engine operating point is involved, but the mass flow is reduced as the ambient pressure is reduced. Both the pressure and temperature decrease with increasing altitude, the latter levelling off at 11 000 m. Because of the dependence of the thrust on the first power of the pressure, the decrease in thrust due to the decrease in $p_a$ more than outweighs the increase due to the reduction in $T_a$. Thus the *thrust* of an engine *decreases with increase in altitude*. Airports at high altitudes in the tropics are often critical with regard to take-off performance, and it may be necessary for aircraft using them to accept significantly reduced payloads. Mexico City is a well known example.

## Variation of fuel consumption and s.f.c. with rotational speed, forward speed and altitude

The fuel consumption of a jet engine together with the fuel capacity of the aircraft determine the range, and the specific fuel consumption (fuel flow per unit thrust for a jet engine) is a convenient indication of the economy of the unit. From the arguments of the previous section it is apparent that both fuel consumption and specific fuel consumption can be evaluated as functions of $N/\sqrt{T_{01}}$ (or $N/\sqrt{T_a}$) and $M_a$.

When a value of the combustion efficiency is assumed, the fuel consumption can readily be determined from the air flow, the combustion temperature rise and the fuel/air ratio curves of Fig. 2.15, in the manner described for the other types of gas turbine engine. To do this, values $p_a$ and $T_a$ have to be assumed to obtain $m$, $T_{03}$ and $T_{02}$ from the dimensionless parameters. The fuel flow is therefore a function of $N/\sqrt{T_a}$, $M_a$, $p_a$ and $T_a$. The dependence of the fuel flow on ambient conditions can in fact be virtually eliminated by plotting the results in terms of the non-dimensional fuel flow ($m_f Q_{net,p}/D^2 p_a \sqrt{T_a}$). For a given fuel and engine the calorific value and linear dimension can be dropped, yielding in practice $m_f p_a \sqrt{T_a}$. Figure 8.17(a) shows a typical set of fuel consumption curves. Unlike $F/p_a$, the fuel parameter depends only slightly on $M_a$ because it is due simply to the variation of compressor inlet conditions and not also to momentum drag. Indeed, if $m_f/p_{01}\sqrt{T_{01}}$ is plotted against $N/\sqrt{T_{01}}$ as in Fig. 8.17(b), the curves merge into a single line for the region where the nozzle is choking. It must be mentioned that although the combustion efficiency is high and constant over most of the working range, it can fall drastically at very high altitudes due to low combustion chamber pressures. Such curves as those of Fig. 8.17 may well underestimate the fuel consumption when $p_a$ is low.

Curves of 'non-dimensional' s.f.c. may be obtained by combining the data of Figs. 8.16 and 8.17(a). Figure 8.18 shows s.f.c./$\sqrt{T_a}$ plotted against $N/\sqrt{T_{01}}$

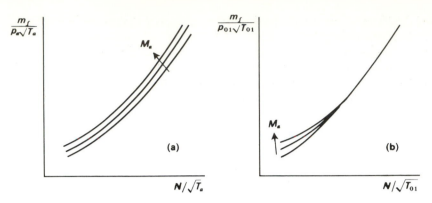

**FIG. 8.17 Fuel consumption curves**

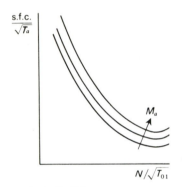

**FIG. 8.18 S.f.c. curves**

for various values of $M_a$. It is apparent that the s.f.c. will improve with increase in altitude due to the decrease in $T_a$. As the s.f.c. is a function only of $\sqrt{T_a}$ and not of $p_a$, however, the effect of altitude is not so marked as in the case of the thrust. (The actual variation in s.f.c. with altitude and Mach number for a simple turbojet, operating at its maximum rotational speed, was shown in Fig. 3.13.) It should be noted that s.f.c. increases with Mach number. Thrust initially drops with increasing Mach number and then rises when the ram effect overcomes the increase of momentum drag; the fuel flow, however, will steadily increase with Mach number because of the increased air flow due to the higher stagnation pressure at entry to the compressor and this effect predominates.

## 8.6    Methods of displacing the equilibrium running line

It was stated earlier that if the equilibrium running line intersects the surge line it will not be possible to bring the engine up to full power without taking some remedial action. As will be shown in the next chapter, even when clear of the surge line, if the running line approaches it too closely the compressor may surge when the engine is accelerated rapidly.

With most modern compressors, surge is likely to be encountered at low values of $N/\sqrt{T_{01}}$ and is less of a problem at high rotational speeds. Many high performance axial compressors exhibit a kink in the surge line as shown in Fig. 8.19. A running line, intersecting the surge line at low speeds and at

FIG. 8.19 Effect of blow-off and increased nozzle area

the kink, is also shown. To overcome the problem it is necessary to lower the running line locally in dangerous regions of operation.

One common method of achieving this is *blow-off*, where air is bled from some intermediate stage of the compressor. Blow-off clearly involves a waste of turbine work, so that the blow-off valve must be designed to operate only over the essential part of the running range. Furthermore, it may be difficult to discharge the air when space around the engine is at a premium as in a nacelle. As an alternative to blow-off a *variable area propelling nozzle* could be used for a jet engine. It will be shown in the next chapter that variable area propelling nozzles serve other useful purposes also.

It can be deduced that either of these methods will produce a reduction in pressure ratio at a given compressor speed and hence will lower the running line. This is most easily demonstrated if we consider operation at the high speed end of the running range, where the constant $N/\sqrt{T_{01}}$ lines on the compressor characteristic are almost vertical (i.e. $m\sqrt{T_{01}}/p_{01} \simeq$ constant) and where both the nozzle and turbine are choked.

If we consider the use of a variable area nozzle first, Fig. 8.20 shows that increasing the nozzle area will cause an increase in turbine pressure ratio and hence non-dimensional temperature drop $\Delta T_{034}/T_{03}$. From equation (8.2),

FIG. 8.20 Effect of variable-area propelling nozzle

satisfying compatibility of flow,

$$\frac{p_{02}}{p_{01}} = \frac{m\sqrt{T_{01}}}{p_{01}} \times \frac{p_{02}}{p_{03}} \times \sqrt{\frac{T_{03}}{T_{01}}} \times \frac{p_{03}}{m\sqrt{T_{03}}} = K_1 \sqrt{\frac{T_{03}}{T_{01}}} \tag{8.23}$$

where $K_1$ is a constant because under the assumed conditions the values of $m\sqrt{T_{01}}/p_{01}$, $p_{02}/p_{03}$ and $m\sqrt{T_{03}}/p_{03}$ are all fixed. From equation (8.6), which satisfies compatibility of work,

$$\frac{T_{03}}{T_{01}} = \frac{\Delta T_{012}}{T_{01}} \times \frac{T_{03}}{\Delta T_{034}} \times \frac{c_{pa}}{c_{pg}\eta_m}$$

Now, at a fixed value of compressor speed $N$ the compressor temperature rise is approximately constant; Fig. 8.21 shows a typical compressor characteristic replotted on a basis of temperature rise vs. pressure ratio. The wide range of pressure ratio obtained at almost constant temperature rise is due to the significant variation of efficiency when operating at conditions away from the design point. Thus if we assume $\Delta T_{012}/T_{01}$ is constant,

$$\frac{T_{03}}{T_{01}} = \frac{K_2}{(\Delta T_{034}/T_{03})} \tag{8.24}$$

where $K_2$ is another constant. Combining equations (8.23) and (8.24) gives

$$\frac{p_{02}}{p_{01}} = \frac{K_3}{\sqrt{(\Delta T_{034}/T_{03})}} \tag{8.25}$$

where $K_3$ is a constant, obtained from $K_1$ and $K_2$.

If the engine speed $N/\sqrt{T_{01}}$ is held constant and the nozzle area is increased, $\Delta T_{034}/T_{03}$ will be increased and equation (8.25) shows that $p_{02}/p_{01}$ will be decreased, hence the running line will be moved away from the surge line. In order to maintain the speed at the required level, it is necessary to reduce the fuel flow; opening the nozzle without reducing the fuel flow would cause the engine to accelerate to a higher speed, as discussed in Chapter 9. It should

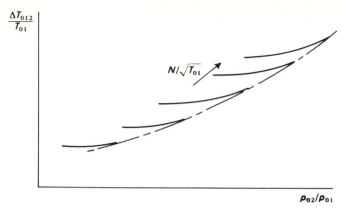

**FIG. 8.21  Compressor characteristics**

also be realized that decreasing the nozzle area will move the operating line towards the surge line; reasons for doing this will be discussed in the next chapter.

When considering the effect of bleed, $m_1$ is no longer equal to $m_3$ and equation (8.23) is modified to

$$\frac{p_{02}}{p_{01}} = K_1 \sqrt{\frac{T_{03}}{T_{01}} \times \frac{m_3}{m_1}} \qquad (8.26)$$

With the nozzle, or power turbine in the case of a shaft power unit, remaining choked, the turbine operating point remains unchanged and equation (8.6) becomes

$$\frac{T_{03}}{T_{01}} = \frac{m_1}{m_3} \times \frac{\Delta T_{012}}{T_{01}} \times \frac{T_{03}}{\Delta T_{034}} \times \frac{c_{pa}}{c_{pg}\eta_m} = K_4 \frac{m_1}{m_3} \qquad (8.27)$$

where $K_4$ is a constant, assuming as before that $\Delta T_{012}/T_{01}$ is constant. Combining equations (8.26) and (8.27)

$$\frac{p_{02}}{p_{01}} = K_5 \sqrt{\frac{m_3}{m_1}} \qquad (8.28)$$

where $K_5$ is a constant.

With the use of bleed, $m_3$ is always less than $m_1$, and the result will once again be to reduce the pressure ratio and lower the operating line. Equation (8.27) also shows that the use of bleed will increase the turbine inlet temperature. The reduction in turbine mass flow necessitates an increase in temperature drop to provide the compressor work, and with the non-dimensional temperature drop fixed this implies an increase in turbine inlet temperature. Both the effects of bleed and increase in nozzle area can be interpreted physically in terms of a decrease in the restriction to the flow, permitting the compressor to operate at a lower pressure ratio for any given rotational speed.

## 8.7   Incorporation of variable pressure losses

The size of the ducting for components such as the intake and exhaust of a gas turbine will be determined primarily by the requirement for low pressure losses at maximum power, which will normally correspond to maximum mass flow. The designer must compromise between low pressure losses and small duct sizes, and the choice will be largely determined by the application envisaged. It can be shown by cycle calculations that the effect of a given pressure loss is dependent on the pressure level at which it occurs, and it is essential to keep the terms $\Delta p_0/p_0$ at all stations as low as possible. With pressure losses in the intake and exhaust of stationary gas turbines occurring at essentially atmospheric pressure, it will be realized that these losses are much more critical than pressure losses in the combustion chamber or the air side of a heat exchanger where the air is entering at the compressor delivery pressure. In a typical installation a pressure loss of 2·5 cm $H_2O$ in the intake may cause about 1 per cent reduction in power output.

Having decided on the size of ducting required to give the desired pressure loss at the design point, the variation of loss with changing operating conditions can be predicted for incorporation in the off-design calculation procedures outlined in this chapter. The velocities in such components as the intake and exhaust ducts will be sufficiently low for the flow to be treated as incompressible, so that $\Delta p_0$ will be proportional to the inlet dynamic head. Expressed in non-dimensional terms this implies, as shown by equation (6.2), that

$$\frac{\Delta p_0}{p_0} \propto \left( \frac{m\sqrt{T_0}}{p_0} \right)^2$$

Assuming the appropriate values at the design condition to be $(\Delta p_0/p_0)_D$ and $(m\sqrt{T_0}/p_0)_D$, then at any off-design condition

$$\frac{\Delta p_0}{p_0} = \left( \frac{\Delta p_0}{p_0} \right)_D \left[ \left( \frac{m\sqrt{T_0}}{p_0} \right) \Big/ \left( \frac{m\sqrt{T_0}}{p_0} \right)_D \right]^2 \qquad (8.29)$$

Values of $m\sqrt{T_0}/p_0$ for each component are obtained in the course of the off-design calculations. The pressure loss from equation (8.29) can be incorporated in the same manner as was the combustion pressure loss $\Delta p_b/p_{02}$, namely by making use of such equations as

$$\frac{p_{03}}{p_{02}} = 1 - \frac{\Delta p_b}{p_{02}}$$

Now, however, the pressure ratio across the component is not constant but changes with $m\sqrt{T_0}/p_0$.

In the particular case of the combustion chamber, the more accurate expression of equation (6.1) can be used. This includes the fundamental loss due to heat addition. Combining (6.1) and (6.2) we would have

$$\frac{\Delta p_b}{p_{02}} \propto \left(\frac{m\sqrt{T_{02}}}{p_{02}}\right)^2 \left[K_1 + K_2\left(\frac{T_{03}}{T_{02}} - 1\right)\right]$$

Such refinement would not normally be worthwhile because the fundamental loss is a small proportion of the total.

# 9

## Prediction of performance—further topics

Evaluation of the off-design performance of complex plant incorporating intercooling, heat-exchange and reheat is inevitably more complicated than for the simple gas turbine, although the basic principles described in Chapter 8 are still applicable. The components of such gas turbines can be arranged in a wide variety of ways, about forty different layouts being possible even with no more than two compressor rotors. A comprehensive survey of the part-load performance and operating stability of many of these possibilities can be found in Ref. (1), wherein use was made of stylized component characteristics. It is as a result of such calculations that it was possible to dismiss certain arrangements as impractical because the running line runs into the surge line under some operating conditions. No attempt will be made to repeat this kind of survey here, and we shall restrict our attention mainly to discussing the prediction of off-design performance in the practical cases of (*a*) high pressure ratio twin-spool engines and (*b*) turbofans. Before discussing the matching procedures for twin-spool engines and turbofans, however, methods of improving the part-load performance of gas turbines will be considered briefly.

The chapter ends with an introduction to transient performance, including a brief description of methods for predicting the acceleration or deceleration of rotor systems. Acceleration rates of gas turbine rotors are obviously dependent on mechanical considerations such as the polar moment of inertia and the maximum temperature which the turbine blades can withstand for short periods, but the limiting factor is usually the proximity of the equilibrium running line to the surge line. Thus a thorough understanding of off-design performance is essential before transient behaviour can be investigated and a suitable control system can be designed.

## 9.1 Methods of improving part-load performance

It was pointed out in the Introduction that the part-load performance of gas turbines intended for vehicular or naval use was of great importance because of the considerable portion of the running time spent at low power. Early studies for both applications resulted in the consideration of complex arrangements incorporating intercooling, heat-exchange and reheat. The sole justification for the marked increase in complexity was the great improvement in part-load specific fuel consumption indicated in Fig. 9.1. Further details regarding the choice of cycle parameters, mechanical layout and development problems can be found in Refs. (2) and (3). Such complex arrangements did not prove successful for either of these applications despite their undoubted thermodynamic merit, the main reason being the mechanical complexity involved. It is interesting to note, nevertheless, that the Ford engine described in Ref. (3) was an extremely compact unit of about 225 kW. As a result of the problems with the complex cycle the navies of the world have concentrated their attention on simple gas turbines, using such arrangements as CODOG, COSAG and COGOG to overcome the part-load problems as discussed in the Introduction. Virtually all proponents of the vehicular gas turbine have now reverted to a low pressure ratio unit with a heat-exchanger: Ref. (4) discusses performance of vehicular gas turbines.

FIG. 9.1  **Part load s.f.c. for simple and complex cycles**

In practice it is found that the vast majority of applications make use of either the high pressure ratio simple cycle or the low pressure ratio heat-exchange cycle. Detailed off-design performance calculations for gas turbines with heat-exchangers would have to account for variation in heat-exchanger effectiveness with engine operating conditions. Heat-exchanger effectiveness depends on the fixed parameters of heat transfer area and configuration (e.g. counter-flow, cross-flow or parallel-flow), and the parameters varying with engine operating conditions such as overall heat transfer coefficient between the two fluids and their thermal capacities ($mc_p$). Methods for estimating the effectiveness of gas turbine heat-exchangers (and intercoolers) are described in Ref. (5) and will not be dealt with here.

The incorporation of a heat-exchanger will cause an increase in pressure loss between compressor delivery and turbine inlet, and also an increase in turbine outlet pressure. Although the additional pressure losses will result in a reduction in power, they will have little effect on the equilibrium running line and the part load behaviour of engines with and without heat-exchanger will be similar. A typical running line for a free turbine engine was shown in Fig. 8.7; the variation of $T_{03}/T_{01}$ with $N/\sqrt{T_{01}}$ can be deduced from this but is more clearly indicated in Fig. 9.2. It was shown in Chapter 2 that the

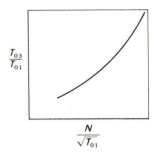

**FIG. 9.2**

thermal efficiency of a real gas turbine cycle was dependent on the turbine inlet temperature, and the rapid drop in $T_{03}$ with decreasing power is the basic cause of the poor part-load performance of the gas turbine. The variation of thermal efficiency with power output for a hypothetical engine with and without heat-exchanger is shown in Fig. 9.3. It can be seen that although the design point thermal efficiency is significantly improved by incorporating

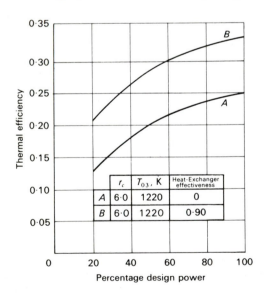

**FIG. 9.3 Variation of thermal efficiency with power output**

a heat-exchanger, the shape of the efficiency curve is not basically altered.†
This is fundamentally due to the fact that in each case there is a similar drop
in turbine inlet temperature as power is reduced.

To improve the part-load efficiency of gas turbines, therefore, some means
must be found of raising the turbine inlet temperature at low powers. In the
majority of applications where good part-load economy is required, e.g.
vehicular and marine, a free turbine would be used. If we restrict our attention
to the free turbine engine, two alternative methods of increasing the part-load
temperature are available. These are, firstly, the use of variable area power
turbine stators and, secondly, a system by which power can be transferred
between the two independent turbine shafts.

## Variable area power turbine stators

Variation of area is accomplished by rotating the nozzle blades, and this
permits the effective throat area to be reduced or increased as shown in
Fig. 9.4. In Chapter 8 it was emphasized that free turbine engines and turbojets were thermodynamically similar because the flow characteristics of a
free turbine and a propelling nozzle impose the same operating restrictions
on the gas generator; thus variable power turbine stators will have the same
effect as a variable area propelling nozzle. Section 8.6 showed that increasing
the propelling nozzle area of a turbojet moved the running line away from
the surge line, while decreasing the area displaced the running line towards
surge. Reference to Fig. 8.7 shows that the latter will cause an increase in
turbine inlet temperature at low powers; it is also likely that the compressor
efficiency will be improved as the surge line is approached. Both of these
effects will improve the part-load s.f.c.

Ideally, the area variation of the power turbine stators can be controlled

**FIG. 9.4 Variable geometry power turbine stators**

† It is possible that for certain values of the cycle parameters the incorporation of a heat-
exchanger will alter the shape of the efficiency curve slightly, but it is unlikely that addition of
a heat-exchanger alone will give any significant improvement in part-load economy. Curve *B*
in Fig. 9.3 was obtained on the assumption of a constant heat-exchanger effectiveness but estimates have shown that the effectiveness does not increase much with reduction in engine power
output.

so that the turbine inlet temperature is maintained at its maximum value as power is reduced as indicated in Fig. 9.5. If the running line at maximum temperature moves to intersect the surge line as shown, it then becomes necessary to reopen the power turbine nozzles for this part of the running range. It should be noted that operation at constant gas generator turbine inlet temperature with reducing power will cause an increase in temperature at entry to the power turbine, because of the reduced compressor power, and the temperature of the hot gases entering the heat-exchanger will also be raised. Temperature limitations in either of these components may restrict operation in this mode. The use of a variable geometry power turbine is particularly advantageous when combined with a heat-exchanger, because the increased turbine outlet temperature is utilized.

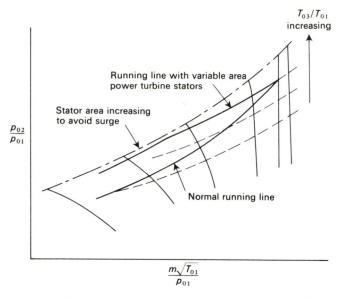

**FIG. 9.5 Effect of variable power turbine stators on running line**

The efficiency of the power turbine will obviously be affected by the position of the variable stators, but with careful design the drop in turbine efficiency can be more than offset by maintaining a higher turbine inlet temperature at part load. Area variations of $\pm 20$ per cent can be obtained with acceptable losses in turbine efficiency, see Ref. (6). The variable stators are liable to be exposed to temperatures in the 1000–1100 K region, but the development problems associated with this do not appear to have been too difficult. A facility for increasing the stator area is also advantageous with respect to starting and accelerating the gas generator. If the stators are rotated still further, the gas generator flow can be directed against the direction of rotation so that the flow impinges on the back of the power turbine blades. This can result in a substantial degree of engine braking which is extremely important for heavy vehicles.

## Power transfer

The concept of transferring controlled amounts of power between the gas generator and power turbine shafts has been patented by General Motors under the name 'Power Transfer'. The primary function of this scheme is to improve the part-load s.f.c. of the engine, but it also provides several other important features such as engine braking: details are given in Ref. (7).

The system is hydromechanical, incorporating a hydraulically actuated clutch. One set of plates of this clutch is driven through suitable gearing from the gas generator rotor and the other set is geared to the power turbine rotor. The clutch torque can be controlled by varying the hydraulic pressure, and the gear ratios are chosen so that under all normal running conditions the plates driven by the power turbine rotate more slowly than those driven by the gas generator.

When a slight hydraulic pressure is applied to the clutch cylinder a torque is exerted between the clutch elements rotating at different speeds. The torque acting on the faster turning plates driven by the gas generator tend to slow this component down; simultaneously, a governor is attempting to hold the speed of the gas generator at the preset level causing the fuel flow to increase and so maintaining the turbine inlet temperature at a higher level. By suitable variation of clutch torque the turbine temperature can be held at its maximum value as the power is reduced, subject to surge limitations, so giving a substantial improvement in part-load economy.

With this arrangement it is also possible to transfer power in the opposite direction. The output shaft can be locked to the gas generator enabling the compressor to absorb power and so provide a large amount of engine braking. Over-speed protection for the power turbine can also be provided by locking the output shaft to the gas generator when a limiting speed is exceeded. In other words, the engine can be operated in the single-shaft mode if required (see under heading 'Single-shaft versus free turbine engines', section 9.5).

'Power Transfer' is an extremely attractive concept which offers substantial gains in part-load economy combined with significant mechanical advantages. The turbine design is not compromised as in the case of the variable geometry power turbine, although a small amount of power is lost in the clutch, and the complexity of the hydraulic clutch arrangement is little more than that of a system for rotating the power turbine stators.

## 9.2  Matching procedures for twin-spool engines

The basic methods used for twin-spool matching are similar to those described in Chapter 8. They differ only from those for single-spool engines in that it is also necessary to satisfy compatibility of flow between the spools. This compatibility requirement gives rise to the phenomenon of *aerodynamic coupling* which determines the ratio of the rotor speeds even though the rotors are mechanically independent of each other. The station numbering used for twin-spool turbojets is shown in Fig. 9.6 and the rotational speeds will be referred to as $N_L$ and $N_H$ for the LP and HP rotors respectively. The corresponding shaft power unit would have a power turbine between 6 and 7 in place of the propelling nozzle.

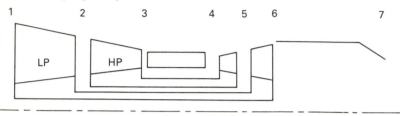

**FIG. 9.6 Station numbering for twin-spool turbojet**

In outlining the procedures we will assume single line turbine characteristics, constant turbine efficiencies and constant percentage combustion pressure loss; these are often close approximations in practice and their use makes it easier to understand what is happening in physical terms. Considering work compatibility, the equations for the LP and HP rotors are

$$\eta_{mL} m c_{pg} \Delta T_{056} = m c_{pa} \Delta T_{012}$$

and

$$\eta_{mH} m c_{pg} \Delta T_{045} = m c_{pa} \Delta T_{023}$$

Flow compatibility must be satisfied between the compressors, the turbines and finally between the LP turbine and nozzle (or power turbine). The problem of flow compatibility between two turbines was dealt with under heading 'Matching of two turbines in series', section 8.4, and Fig. 9.7 illustrates the further restriction in operating conditions resulting from the introduction of a nozzle (or power turbine).

**FIG. 9.7 Flow compatibility for twin-spool engine**

Calculations for a twin-spool engine can be lengthy but they may be considerably simplified once it is appreciated that the HP rotor of a twin-spool engine is equivalent to a single-spool turbojet with a fixed nozzle the area of which is defined by the throat area of the LP turbine stators. It was shown in Chapter 8 that for a single-spool turbojet a unique running line is defined when the nozzle is choked, and it follows that a similarly unique running line will be defined on the HP compressor characteristic of a twin-spool engine *when the LP turbine stators are choked.* (It should be noted that in practice the LP turbine of a twin-spool unit will be choked over most of the useful running range, but may become unchoked at idle conditions.) The position of the unique running line *on the HP characteristic* can be determined by considering the HP rotor alone, for which compatibility of flow and work yields

$$\frac{m\sqrt{T_{04}}}{p_{04}} = \frac{m\sqrt{T_{02}}}{p_{02}} \times \frac{p_{02}}{p_{03}} \times \frac{p_{03}}{p_{04}} \times \sqrt{\frac{T_{04}}{T_{02}}} \qquad (9.1)$$

$$\frac{\Delta T_{045}}{T_{04}} = \frac{\Delta T_{023}}{T_{02}} \times \frac{T_{02}}{T_{04}} \times \frac{c_{pa}}{c_{pg}\eta_{mH}} \qquad (9.2)$$

It can be seen that equations (9.1) and (9.2) are the same as equations (8.2) and (8.6) apart from the difference in station numbering, and they can be solved by the trial and error procedure described in section 8.3 to yield a value of $T_{04}/T_{02}$ for each point on a constant $N_H/\sqrt{T_{02}}$ line on the HP compressor characteristic. For the particular case when the LP turbine is choked, the HP turbine will be restricted to operation at a fixed non-dimensional point with the values of $p_{04}/p_{05}$, $m\sqrt{T_{04}}/p_{04}$ and $\Delta T_{045}/T_{04}$ all fixed. Substituting the relevant values of $m\sqrt{T_{04}}/p_{04}$ and $\Delta T_{045}/T_{04}$ in equations (9.1) and (9.2), the running line defined by LP turbine stator choking can be established on the HP characteristic, and it will be similar to that shown in Fig. 8.7. *The importance of establishing the LP choking line is that it greatly facilitates satisfying the requirement for compatibility of flow between the two compressors* as will be apparent in what follows.

Although much of the procedure to be described in this section is equally applicable to turbojets and shaft power units with a separate power turbine, it is most easily explained for the case of a turbojet with a fully variable propelling nozzle. Any effect resulting from variation of nozzle area will directly influence the LP turbine, and hence the LP compressor; the HP turbine, however, is separated from the nozzle by the LP turbine and, if the LP turbine is choked, the HP rotor will be shielded from disturbances caused by the variable nozzle. The use of a fully variable nozzle permits operation over a wide area of the LP compressor characteristic as in the case of a single-spool engine, even though the operating region on the HP characteristic is a single line determined by choking of the LP turbine. Thus if we assume a fully variable nozzle we can start with any point on the LP characteristic, and the final step in the calculations will be the determination of the required nozzle area. No iteration is required in this procedure, which can be summarized as follows.

(a) Determine the inlet conditions $T_{01}$ and $p_{01}$ from the ambient and flight conditions; for a land based unit $T_{01} = T_a$ and $p_{01} = p_a$.

(b)   Select any point on a line of constant $N_L/T_{01}$ on the LP compressor characteristic, which will specify values of $m\sqrt{T_{01}}/p_{01}$, $p_{02}/p_{01}$ and $\eta_{cL}$. The compressor temperature rise $\Delta T_{012}$ and temperature ratio $T_{02}/T_{01}$ can then be calculated using equation (8.4).

(c)   The non-dimensional flow at LP compressor exit, $m\sqrt{T_{02}}/p_{02}$, is obtained from the identity

$$\frac{m\sqrt{T_{02}}}{p_{02}} = \frac{m\sqrt{T_{01}}}{p_{01}} \times \frac{p_{01}}{p_{02}} \times \sqrt{\frac{T_{02}}{T_{01}}}$$

and for compatibility of flow between the compressors this is the non-dimensional flow at entry to the HP compressor.

(d)   With the operating line on the HP compressor characteristic determined by LP turbine choking, the known value of $m\sqrt{T_{02}}/p_{02}$ defines the operating point on the HP compressor characteristic. This gives the values of $p_{03}/p_{02}$, $N_H/\sqrt{T_{02}}$ and $\eta_{cH}$, and the values of $\Delta T_{023}$ and $T_{03}/T_{02}$ can be found.

(e)   The overall compressor pressure ratio is now given by

$$\frac{p_{03}}{p_{01}} = \frac{p_{02}}{p_{01}} \times \frac{p_{03}}{p_{02}}$$

(f)   The turbine inlet pressure $p_{04}$ can be found from

$$p_{04} = \frac{p_{03}}{p_{01}} \times \frac{p_{04}}{p_{03}} \times p_{01}$$

where $p_{04}/p_{03}$ is obtained from the combustion pressure loss.

(g)   The value of $m\sqrt{T_{04}}/p_{04}$ is known, having been determined by flow compatibility between the two turbines, and $m$ can be found from $m\sqrt{T_{01}}/p_{01}$, $T_{01}$ and $p_{01}$.

(h)   The turbine inlet temperature $T_{04}$ can now be found from

$$T_{04} = \left( \frac{m\sqrt{T_{04}}}{p_{04}} \times \frac{p_{04}}{m} \right)^2$$

(i)   The HP turbine is operating at a fixed non-dimensional point with $p_{04}/p_{05}$ and $\Delta T_{045}/T_{04}$ determined by LP turbine choking. Thus conditions at entry to the LP turbine, $p_{05}$ and $T_{05}$, can be obtained from

$$p_{05} = p_{04} \times \frac{p_{05}}{p_{04}}$$

and $T_{05} = T_{04} - \Delta T_{045}$

(j)   We must now satisfy the work requirement for the LP compressor. This is given by

$$\eta_{mL} m c_{pg} \Delta T_{056} = m c_{pa} \Delta T_{012}$$

$\Delta T_{012}$ is known, so $\Delta T_{056}$ can be calculated.

(k)  With $\Delta T_{056}$ and $T_{05}$ known, the temperature at exit from the LP turbine and entry to the nozzle is given by

$$T_{06} = T_{05} - \Delta T_{056}$$

(l)  The LP turbine pressure ratio $p_{05}/p_{06}$ is found from

$$\Delta T_{056} = \eta_{tL} T_{05}\left[ 1 - \left(\frac{1}{p_{05}/p_{06}}\right)^{(\gamma-1)/\gamma}\right]$$

Once $p_{05}/p_{06}$ is known $p_{06}$ is given by

$$p_{06} = p_{05} \times \frac{p_{06}}{p_{05}}$$

(m)  $m$, $T_{06}$ and $p_{06}$ are now established. The overall nozzle pressure ratio $p_{06}/p_a$ is then known, and the calculation of the nozzle area given the inlet conditions, pressure ratio and mass flow is quite straightforward (see Chapter 3). Once the required nozzle area has been calculated, the matching procedure is complete and an equilibrium running point has been obtained for that particular nozzle area and initial value of $N_L/\sqrt{T_{01}}$.

(n)  The procedure can be repeated for points on other $N_L/\sqrt{T_{01}}$ lines. All the data required for a complete performance calculation, e.g. of the thrust, fuel flow and s.f.c. at each running point, are now available.

Note that if a fixed nozzle were used, the area calculated in step (m) would not in general be equal to the specified area; it would then be necessary to return to step (b), try another point on the same $N_L/\sqrt{T_{01}}$ line on the LP characteristic and iterate until the correct nozzle area was obtained.

The effect of the LP turbine operating on the unchoked part of its characteristic is to raise the running line on the HP characteristic (i.e. displace it towards surge), because of the reduced non-dimensional flow at exit from the HP turbine. The calculation described above must be modified somewhat to deal with an unchoked LP turbine, but this condition is only liable to occur at low powers and the modification will not be dealt with here.

Finally, if a shaft power unit were being considered, step (m) yields $m\sqrt{T_{06}}/p_{06}$ at inlet to the power turbine and also the power turbine pressure ratio $p_{06}/p_a$. If this value of $m\sqrt{T_{06}}/p_{06}$ does not agree with the value from the power turbine characteristic for the known pressure ratio (see Fig. 9.7), again it is necessary to return to step (b) and repeat the calculation.

## 9.3  Some notes on the behaviour of twin-spool engines

Complete off-design performance calculations for twin-spool engines are obviously time consuming and in practice would be carried out using digital computers. Some aspects of twin-spool behaviour, however, can be deduced from an understanding of the matching procedure and the more important of these will be briefly described.

## Aerodynamic coupling of rotor speeds

From step (*c*) of the matching procedure described in the previous section it can be seen that once an operating point on the LP characteristic has been chosen, the corresponding operating point on the HP characteristic is fixed. Thus for a fixed value of $N_L/\sqrt{T_{01}}$ and nozzle area, the value of $N_H/\sqrt{T_{02}}$ is determined by flow compatibility between the compressors. The HP compressor non-dimensional speed can be expressed in terms of the temperature at engine inlet by

$$\frac{N_H}{\sqrt{T_{01}}} = \frac{N_H}{\sqrt{T_{02}}} \times \sqrt{\frac{T_{02}}{T_{01}}}$$

$N_L/\sqrt{T_{01}}$ and $N_H/\sqrt{T_{01}}$ are then directly proportional† to the actual mechanical speeds of the two rotors.

A typical variation of $N_L/\sqrt{T_{01}}$ with $N_H/\sqrt{T_{01}}$ for a turbojet is shown in Fig. 9.8; note that the relation between the speeds is dependent on nozzle

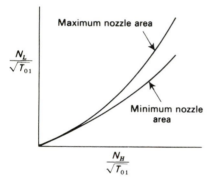

**FIG. 9.8  Speed relationship for twin-spool engine**

area, and only for a fixed nozzle will it be unique. At a fixed value of $N_H$ any increase in nozzle area will cause an increase in $N_L$, the physical reason being that opening the nozzle will increase the pressure ratio across the LP turbine so causing an increase in LP rotor torque.

## Effect of variable area propelling nozzle

We have seen that the use of a fully variable nozzle permits operation over a wide range of the LP compressor characteristic, although the HP running line is not affected by nozzle variation as long as the LP turbine is choked. Typical running lines are shown in Fig. 9.9. It is important to note that increasing the nozzle area moves the LP running line *towards* surge, which is the opposite effect to that obtained on a single-spool engine. The reason for this is the increase in LP turbine power resulting from the redistribution

---

† If the 'equivalent speed' $N/\sqrt{\theta_0}$ were used instead of $N/\sqrt{T_0}$ (see section 4.6) the numbers will be equal to the mechanical speeds under reference entry conditions.

of pressure ratio between the LP turbine and the nozzle. One of the major advantages of the variable nozzle is that it permits selection of rotor speed and turbine inlet temperature independently; or, in terms more relevant to an aircraft, of air flow and thrust. This is a valuable feature for an engine that has to operate over a wide range of intake temperature which produces significant changes in non-dimensional speed and mass flow. Reference (8) gives an excellent picture of the problems involved in matching the engine and intake for a supersonic transport.

(a)

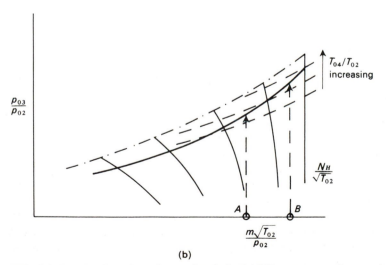

(b)

**FIG. 9.9  Running lines for twin-spool turbojet (a) LP compressor characteristic, (b) HP compressor characteristics**

It was shown in Chapter 8 that lines of constant $T_{03}/T_{01}$ can be drawn on the compressor characteristic of a single-spool engine and these represent possible equilibrium running lines when a fully variable nozzle is used. As can be seen from Fig. 8.7, the value of $T_{03}/T_{01}$ increases as lines of constant temperature ratio move towards surge. For a twin-spool engine lines of constant $T_{04}/T_{02}$ can be drawn on the HP compressor characteristic and they behave similarly. If we consider the LP compressor characteristic, however, and plot lines of constant $T_{04}/T_{01}$ as shown in Fig. 9.9(a), it will be seen that they move *away* from surge as the value of $T_{04}/T_{01}$ is increased. Thus again we have behaviour opposite to that found for the single-spool turbojet. This can be explained if we consider operation at a constant value of $N_L/\sqrt{T_{01}}$. With the nozzle fully open the LP compressor operates at point $A$ on Fig. 9.9(a) and closing the nozzle moves the operating point to $B$. The value of $m\sqrt{T_{01}}/p_{01}$ is constant for the vertical $N_L/\sqrt{T_{01}}$ line shown and the temperature rise across the LP compressor will be approximately fixed by the rotational speed. Thus if the LP compressor pressure ratio is decreased from $A$ to $B$ the value of $m\sqrt{T_{02}}/p_{02}$ will increase because

$$\frac{m\sqrt{T_{02}}}{p_{02}} = \frac{m\sqrt{T_{01}}}{p_{01}} \times \frac{p_{01}}{p_{02}} \times \sqrt{\frac{T_{02}}{T_{01}}}$$

The increase in $m\sqrt{T_{02}}/p_{02}$ results in an increase in HP compressor non-dimensional speed, pressure ratio and $T_{04}/T_{02}$ as shown in Fig. 9.9(b). With the value of $T_{02}$ approximately constant for a fixed rotational speed, it can be seen that $T_{04}$, and hence $T_{04}/T_{01}$, must increase as we move the LP operating point away from surge. It is also evident that a decrease in LP pressure ratio due to nozzle variation will be compensated by an increase in HP pressure ratio.

## Presentation of performance

In Chapter 8 it was shown that the performance of a single-spool engine could be presented as a function of $N/\sqrt{T_{01}}$. The performance of a twin-spool fixed geometry turbojet can be presented in terms of either the LP or HP rotor speeds because of the fixed relation between the speeds. For an engine with variable nozzle the performance would have to be presented in terms of any two of $N_L$, $N_H$ and $A_7$. An example of a 'carpet' plot showing the variation of thrust with $N_L$ and $N_H$ for a variable nozzle turbojet is shown in Fig. 9.10; other quantities such as fuel flow and s.f.c. could be presented in a similar manner.

For fixed geometry engines the variation of turbine inlet temperature and fuel flow could be plotted versus either $N_L$ or $N_H$, but because of the rapid variation of both with $N_H$ the latter may be more suitable as abscissa. It should be noted that although the HP rotor speed will always be higher than the LP rotor speed and the HP turbine blades are at the higher temperature, they may not be the critical components because the LP turbine blades are usually substantially longer.

The mass flow through a twin-spool engine is primarily determined by the LP compressor speed, and consequently the performance of a twin-spool shaft power unit with a free power turbine is conveniently presented in a manner similar to that of Fig. 8.10 with lines of constant LP compressor speed replacing lines of constant gas generator speed.

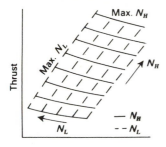

**FIG. 9.10**

## 9.4    Matching procedures for turbofan engines

The approach described in section 9.2 is also applicable to turbofans, but in this case we must take into account the division of flow between the by-pass duct and the gas generator, which will vary with off-design operating conditions. Only the simplest case, that of the twin-spool turbofan with separate exhausts shown in Fig. 3.14, will be considered here. The additional information required is the by-pass or 'cold' nozzle characteristic, as indicated in Fig. 9.11; the turbine characteristics and the hot stream nozzle characteristics will be similar to those of the twin-spool turbojet.

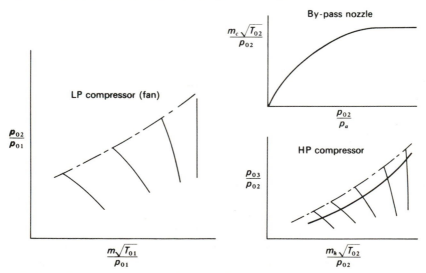

**FIG. 9.11  Characteristics required for turbofan matching**

Let us consider a turbofan with both nozzles fixed. The running line corresponding to LP turbine choking can be established on the HP characteristic in exactly the same manner as for the twin-spool turbojet. The problem now is to establish the proportion of the total flow ($m$) passing through the by-pass duct ($m_c$) and the gas generator ($m_h$). Suffixes $h$ and $c$ refer to 'hot' and 'cold' streams respectively. One possible procedure is as follows.

(a)  Select ambient and flight conditions, giving values of $p_{01}$ and $T_{01}$.
(b)  Select a value of $N_L/\sqrt{T_{01}}$ and guess any point on this line. The value of $m$ entering the fan is then known, and the value of $m\sqrt{T_{02}}/p_{02}$ at exit from the fan can be calculated.
(c)  From the known value of $p_{02}$, the pressure ratio across the by-pass nozzle, $p_{02}/p_a$, can be calculated and the value of $m_c\sqrt{T_{02}}/p_{02}$ can be found from the by-pass nozzle characteristic. The value of $m_c$ can also be found, and $m_h = m - m_c$.
(d)  The gas generator non-dimensional flow can then be found from

$$\frac{m_h\sqrt{T_{02}}}{p_{02}} = \frac{m\sqrt{T_{02}}}{p_{02}} - \frac{m_c\sqrt{T_{02}}}{p_{02}}$$

(f)  Knowing the gas generator non-dimensional flow we can enter the HP characteristic, and establish the overall pressure ratio and turbine inlet temperature as for the twin-spool turbojet [see steps (e) to (h) of section 9.2]. The HP turbine pressure ratio and temperature drop are known as a result of flow compatibility between the two turbines, so that the pressure and temperature at entry to the LP turbine, $p_{05}$ and $T_{05}$, can be determined as before.
(g)  The LP turbine temperature drop is given by

$$\eta_{mL}m_h c_{pg}\Delta T_{056} = mc_{pa}\Delta T_{012}$$

(h)  The LP turbine pressure ratio is then found from $T_{05}$, $\Delta T_{056}$ and $\eta_{tL}$. We now know $p_{06}$ and $T_{06}$.
(i)  Knowing $m_h$, $p_{06}$, $T_{06}$ and $p_{06}/p_a$ the hot stream nozzle area can be calculated; in general this will not agree with the specified value and it is necessary to return to step (b), select a new point on the LP characteristic, and repeat the process until agreement is reached.

The close similarity between the twin-spool and turbofan matching procedures can be seen. The method described may readily be extended to deal with three-spool engines, and the configuration shown in Fig. 3.18(c) is equivalent to a simple turbofan with a twin-spool gas generator. The use of mixed exhausts presents a further complication because it is then necessary to include equations satisfying conservations of energy and momentum for the mixing process. Neither of these topics will be pursued further here.

## 9.5   Transient behaviour of gas turbines

In certain applications the transient response of gas turbines following a demand for a change in output can be critical; in other applications good response may merely be desirable. Aircraft engines are an obvious example of an application where the transient behaviour is critical; the prime requirement for civil aircraft is for a rapid thrust response to cope with a baulked landing when the aircraft is close to touchdown but is forced to overshoot the runway. Lifting engines for VTOL aircraft provide another example, where an engine failure on a multi-engined installation may result in serious unbalanced forces which have to be adjusted rapidly. Rapid response is also required from gas turbines used for emergency generation of electricity which must often deliver their rated output within two minutes of the start signal. The response rate of the gas turbine itself from idle to maximum power must be less than about ten seconds to allow time for the starting and synchronization sequences. The transient response of a vehicle gas turbine to changes in throttle setting is not critical. Although the relatively slow response of a gas turbine has frequently been regarded as a major disadvantage, for heavy vehicles where it is likely to be competitive the vehicle acceleration will be largely independent of the engine acceleration.

Until recently, relatively little attention has been paid to the prediction of the transient behaviour of gas turbines, and the response rate of an engine was established empirically during development testing. Now, however, the transient behaviour is often predicted from a knowledge of the off-design performance. A detailed knowledge of the dynamic response at the design stage is becoming increasingly important for the design and development of control systems for advanced gas turbines, and it is common practice for manufacturers to be asked to guarantee the response rate of their products when negotiating a contract.

The acceleration of gas turbines is obviously dependent on such factors as the polar moment of inertia of the rotor system and the maximum temperature which the turbine blades can withstand for short periods. Usually the limiting factor on acceleration is the proximity of the surge line to the equilibrium running line, and this is particularly critical at the start of an acceleration from low powers. The configuration of the gas turbine will have a major effect on its transient behaviour; for example, the behaviour of single-shaft and free turbine engines delivering shaft power will be quite different, and a twin-spool engine will respond very differently from a single-spool engine. Only a brief introduction to the problem of engine transients will be given in this book, principally via the single-spool engine with and without free power turbine, and the interested reader must then turn to the specialized literature.

### Prediction of transient performance

It will by now be clear that all off-design equilibrium running calculations are based on satisfying the requirements for compatibility of flow and work between the components. During transient operation a gas turbine can be considered to satisfy compatibility of flow† but not of work, and the excess or deficiency of power applied to a rotor can be used to calculate its acceleration or deceleration. The problem then becomes one of calculating increments of net torque associated with increments of fuel flow and integrating to find the change of rotor speed.

The acceleration of the compressor rotor and the excess torque $\Delta G$ are related by Newton's Second Law of Motion, namely

$$\Delta G = J\dot{\omega}$$

where $J$ is the polar moment of inertia of the rotor and $\dot{\omega}$ is its angular acceleration. The excess torque $\Delta G$ is given by

$$\Delta G = G_t - G_c \text{ (free turbine engine)}$$

or

$$\Delta G = G_t - (G_c + G_l) \text{ (single-shaft engine)}$$

where suffixes $t$, $c$ and $l$ refer to turbine, compressor and load respectively. *It should be noted that the net torque is the difference between two quantities of similar magnitude, and a small change in either may result in a much larger change in the torque available for acceleration.* The problem now is that of obtaining the torques during transient operation. The turbine torque may be greater or less than the compressor torque, corresponding to acceleration or deceleration of the rotor.

To obtain the torques we may use the methods for off-design performance described in Chapter 8. If we consider *any* point on the compressor characteristic, which will not in general be an equilibrium running point, we can satisfy flow compatibility between compressor and turbine by using the identity

$$\frac{m\sqrt{T_{03}}}{p_{03}} = \frac{m\sqrt{T_{01}}}{p_{01}} \times \frac{p_{01}}{p_{02}} \times \frac{p_{02}}{p_{03}} \times \sqrt{\frac{T_{03}}{T_{01}}}$$

Consider first the turbine non-dimensional flow $m\sqrt{T_{03}}/p_{03}$: it can normally be assumed to be independent of speed and a function of the pressure ratio alone. The value of the gas generator turbine pressure ratio $p_{03}/p_{04}$ corresponding to the selected value of $p_{02}/p_{01}$ on the compressor characteristic can be easily obtained for both single-shaft and free turbine engines; for a single-shaft engine it is fixed by the compressor pressure ratio and the combustion pressure loss, and for a free turbine engine it can be found from

† During rapid transients, pressures cannot change instantaneously because of the finite volume between the components, and the assumption of flow compatibility at all times is not exactly true although it is a good approximation. Methods for dealing with non-instantaneous pressure changes are discussed in Ref. (11), and are subject to further research; the effects are confined to a very short period immediately following the start of a transient.

Fig. 8.9. Thus it can be seen that $m\sqrt{T_{03}}/p_{03}$ is a function of $p_{02}/p_{01}$ for both types of engine. Rearranging the equation for flow compatibility given above

$$\sqrt{\frac{T_{03}}{T_{01}}} = \frac{m\sqrt{T_{03}}}{p_{03}} \times \frac{p_{02}}{p_{01}} \times \frac{p_{03}}{p_{02}} \bigg/ \frac{m\sqrt{T_{01}}}{p_{01}}$$

With constant percentage combustion pressure loss $p_{03}/p_{02}$ will be constant, and it can be seen that the value of $T_{03}/T_{01}$ satisfying flow compatibility can be obtained for *any* point on the compressor characteristic without reference to work compatibility. The fuel flow required to produce this temperature ratio can be calculated because the air flow, compressor delivery temperature and combustion temperature rise are all known. If the process is repeated for a series of points, lines of constant $T_{03}/T_{01}$ can be drawn on the compressor characteristic. These lines will have a similar form to the constant temperature lines shown for a free turbine engine in Fig. 8.7, but it must be emphasized they are not the *same* lines, because they represent non-equilibrium running conditions. (Constant temperature lines of this kind drawn on the compressor characteristic for a single-shaft engine would, however, represent a series of equilibrium running points for an engine driving a load which can be set independently of speed, e.g. a hydraulic dynamometer.)

With the turbine operating point and turbine inlet temperature known the power developed by the turbine can be calculated. Thus

$$\text{turbine power} = \eta_m m c_{pg} \Delta T_{034}$$

where

$$\Delta T_{034} = \frac{\Delta T_{034}}{T_{03}} \times \frac{T_{03}}{T_{01}} \times T_{01}$$

The compressor power is given by $m c_{pa} \Delta T_{012}$ and for an engine with a free turbine the net torque on the compressor rotor is given by

$$\Delta G = (\eta_m m c_{pg} \Delta T_{034} - m c_{pa} \Delta T_{012})/2\pi N$$

For a single-shaft engine the load torque would also have to be included.

Once the net torque has been obtained, the angular acceleration of the compressor rotor can be calculated; it may be assumed that this will be constant for a small interval of time and the resulting change in speed can be found. The process can then be repeated many times to provide a transient running line starting from some convenient equilibrium running point. The shape of this transient running line will be determined either by the choice of a limiting $T_{03}$ (which might be 50 K higher than the design point value) or the location of the surge line. The fuel flow required for operation along the desired running line can be calculated and the provision of this fuel during the transient is the responsibility of the designer of the fuel control system. Typical acceleration and deceleration trajectories on the compressor characteristic for a free turbine engine or a single-shaft engine with a propeller load are shown in Fig. 9.12. With an acceleration the initial movement towards surge shown in Fig. 9.12 is due to the rise in temperature following an increase in fuel flow, which takes place before the rotor has had time to increase its speed and provide an increase in mass flow. Too large an initial increase in fuel flow will cause the compressor to surge, resulting in very high temperatures which could destroy the turbine. The acceleration procedure for a single-shaft engine driving an electric generator would be to bring the rotor system up to full speed before applying the load and there should be no problem with regard to surge.

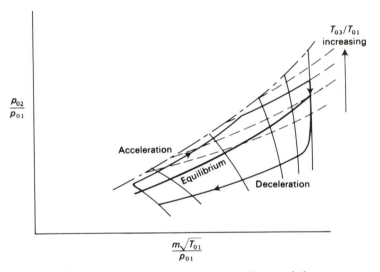

**FIG. 9.12  Transient trajectories on compressor characteristic**

During decelerations the operating point is moving away from surge as shown and the turbine inlet temperature is decreasing; the only problem that may arise is 'flame-out' of the combustion chamber because of very weak mixtures. This can be overcome by scheduling the deceleration fuel flow as a function of rotor speed to prevent too rapid reduction of fuel flow. In some engines decelerations are carried out by cutting off the fuel flow completely but it then becomes necessary to restart the combustion process before power can be restored.

We have seen in section 8.6 that if the equilibrium running line intersects the surge line, either blow-off or variable geometry can be used to lower the running line, but it should be noted that neither would have much effect on the *transient* running line. Rates of acceleration will be slowed down by blow-off, because of the reduction in mass flow through the turbine, but can be improved by a facility for increasing the area of the propelling nozzle (or power turbine stators).

We pointed out under the heading 'Variable area power turbine stators', that gas turbines for vehicular applications may well incorporate a variable geometry power turbine to improve the part-load economy, achieving this by *decreasing* the stator area as power is reduced. Conversely, on starting, the effect of *increasing* the stator area is to decrease the pressure ratio across the power turbine and increase that across the gas generator turbine. The resulting transient increase in gas generator turbine torque provides a substantially greater increase in *net* torque available for acceleration.

## Single-shaft versus free turbine engines

The choice of a single-shaft or free turbine configuration will normally be made on the basis of some primary requirement such as low cost or good low speed torque characteristics, but the transient behaviour might also be a deciding factor.

With a free turbine engine, power reduction must be obtained by reducing the gas generator speed, because of the fixed running line determined by the flow characteristic of the power turbine. Thus to restore power it is necessary to accelerate the gas generator up to its maximum speed. With a single-shaft engine, however, the load may be varied at constant speed, e.g. as with an electric generator or a variable pitch propeller. This means that there is no need to accelerate the gas generator and power may be restored very quickly by increasing the fuel flow. Such a feature can be very useful for turboprops where the power can be altered by changing the pitch of the propeller. The single-shaft engine has the obvious additional advantage that in the event of load being shed the compressor acts as a very efficient brake and for this reason regulation of output speed is easier than with a free turbine engine.

The starting power requirements for large gas turbines must be carefully considered; this is particularly important for emergency generating plant where one of the prime requirements is to achieve maximum output quickly. With a single-shaft unit the entire rotating assembly must be brought up to speed, and this may require the use of a steam turbine or diesel engine with a power of the order of 3 MW for a 100 MW unit. The free turbine engine is in a much more advantageous position, because neither the power turbine nor the load are driven by the starter. A twin-spool gas generator is even more favourable: only the HP rotor need be motored and the starter power will be less than 100 kW even for a large unit.

At low powers, with a generator as load, the part-load fuel consumption of a free turbine engine will be superior to that of a single-shaft engine. Figs. 8.7 and 8.5 show the equilibrium running lines in each case. The running line of the free turbine engine follows the locus of maximum compressor efficiency more closely than that of the single-shaft engine, and in general $T_{03}$ does not fall off so rapidly.

Finally, free turbine units can readily provide power at either 50 or 60 Hz for the European or North American markets, by minor changes in the speed of the power turbine with very small change in efficiency. To do this with a single-shaft unit would require modifications to the reduction gearbox.

## Twin-spool transient running lines

A full discussion of twin-spool transient behaviour cannot be covered in a book of this nature, but some important differences from single-spool behaviour must be mentioned. The transient behaviour of the HP rotor is similar to that of a simple jet engine and the trajectories on the HP characteristic will be similar to those shown in Fig. 9.12. Consideration of the lines of constant $T_{04}/T_{01}$ on the LP characteristic would suggest that trajectories for accelerations and decelerations would be as shown in Fig. 9.13, with the

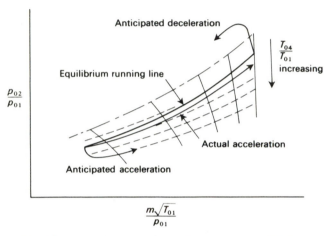

FIG. 9.13  Transient trajectories on LP compressor of twin-spool unit

likelihood of LP surge following a rapid deceleration. The lines of $T_{04}/T_{01}$, however, apply only to steady state operation and are irrelevant to transient behaviour. Detailed calculations and experimental investigations, described in Refs. (9) and (10), have shown that the LP transient running lines follow the equilibrium running line very closely, both for accelerations and decelerations.

We have noted that the acceleration of a single-spool turbojet engine can be improved by increasing the nozzle area. Although care must be taken when extending this idea to twin-spool turbojets, considerable advantage can result from judicious manipulation of the nozzle area during transients. Increasing the nozzle area during an acceleration will improve the LP rotor acceleration, because of the increase in LP turbine torque, but it should be realized that the LP running line will be moved towards surge. The mass flow is increased more rapidly because of the improved LP rotor acceleration, and this in turn permits the fuel flow to be increased more rapidly. In practice it is found that the thrust response is primarily determined by the rate of increase of fuel flow, so that the thrust response is improved at the expense of LP surge margin. Detailed calculations have shown that the HP surge margin is improved, however, and it is possible to trade surge margins between the two compressors to get the best response.

When operating at high forward speeds the nozzle will usually be open and the LP operating point will be close to surge. Emergency decelerations from this condition could result in LP surge, which may lead to a flame-out. A study of emergency decelerations in Refs. (9) and (10) showed that the best method of avoiding surge was to close the nozzle fully before reducing the fuel flow; this moved the LP operating point down to the running line associated with minimum nozzle area and also reduced the LP rotor speed. Immediate reduction of fuel flow, on the other hand, resulted in a running line very close to surge, as did simultaneous reduction of fuel flow and closing of the nozzle.

## Simulation of transient response

The development of a suitable control system requires a deep understanding of the transient behaviour of the gas turbine to be controlled, particularly for new types of engine where no previous experience exists. If a mathematical model or 'simulation' describing the engine dynamics is constructed and stored in a suitable computer, it can provide designers of gas turbines with an extremely versatile tool with which to investigate a wide variety of problems. A prime requirement for the simulation is that it should be capable of covering the entire running range of the engine and also of incorporating such modifications as blow-off and variable geometry as necessary. This can best be achieved by basing the simulation on the methods of off-design performance calculation presented in Chapters 8 and 9; the use of the normal component characteristics makes the simulation flexible in use and easy to understand, and has the further advantage that the component characteristics can be modified in the light of rig testing as the development programme proceeds. Typical problems which can readily be investigated include optimization of acceleration fuel schedules, operation of variable geometry devices and overspeed protection. Incorrectly chosen operating procedures, particularly under emergency conditions, may be hazardous to both engine and operator.

The information flow chart required for a mathematical model of a simple turbojet engine is shown in Fig. 9.14. It can be seen that the pressure ratios across each component are governed by flow compatibility, the pressure ratios in turn determine the temperature ratios, and the turbine temperature drop (proportional to the power developed by the turbine) is determined by the pressure ratio and turbine inlet temperature (controlled by the fuel flow).

The simulation may be carried out using analogue, digital or hybrid computers and appropriate techniques are described in Refs. (9) to (12). Their judicious use can give an invaluable insight into transient response without

**FIG. 9.14 Information flow chart for single-spool turbojet**

endangering an engine, and can be expected to result in significant savings in development time and hence cost.

## 9.6   Principles of control systems

Figure 9.15 shows the main features of a gas turbine control system. The fundamental requirement is to maintain the safety of the engine, regardless of how the operator moves the throttle lever or how the inlet conditions (e.g. altitude) are changing. The control system must ensure that the critical operating limits of rotational speed and turbine inlet temperature are never exceeded, and that compressor surge is avoided. The sensing of rotational speed presents no problems and a variety of frequency measuring devices are available. Turbine temperature, on the other hand, is very difficult to measure. It is not normally practicable to locate temperature probes at inlet to the

**FIG. 9.15  Control system components**

turbine, where the temperature is very high, and the temperature is measured at some downstream location, typically at inlet to the power turbine on shaft power units and in the jet pipe on jet engines and turbofans. Thus the temperatures used to protect the turbines are *indirect* measures of the critical temperatures. The temperatures are usually measured by thermocouples spaced around the annulus: typically six to eight probes are used. In the case of turbines with highly cooled blades, it is the actual *blade* temperature which is important and this may be sensed by radiation pyrometry in advanced engines.

The methods described in Chapters 8 and 9 can be used to provide the control system designer with information about the fuel flow required for steady-state operation over the entire range of operating conditions. Analysis of transient performance can predict the maximum fuel flow which can be used for acceleration without encountering surge or exceeding temperature limits. A typical schedule for fuel flow for a simple jet engine is shown in Fig. 9.16. Mathematical models for simulating the transient behaviour are an essential tool for optimization of fuel schedules, which must be experimentally verified during the engine development programme.

The performance calculations described in Chapters 8 and 9 showed that the variations of all the key parameters are wholly determined by the matching of the compressor, turbine and nozzle characteristics. It is important to realize that if the engine has *fixed geometry*, the steady-state performance cannot be altered in any way by the control system. If, however, devices such as variable IGVs, variable compressor stators or variable nozzles (turbine or propelling) are included, the operation of these can be integrated with the control system to modify the performance. The level of sophistication required of the control system is strongly dependent on the complexity of the engine.

Figure 9.15 shows that the control system must incorporate both a computing section and a fuel metering section. For many years both of these

**FIG. 9.16 Typical fuel schedule**

functions were met by *hydromechanical* systems, in which fuel passing through the unit provided the necessary hydraulic actuation of a variety of pistons, bellows and levers which metered the required fuel to the combustion system. In recent years much development of *digital* control systems has taken place; the increasing computational capacity and rapidly decreasing cost of small digital computers has made them quite feasible for use in control systems. It should be realized, however, that a fuel metering system is still required; the function of the digital computer is limited to computation and it is still necessary to open and close a tap to control fuel flow. Digital control systems will increasingly be used for data acquisition, using the data measured by the sensors, and unusual changes may be used in *diagnostic* systems for examining the mechanical condition of the engine; thermodynamic measurements, combined with vibration analysis and chemical analysis of the lubricating oil, can all be used for Engine Health Monitoring (EHM). The successful development of EHM systems will result in gains in maintenance cost and overhaul life and is the subject of intensive research.

To close this section, a very simple example of the use of the control system for maintaining an engine within operating limits will be considered. Let us consider a turboshaft engine with a free power turbine, used in a helicopter; the gearbox will be very highly stressed to keep weight to a minimum and the maximum power permissible will be strictly limited. The worked example in Chapter 8 showed the significant increase in power on a cold day, and conversely the power will decrease on a hot day. The variation in power would be as shown by the full line in Fig. 9.17; the maximum power limit is shown dotted, and at high ambient temperatures it will also be necessary to limit turbine temperature causing an even more rapid decline in power, as

**FIG. 9.17 Power and temperature limiting**

previously discussed with respect to jet engines. One method of achieving these limits would be by controlling gas generator speed as shown; the gas generator could only be run at maximum speed where neither limit was exceeded and at both low and high temperatures its value would be reduced.

Control system design is a specialized field which is changing rapidly and the interested reader must turn to the current literature. The control designer, in turn, must have a full understanding of the system to be controlled which necessitates an appreciation of gas turbine performance.

# Appendix A
## *Some notes on gas dynamics*

Owing to the increasing tendency towards specialization even at first degree and diploma level, it may be that some readers will not have been exposed to a course in gas dynamics. It is hoped that this Appendix will provide them with an adequate summary of those aspects which are relevant to gas turbine theory, and that it will serve others as useful revision material.

## A.1 Compressibility effects (qualitative treatment)

It is well known that when the relative velocity between a gas and a solid body reaches a certain value, the flow behaves in a quite different manner to that expected from a study of hydrodynamics. The effects produced, which manifest themselves as additional loss of stagnation pressure in the stream, do not arise when the fluid is a liquid. This suggests that the phenomena are due to the change in density which accompanies a change in pressure of a gas. The idea is strengthened by the fact that the phenomena only occur at high speeds when the pressure changes set up by the relative motion, and therefore the density changes, become considerable. In consequence, the phenomena here described are known as compressibility effects.

When, in a mass of gas at rest, a small disturbance results in a slight local rise of pressure, it can be shown that a pressure wave is propagated throughout the gas with a velocity which depends upon the pressure and density of the gas. This velocity is the speed of sound in the gas, or *sonic velocity a*, given by

$$a = \sqrt{(\gamma p/\rho)} \text{ or } \sqrt{(\gamma RT)}$$

where $p$, $\rho$ and $T$ are the local pressure, density and temperature of the gas.

In all processes related to the propagation of pressure waves, the changes take place so rapidly that there is no time for any heat transfer between adjacent layers of fluid; the processes are therefore adiabatic. Also, when the amplitude of the pressure wave is small and there is no material alteration in the pressure and temperature of the gas, as is true of an ordinary sound wave, there is no increase of entropy. The propagation of a sound wave is therefore not only adiabatic but isentropic.

Now consider what happens when a similar disturbance occurs in a gas flowing in one direction with a velocity $C$. The velocity of propagation of the pressure wave relative to the gas will still be equal to the speed of sound, $a$. Relative to a fixed point, however, say the walls of the passage confining the gas, the speed of propagation will be $(a+C)$ downstream and $(a-C)$ upstream. It follows that if the velocity of the gas is greater than the sonic velocity, i.e. supersonic, there can be no propagation of the pressure wave upstream at all. This is the usual physical explanation given for the existence of a critical condition in nozzle flow. When once the pressure drop across a nozzle is great enough to cause the gas velocity to reach the local sonic value, no further decrease in outlet pressure will be propagated upstream and no further increase in mass flow will be obtained.

Figure A1 illustrates the effects just described, and a useful picture to have in mind is that of the ever-widening circles of ripples formed by a stone thrown into a pond. When a disturbance, such as an intermittent electric spark, is placed in a gas stream moving with subsonic velocity $(C < a)$, the radius of a spherical pressure wave after time $t$ will be $at$, while the centre of this wave will have moved downstream a distance $Ct$. All waves emitted subsequently will lie within the spherical wave front of this wave, as shown in Fig. A1(a). On the other hand, when $C > a$ as in Fig. A1(b), the spherical wave

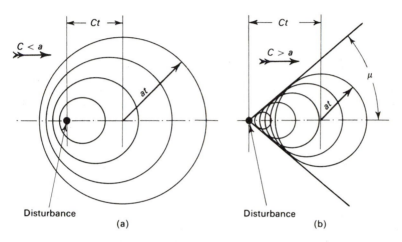

FIG. A1 Sound waves in a moving stream

fronts will move downstream at a greater rate than the radii of the waves increase. All the spherical waves will therefore lie within a cone having its apex at the point of the disturbance.

The effect of a small solid particle placed in a stream of gas is that of a disturbance emitting pressure waves continuously, so that the spherical wave fronts of Fig. A1(b) appear as a single conical wave front of semi-angle $\mu$, where $\mu$ is given by

$$\mu = \sin^{-1}\frac{at}{Ct} = \sin^{-1}\frac{a}{C}$$

The ratio $C/a$ frequently arises in the mathematical treatment of compressible flow, and it is the well known *Mach number M*. The angle $\mu$ is commonly called the *Mach angle*, and the conical wave front a *Mach wave*. Thus

Mach angle $\mu = \sin^{-1}(1/M)$

So far we have been considering pressure impulses of very small amplitude, such that there is no permanent change in the pressure and temperature of the gas as the wave moves through it, and consequently such that there is no change in entropy. In many practical cases of gas flow relative to a solid body these conditions are not fulfilled; there is a marked pressure and temperature difference across the wave, and there is an increase in entropy indicating an irreversible dissipation of kinetic energy which is manifested by a loss of stagnation pressure. The wave front represents a discontinuity in the flow, and as the change of pressure is to all intents and purposes instantaneous, the wave is termed a *shock wave*. The Mach wave previously discussed can be regarded as the weakest possible form of shock wave. The shock wave formed by a projectile travelling at supersonic speed, for example, is analogous to the bow wave set up by a ship: the water, unable to escape rapidly enough past the sides of the ship piles up to form a vee-shaped wave which travels along with the ship. In the case of the projectile, the air outside the region enclosed by the conical wave front does not receive a signal warning it of the approach of the solid object creating the disturbance, and hence the formation of the shock wave at the nose of the projectile. It must be stressed that it is the relative motion which is important; it does not matter whether the body or the fluid or both are moving.

We have said that there is a pressure difference across a shock wave. We must now ask whether it is a pressure rise or pressure drop in the direction of gas flow relative to the body, that is, through the shock wave. Both experiment and theory indicate that a shock wave can only be formed when a supersonic flow is decelerated. The velocities in the divergent part of a convergent-divergent nozzle are supersonic, but if the nozzle is operating at the pressure ratio for which it is designed no shock waves will be formed because the flow is accelerating under the influence of the pressure drop. Consider, on the other hand, what happens when the outlet pressure is appreciably above the value which would give just the right amount of expansion to suit the outlet area of the nozzle. Under these conditions the nozzle over-expands the gas so that before the gas can discharge into the surroundings some re-compression and deceleration of the gas must occur. This re-compression can only be brought about by a shock wave in the divergent part of the nozzle, because a convergent duct is necessary for isentropic diffusion of a supersonic stream. Figure A2 shows typical pressure distributions along a nozzle when

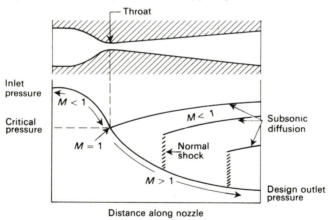

FIG. A2  **Shock wave in nozzle flow**

the outlet pressure is above the design value. As the outlet pressure is reduced, the plane normal shock wave moves towards the exit, and further reduction towards the design outlet pressure is accompanied by a sudden change to a complex system of oblique shock waves downstream of the exit.

To sum up so far:

(a)  A shock wave only occurs at supersonic speeds in a decelerating flow.

(b)  There is a *rise* in *static* pressure and temperature through the wave in the direction of relative motion of the gas, i.e. a shock wave is a compression wave.

(c)  There is a *drop* in *stagnation* pressure through the wave, some of the kinetic energy being dissipated with a consequent increase in entropy.

Finally, we may now turn to consider the effect of shock waves on the flow of air over an aerofoil. This will be directly applicable to problems associated with the design of axial compressor and turbine blading which are simply rows of aerofoils. First, consider the changes of pressure and velocity which occur around the aerofoil of Fig. A3,

Region of high velocity
and low pressure

**FIG. A3 Streamline flow over aerofoil**

assuming that an air stream is flowing past it with subsonic velocity. As indicated by the convergence and divergence of the streamlines, the air moving over the curved top surface must first accelerate and then decelerate, the outer streamlines effectively acting as the wall of a duct. This produces a region of increased velocity and reduced static pressure on the top surface. It is this region of low pressure or suction on the top surface which produces most of the lift of an aerofoil. The amount by which the velocity is increased along the top surface will depend upon the shape and camber of the aerofoil: it is quite possible for the local velocity, somewhere near the point of maximum camber, to be 1·5 times the velocity of the main stream. Furthermore, the acceleration will be accompanied by a fall in static temperature, so that the local sonic velocity $\sqrt{(\gamma RT)}$ will be reduced below that of the free stream. Both effects contribute to an increase in Mach number, and there may well be a region on the top surface where the flow is supersonic when the Mach number of the main stream is only about 0·7. But we have seen that when a deceleration occurs at supersonic speeds in a diverging passage a shock wave will be formed. The effect of this is illustrated in Fig. A4. If under these conditions a pitot tube

Shock wave     Loss of stagnation
pressure

**FIG. A4 Shock wave on aerofoil**

is traversed across the flow in front of and behind the aerofoil, the loss of stagnation pressure will be found to vary in the manner shown. The loss of head due to the shock wave itself is represented by the part of the curve $AB$; the loss of head represented by the peak of the curve may be very much greater than that across the shock wave, however, and requires some explanation.

When a fluid flows along a surface there is a thin layer of fluid, known as a boundary layer, in which there is a steep velocity gradient due to viscous friction, the velocity dropping to zero at the solid surface. Now the pressure gradient across the shock wave opposes the direction of flow and consequently, in the boundary layer where the kinetic energy is small, the shock wave may arrest the motion altogether. The boundary layer will thicken just in front of the shock wave, and may break away from the surface at the rear of it. If this breakaway of the boundary layer occurs, it will result in the initiation of a vortex trail involving considerable dissipation of energy. This, then, is the reason for the large loss of stagnation pressure in the wake of the aerofoil, and the reason why the Mach number of the main stream should be kept below the value likely to cause the formation of shock waves with this shape of aerofoil.

We may now turn to the mathematical analysis of compressible flow in a few simple, classical, flow situations. Much of the algebra is too lengthy to be given here, but by its omission we hope to enable the reader to see the wood: for the trees he can turn to the many excellent standard texts on gas dynamics, e.g. Refs. (3) and (4).

## A.2  Basic equations for steady one-dimensional compressible flow of a perfect gas in a duct

A flow can be regarded as *one-dimensional* if

(a)  changes in flow area and curvature of the axis are gradual,
(b)  all properties are uniform across planes normal to the axis,
(c)  any heat transfer per unit mass flow (dQ), across surface area of duct (dS), changes the properties uniformly over the cross-section,
(d)  the effect of friction can be represented by a shear stress $\tau$ at the wall.

The flow is *steady* if there is no change in the mass flowing per unit time at successive planes along the duct, and if the properties of the gas at any plane do not change with time.

Firstly, because we are dealing with a perfect gas we have the *equation of state*

$$\frac{p}{\rho} = RT \text{ or } \frac{dp}{p} = \frac{d\rho}{\rho} + \frac{dT}{T} \tag{1}$$

Secondly, application of the *conservation laws* yields the following equations in integral and differential form (see Fig. A5):

**FIG. A5  One-dimensional flow**

*Conservation of mass* (continuity equation)

$$m = \rho_1 A_1 C_1 = \rho_2 A_2 C_2 \text{ or } \frac{d\rho}{\rho} + \frac{dC}{C} + \frac{dA}{A} = 0 \tag{2}$$

*Conservation of momentum*

$$(\rho_2 C_2^2 A_2 - \rho_1 C_1^2 A_1) + (p_2 A_2 - p_1 A_1) - \tfrac{1}{2}(p_2 + p_1)(A_2 - A_1) + \tau S = 0$$
$$\rho A C \, dC + A \, dp + \tau \, dS = 0 \left.\begin{array}{c}\\\\\end{array}\right\} \tag{3}$$

*Conservation of energy*

$$Q = c_p(T_2 - T_1) + \tfrac{1}{2}(C_2^2 - C_1^2) = c_p(T_{02} - T_{01})$$
$$dQ = c_p \, dT + C \, dC = c_p \, dT_0 \left.\begin{array}{c}\\\\\end{array}\right\} \tag{4}$$

Thirdly, the *Second Law of Thermodynamics* must be satisfied, i.e.

for adiabatic processes $\Delta s \geqslant 0$     (5)

For a perfect gas with constant specific heats we have the specific entropy $s = f(p, T)$ given by

$$\Delta s = c_p \ln \frac{T_2}{T_1} - R \ln \frac{p_2}{p_1}$$

and for the special case of a reversible adiabatic (isentropic) process we can put $\Delta s = 0$ to give

$$\frac{p}{T^{\gamma/(\gamma-1)}} = \text{constant or } \frac{p}{\rho^\gamma} = \text{constant}$$     (6)

Finally, although no additional physical principle is introduced, the algebra is often simplified if velocities are (a) expressed in terms of Mach number $M$, or (b) taken into account implicitly by making use of the concepts of stagnation pressure $p_0$, temperature $T_0$ and density $\rho_0$. By definition $M = C/a$. For any fluid, the local sonic velocity $a$ is given by $\sqrt{(dp/d\rho)_s}$, while for a perfect gas it can be expressed variously as

$$a \text{ (or } C/M) = \sqrt{(\gamma p/\rho)} = \sqrt{(\gamma RT)} = \sqrt{[(\gamma-1)c_pT]}$$     (7)

By definition (see under heading 'Stagnation properties', p. 42), the stagnation properties in terms of $M$ and the static values become

$$\frac{T_0}{T} = 1 + \frac{C^2}{2c_pT} = \left[1 + \frac{\gamma-1}{2}M^2\right]$$

$$\frac{p_0}{p} = \left(\frac{T_0}{T}\right)^{\gamma/(\gamma-1)} = \left[1 + \frac{\gamma-1}{2}M^2\right]^{\gamma/(\gamma-1)}$$     (8)

$$\frac{\rho_0}{\rho} = \left(\frac{p_0}{p}\right)^{1/\gamma} = \left[1 + \frac{\gamma-1}{2}M^2\right]^{1/(\gamma-1)}$$

We may obtain an important flow equation by combining the differential forms of equations (1) to (4) and relations (7). Thus combining (1) and (2) to eliminate $d\rho/\rho$ we have

$$\frac{dC}{C} + \frac{dA}{A} + \frac{dp}{p} - \frac{dT}{T} = 0$$

From (3) and (7) we get

$$\frac{dp}{p} = -\gamma M^2 \frac{dC}{C} - \frac{\gamma \tau S}{\rho A a^2}$$

From (4) and (7) we get

$$\frac{dT}{T} = \frac{dQ}{c_pT} - (\gamma-1)M^2\frac{dC}{C}$$

Combining these three equations we have finally

$$(M^2-1)\frac{dC}{C} = \frac{dA}{A} - \frac{dQ}{c_pT} - \frac{\gamma \tau dS}{\rho A a^2}$$     (9)

Equation (9) expresses the effect on the velocity of (a) changes in flow area, (b) heat transferred and (c) viscous friction. Considering each of these factors acting in isolation we can draw the following qualitative conclusions.

*When $M < 1$ (subsonic flow)* the flow will *accelerate* (+ve dC) if (a) the duct converges (−ve dA) or (b) heat is transferred to the gas (+ve dQ). Conversely, the flow will decelerate if the duct diverges or is cooled. The effect of friction is always to accelerate subsonic flow: the physical reason is that the transformation of directed kinetic energy into internal energy is such that the decrease in density, consequent upon the rise in temperature, predominates and the flow velocity increases to satisfy the continuity equation.

When $M > 1$ (*supersonic flow*) the flow will *accelerate* if (*a*) the duct diverges or (*b*) the duct is cooled. Conversely, it will decelerate if the duct converges or is heated. The effect of friction is always to decelerate supersonic flow.

When $M = 1$ we have the condition where the flow velocity $C$ equals the local velocity of sound $a$. Values of all quantities at this condition will be denoted by an asterisk, e.g. $T^*$, $a^*$, $A^*$, $p^*$. Equation (9) shows that $M$ may be unity when the duct has a throat ($dA = 0$); or when sufficient heat has been added in a duct of constant flow area (supersonic flow decelerates towards $M = 1$ and subsonic flow accelerates towards $M = 1$); or when a constant area adiabatic duct is sufficiently long for friction to decelerate a supersonic flow or accelerate a subsonic flow to the condition $M = 1$. Under all such conditions the duct is said to be *choked* because the mass flow is the maximum which the duct can pass with the given inlet conditions: further heat transfer or friction in a constant area duct merely causes a reduction of mass flow in the duct.

Note that the presence of a throat does not necessarily imply $M = 1$ at the throat, because the duct may be acting as a venturi with subsonic expansion followed by subsonic diffusion. But we are here considering only a *continuous* expansion or compression along the duct (i.e. $dC$ is not changing sign during the process). For this reason also, we cannot expect equation (9) to yield information about discontinuities such as shock waves which will be considered later. Before doing so we will summarize the analysis which gives us quantitative information about the behaviour of a gas undergoing a continuous expansion or compression under the separate actions of changes in flow area (section A.3), heat transfer (A.4), and friction (A.5). This will involve algebraic manipulation of the integrated forms of equations (1)–(4) with appropriate terms omitted, together with relations (5)–(8) as appropriate. In what follows we shall often find it convenient to consider changes which occur when a fluid flowing with an arbitrary Mach number $M$ through area $A$ and with properties denoted by $p$, $p_0$, $T$, $T_0$, etc., is brought to a reference state where $M = 1$ and the relevant quantities are denoted by $A^*$, $p^*$, $p_0^*$, $T^*$, etc.

## A.3    Isentropic flow in a duct of varying area

For reversible adiabatic flow $\tau$ and $Q$ are zero, and the isentropic relations (6) are applicable. The relevant equations become:

$$p_1/\rho_1 T_1 = p_2/\rho_2 T_2$$
$$m = \rho_1 A_1 C_1 = \rho_2 A_2 C_2$$

Making use of $p/\rho^\gamma = $ constant, the momentum and energy equations (3) and (4) can be shown to be identical, namely

$$c_p(T_2 - T_1) + \tfrac{1}{2}(C_2^2 - C_1^2) = 0 \text{ or } c_p(T_{02} - T_{01}) = 0$$

We shall consider state 1 to be 'reservoir' or stagnation conditions, and state 2 to be any arbitrary state denoted by suffixless quantities. Then, combining the equations, and making use of relations (6)–(8), we arrive at the well known equations for flow in nozzles (or diffusers):

$$C = \left[ \frac{2\gamma}{\gamma - 1} RT_0 \left\{ 1 - \left( \frac{p}{p_0} \right)^{(\gamma - 1)/\gamma} \right\} \right]^{\frac{1}{2}} \tag{10}$$

$$\frac{m\sqrt{T_0}}{A p_0} = \left[ \frac{2\gamma}{\gamma - 1} \frac{1}{R} \left( \frac{p}{p_0} \right)^{2/\gamma} \left\{ 1 - \left( \frac{p}{p_0} \right)^{(\gamma - 1)/\gamma} \right\} \right]^{\frac{1}{2}} \tag{11}$$

$m\sqrt{T_0}/Ap_0$ is a maximum at the throat, where $C = a$ (i.e. $M = 1$) and the other quantities are denoted by asterisks. Differentiating (11) with respect to $p$ and equating to zero gives

$$\frac{p^*}{p_0} = \left(\frac{2}{\gamma+1}\right)^{\gamma/(\gamma-1)} \tag{12}$$

$$\frac{T^*}{T_0} = \left(\frac{2}{\gamma+1}\right) \tag{13}$$

Substitution of (12) in (11) then yields

$$\frac{m\sqrt{T_0}}{A^*p_0} = \left[\frac{\gamma}{R}\left(\frac{2}{\gamma+1}\right)^{(\gamma+1)/(\gamma-1)}\right]^{\frac{1}{2}} \tag{14}$$

Dividing (14) by (11), and introducing $p/p_0 = f(M)$ from (8), we obtain the value of $A$ at any $M$ in terms of the throat area $A^*$, namely

$$\frac{A}{A^*} = \frac{1}{M}\left[\frac{2+(\gamma-1)M^2}{(\gamma+1)}\right]^{(\gamma+1)/2(\gamma-1)} \tag{15}$$

Values of $p_0/p$, $T_0/T$, $\rho_0/\rho$ from relations (8) and $A/A^*$ from equation (15) are tabulated in the isentropic flow table of Ref. (1), with $M$ as the argument. The table covers the range $M = 0.01$–4, and is for dry air with $\gamma = 1.403$. Reference (2) is a much fuller set of tables covering a range of values of $\gamma$ and a wider range of $M$. Given the passage shape, i.e. values of $A/A^*$ at successive distances along the duct, corresponding values of $M$ and thence $p/p_0$ can be read from the isentropic flow table. Two values of $M$ and $p/p_0$ will be obtained from the table for each value of $A/A^*$, corresponding to the two real roots of equation (15). The pressure distribution can then be plotted as in Fig. A6. Curve

Distance along axis

$A/A^* > 1 \longleftarrow\!\!\!\mid\!\!\!\longrightarrow A/A^* > 1$

**FIG. A6**

BCD refers to the continuous expansion (or compression depending on the direction of flow), while BCE represents the limiting pressure distribution above which the passage is acting as a venturi.

## A.4   Frictionless flow in a constant area duct with heat transfer

We have seen that when $A$ is constant and $\tau$ is zero, heat transfer *to* the gas causes a subsonic flow to accelerate towards $M = 1$ and a supersonic flow to decelerate towards $M = 1$. This idealized flow is referred to as *Rayleigh flow*. One important effect, to which reference was made in Chapter 6, is that heat transfer to a subsonic flow in a duct of constant area must be accompanied by a fall in pressure. The pressure difference is necessary to provide the force required to accelerate the flow, i.e. to satisfy the momentum equation.

The relevant equations for a Rayleigh flow from some arbitrary state $M$, $T$, etc., to the state where $M = 1$ and quantities are denoted by asterisks, are as follows:

$$p/\rho T = p^*/\rho^* T^*$$
$$\rho C = \rho^* C^* (= \rho^* a^*)$$
$$(\rho^* C^{*2} - \rho C^2) + (p^* - p) = 0$$
$$Q = c_p(T^* - T) + \tfrac{1}{2}(C^{*2} - C^2) = c_p(T_0^* - T_0)$$

From the momentum equation, in conjunction with relations (7), we have

$$\frac{p}{p^*} = \frac{1+\gamma}{1+\gamma M^2} \tag{16}$$

From this, together with the continuity equation and equation of state, we obtain

$$\frac{T}{T^*} = M^2 \left[\frac{1+\gamma}{1+\gamma M^2}\right]^2 \tag{17}$$

From (16) and (17) and relations (8), by writing

$$\frac{p_0}{p_0^*} = \left(\frac{p_0}{p}\right)\left(\frac{p}{p^*}\right)\left(\frac{p^*}{p_0^*}\right), \text{ and likewise for } \frac{T_0}{T_0^*}$$

we obtain

$$\frac{p_0}{p_0^*} = \left[\frac{1+\gamma}{1+\gamma M^2}\right]\left[\frac{2+(\gamma-1)M^2}{\gamma+1}\right]^{\gamma/(\gamma-1)} \tag{18}$$

$$\frac{T_0}{T_0^*} = \frac{(1+\gamma)M^2}{(1+\gamma M^2)^2}\left[2+(\gamma-1)\,M^2\right] \tag{19}$$

Finally, from the energy equation we have

$$\frac{Q}{c_p T_0^*} = 1 - \frac{T_0}{T_0^*} \tag{20}$$

Equations (16)–(19) are the Rayleigh functions, and values of $p/p^*$, $T/T^*$, $p_0/p_0^*$, $T_0/T_0^*$ are given in the tables referred to in the previous section. To obtain values of a property at state 2, knowing the value in state 1, we merely apply such equations as:

$$\frac{p_2}{p_1} = \frac{p_2/p^*}{p_1/p^*} \text{ or } \frac{T_{02}}{T_{01}} = \frac{T_{02}/T_0^*}{T_{01}/T_0^*}$$

For example, given the inlet conditions $p_{01}$, $T_{01}$ and $M_1$, we may obtain $p_{02}$ and $M_2$ for any given final temperature $T_{02}$ (or temperature ratio $T_{02}/T_{01}$) as follows.

> For given $M_1$ read $p_{01}/p_0^*$ and $T_{01}/T_0^*$
> Evaluate $T_{02}/T_0^*$ from $(T_{02}/T_{01})(T_{01}/T_0^*)$
> Read $M_2$ corresponding to $T_{02}/T_0^*$
> Read $p_{02}/p_0^*$ corresponding to $M_2$
> Evaluate $p_{02}$ from $(p_{02}/p_0^*)(p_0^*/p_{01})p_{01}$

Finally, we could read $p_1/p_{01}$ corresponding to $M_1$ from the isentropic flow table, and evaluate the 'fundamental pressure loss factor' referred to in Chapter 6, namely $(p_{01} - p_{02})/(p_{01} - p_1)$. If this were done for a series of subsonic values of $M_1$ and $T_{02}/T_{01} > 1$, the results would appear as in Fig. A7. The chain dotted curve represents the limiting case where the heat transfer is sufficient to accelerate the flow to $M_2 = 1$: this condition is often referred to as *thermal choking*. Note that as the inlet Mach number approaches zero the fundamental pressure loss factor tends to $[(T_{02}/T_{01}) - 1]$: c.f. the incompressible flow result obtained in section 6.3.

**FIG. A7 Stagnation pressure loss in Rayleigh flow**

## A.5    Adiabatic flow in a constant area duct with friction

The idealization considered here is referred to as *Fanno flow*. We have seen in section A.2 that friction accelerates subsonic flow towards $M = 1$ and decelerates supersonic flow towards $M = 1$. The length of duct which is sufficient to change the Mach number from any value $M$ to the value unity will be referred to as $L^*$. Friction data are normally given in terms of relations between a friction factor $f$ and Reynolds number $Re$. The definition of $f$ used here (other definitions are also used) is

$$\tau = f\rho C^2/2$$

The relevant equations for Fanno flow are

$$p/\rho T = p^*/\rho^* T^*$$
$$\rho C = \rho^* C^* (= \rho^* a^*)$$
$$(\rho^* C^{*2} - \rho C^2) + (p^* - p) + \tau S = 0$$
$$c_p(T^* - T) + \tfrac{1}{2}(C^{*2} - C^2) = c_p(T_0^* - T_0) = 0$$

From the energy equation the stagnation temperature is constant, and this together with the $T_0/T$ relation (8) yields

$$\frac{T}{T^*} = \frac{\gamma+1}{2+(\gamma-1)M^2} \tag{21}$$

Using this in conjunction with the continuity equation and equation of state, we get

$$\frac{p}{p^*} = \frac{1}{M}\left[\frac{\gamma+1}{2+(\gamma-1)M^2}\right]^{\frac{1}{2}} \tag{22}$$

And using the $p_0/p$ relation (8) we obtain

$$\frac{p_0}{p_0^*} = \frac{1}{M}\left[\frac{2+(\gamma-1)M^2}{\gamma+1}\right]^{(\gamma+1)/2(\gamma-1)} \tag{23}$$

The equation expressing $L^*$ for a given $M$ and $f$ is best found from the differential form of the momentum equation (3), with $\tau$ replaced by $\tfrac{1}{2}f\rho C^2$. Dividing throughout by $A$, the momentum equation becomes

$$\rho C\,\mathrm{d}C + \mathrm{d}p + \frac{f\rho C^2\,\mathrm{d}S}{2A} = 0$$

For circular ducts of diameter $D$,

$$\frac{dS}{A} = \frac{\pi D\, dL}{\pi D^2/4} = \frac{4 dL}{D}$$

And for non-circular ducts $D$ can be replaced by the *equivalent diameter* ($= 4 \times$ flow area/perimeter). Introducing the equation of state, continuity equation and relations (7) and (8), after much algebra we arrive at a differential equation for $4f\, dL/D$ in terms of $M$. Integrating between $M$ and 1 we have finally

$$\frac{4f L^*}{D} = \frac{1 - M^2}{\gamma M^2} + \frac{\gamma + 1}{2\gamma} \ln \left[ \frac{(\gamma + 1)M^2}{2 + (\gamma - 1)M^2} \right] \tag{24}$$

The Fanno functions (21) to (24) are given in the compressible flow tables. To find a length $L$ over which the flow changes from $M_1$ to $M_2$ it is only necessary to subtract the two values of $4f L^*/D$, namely

$$\frac{4f L}{D} = \left[ \left( \frac{4f L^*}{D} \right)_{M_2} - \left( \frac{4f L^*}{D} \right)_{M_1} \right]$$

It should be appreciated that we have assumed $f$ to be constant, i.e. that it does not change along the duct. Since the Reynolds number is $\rho CD/\mu = GD/\mu$, and the mass velocity $G$ is constant because the area $A$ is constant, $Re$ only changes due to the variation of $\mu$ with $T$. This is small for gases. Furthermore, in turbulent pipe flow $f$ is a weak function of $Re$. Thus little error is incurred by using an appropriate mean value of $f$.

In practical situations variation in area, heat addition, and friction may be present simultaneously. Methods for combining the results of the preceding three sections can be found in any text on gas dynamics.

## A.6   Plane normal shock waves

When shock waves occur normal to the axis of flow, they are discontinuities which occupy a finite but very short length of duct as depicted in Fig. A8(a). For this reason they can be treated as adiabatic frictionless processes in a duct of constant cross-sectional area. In general, shock waves are formed when the conditions are such that the three

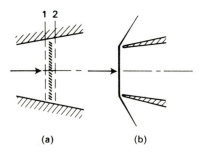

(a)              (b)

FIG. A8

conservation laws cannot be satisfied simultaneously with an assumption of reversible flow. What has to be relinquished is the idealization of reversibility, even though the flow is being regarded as frictionless. Then, if the process is adiabatic, all that the Second Law of Thermodynamics requires is that there should be an increase in entropy in the direction of flow.

The relevant equations relating properties on either side of a shock wave are

$$p_1/\rho_1 T_1 = p_2/\rho_2 T_2$$
$$m = \rho_1 C_1 = \rho_2 C_2$$
$$(\rho_2 C_2^2 - \rho_1 C_1^2) + (p_2 - p_1) = 0$$
$$c_p(T_2 - T_1) + \tfrac{1}{2}(C_2^2 - C_1^2) = c_p(T_{02} - T_{01}) = 0$$

From the momentum and continuity equations, together with relations (7), we obtain the pressure ratio across the shock wave as

$$\frac{p_2}{p_1} = \frac{1 + \gamma M_1^2}{1 + \gamma M_2^2} \tag{25}$$

Also, since $T_{02} = T_{01}$ from the energy equation, we have

$$\frac{T_2}{T_1} = \left(\frac{T_2}{T_{02}}\right)\left(\frac{T_{01}}{T_1}\right) = \frac{2 + (\gamma - 1)M_1^2}{2 + (\gamma - 1)M_2^2} \tag{26}$$

At this stage the value of $M_2$ is unknown but from the continuity equation, equation of state, and relations (7) we can show that $M_2$ is uniquely related to $M_1$ by the equation

$$\frac{M_2}{M_1} = \frac{p_1}{p_2}\sqrt{\frac{T_2}{T_1}}$$

On substitution from (25) and (26) we get

$$M_2^2 = \frac{(\gamma - 1)M_1^2 + 2}{2\gamma M_1^2 - (\gamma - 1)} \tag{27}$$

Finally, substituting for $M_2$ in (25) and (26) we obtain the pressure and temperature ratio in terms of $M_1$ given by

$$\frac{p_2}{p_1} = \frac{2\gamma M_1^2 - (\gamma - 1)}{\gamma + 1} \tag{28}$$

$$\frac{T_2}{T_1} = \left[2\gamma M_1^2 - (\gamma - 1)\right]\left[\frac{2 + (\gamma - 1)M_1^2}{(\gamma + 1)^2 M_1^2}\right] \tag{29}$$

The pressure ratio $p_{02}/p_1$ can be found from $(p_{02}/p_2)(p_2/p_1)$, using (28), (27) and the $p_0/p$ relation (8), the result being

$$\frac{p_{02}}{p_1} = \left[\frac{\gamma + 1}{2} M_1^2\right]^{\gamma/(\gamma - 1)} \bigg/ \left[\frac{2\gamma M_1^2 - (\gamma - 1)}{\gamma + 1}\right]^{1/(\gamma - 1)} \tag{30}$$

The normal shock functions (27) to (30) are tabulated in Ref. (1). If required, the stagnation pressure ratio can be found from

$$\frac{p_{02}}{p_{01}} = \left(\frac{p_{02}}{p_1}\right)\left(\frac{p_1}{p_{01}}\right)$$

remembering that $p_1/p_{01}$ can be found from the isentropic table for the given $M_1$. The fuller tables of Ref. (2) include $p_{02}/p_{01}$ as well as $p_{02}/p_1$.

Evaluation of equation (27) shows that for $M_1 > 1$, $M_2$ is less than 1; while for $M_1 < 1$, $M_2$ is greater than 1. There is nothing in the foregoing analysis to suggest that both these flow situations are not possible. But knowing $p_2/p_1$ and $T_2/T_1$ from (28) and (29), it is a simple matter to calculate the change in entropy of the gas from

$$\frac{\Delta s}{c_v} = \gamma \ln \frac{T_2}{T_1} - (\gamma - 1) \ln \frac{p_2}{p_1}$$

For values of $M_1 > 1$ we find that the values of $T_2/T_1$ and $p_2/p_1$ yield positive values of $\Delta s$, whereas when $M_1 < 1$ the change of entropy is negative. The latter situation is physi-

cally impossible because it contradicts the Second Law of Thermodynamics. Thus the
normal shock functions are only tabulated for values of $M_1 > 1$.

From the foregoing we may deduce that a normal shock wave can only arise in a
supersonic flow, and that it produces a sudden reduction in velocity to a subsonic value.
Thus the normal shock is a compression process and $p_2/p_1 > 1$ for all $M_1 > 1$. Owing to
the irreversibility introduced by the shock wave, however, there is a loss of *stagnation*
pressure, i.e. $p_{02}/p_{01} < 1$. All these characteristics are immediately obvious from the
tables or the graphical representation in Fig. A9. We note also, from the $M_2 = f(M_1)$

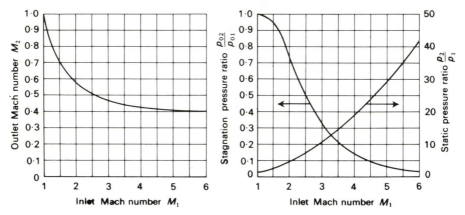

FIG. A9  Changes in Mach number and pressure across a plane normal shock wave

curve, that $M_2$ is asymptotic to a definite value as $M_1$ increases, and that $M_2$ is close
to this lower limit when $M_1$ has increased to about 5. By dividing the numerator and
denominator of equation (27) by $M_1^2$ and letting $M_1$ tend to infinity, the lower limit of
$M_2$ is seen to equal $\sqrt{[(\gamma - 1)/2\gamma]}$ which is approximately 0.38 for air.

It is clearly of interest to see how efficient the normal shock is as a compressor,
because it can be made use of at the front of intakes of turbojet and ramjet engines when
these are operating at supersonic flight speeds. The simple 'pitot' intake is illustrated
in Fig. A8(b). The efficiency of the process can be defined as the ratio of the ideal (isentropic) temperature rise to the actual temperature rise for a given pressure ratio $p_2/p_1$.
In which case we have the isentropic efficiency $\eta$, where

$$\eta = \frac{T_2' - T_1}{T_2 - T_1} = \frac{[(p_2/p_1)^{(\gamma - 1)/\gamma} - 1]}{T_2/T_1 - 1}$$

$p_2/p_1$ and $T_2/T_1$ can be found from the normal shock table for a range of $M_1$, and hence
$\eta$ can be plotted versus $M_1$ as in Fig. A10. The efficiency falls off rapidly as $M_1$ is increased.
Nevertheless, when it is realized that $p_2/p_1 = 4.5$ for $M_1 = 2$ (from Fig. A9), and that
this pressure ratio has been achieved in a negligible length of intake, an efficiency of
78 per cent (from Fig. A10) is not too low to be useful. For this example, $M_2$ is 0.58
(Fig. A9) and after the shock wave a further pressure rise can be obtained by ordinary
subsonic diffusion in a divergent duct. Higher efficiencies can only be obtained by designing supersonic intakes to operate with a system of oblique shocks. This type of shock
forms the subject of the next section.

The main features of Rayleigh flow, Fanno flow and flow through normal shocks
can be summarized neatly by drawing the processes on a $T$–$s$ diagram as in Fig. A11.
Such a diagram is a useful mnemonic. The Fanno and normal shock processes are
shown as dashed lines because they are essentially irreversible processes. The three lines
are all drawn for the same value of mass flow per unit area (i.e. $\rho C$) which we have seen

is constant for all three types of flow. This is why states 1 and 2 on either side of the normal shock coincide with the points of intersection of the Fanno and Rayleigh lines. There are four features which perhaps are not emphasized sufficiently by notes on the figure. (i) A Fanno process can occur only from state 2 towards state 3 or from 1 towards 3. It cannot pass *through* 3 without a decrease in entropy which would contravene the Second Law of Thermodynamics. (ii) A Rayleigh process can occur between 1 and 4 in

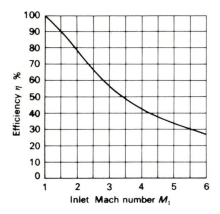

FIG. A10 Isentropic efficiency of plane shock wave

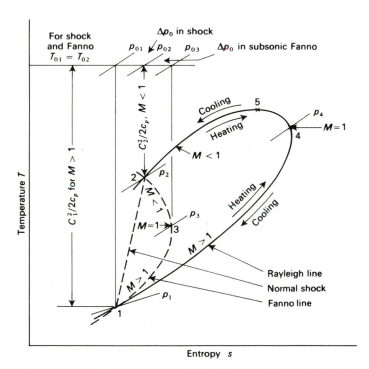

FIG. A11 Main features of Rayleigh, Fanno and normal shock processes

either direction or between 2 and 4 in either direction, but it cannot proceed *through*
state 4 in either direction. In practice, the picture is modified because friction is present
simultaneously with heating or cooling; but also it would be physically difficult to
suddenly change from heating to cooling at the point in the duct where the gas attains
state 4 and, without doing so, passage through state 4 would clearly contravene the
Second Law. (iii) It is interesting to note that when a subsonic flow is heated, the maximum temperature is reached at state 5 (where it can be shown that $M = 1/\sqrt{\gamma}$), although
the stagnation temperature must continue to rise with energy input until the gas attains
state 4. In other words, between 5 and 4 the density decreases, and hence the velocity
increases, at such a rate that the static temperature falls. (iv) when point 3 on the Fanno
line or point 4 on the Rayleigh line is reached the flow is choked, and further friction (i.e.
additional length of pipe), or heating in the case of Rayleigh flow, causes the state to move
on to another Fanno line or another Rayleigh line respectively. $M$ remains equal to
unity. These would be lines appropriate to a reduced mass flow per unit area, and would
lie to the right of those shown because of the increase in entropy due to additional
friction in one case and heating in the other.

## A.7   Oblique shock waves

In certain types of supersonic flow, shock waves are formed at an inclination to the
direction of flow as in Fig. A12. Strictly speaking the analysis of such phenomena involves

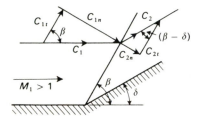

**FIG. A12  Oblique shock**

a study of two-dimensional flow although, as will be made apparent, some of the equations obtained for plane normal shock waves can be used with a little modification. The
analysis of *plane* oblique shock waves shows that for any given incident Mach number,

(a)   there is a limiting flow deflection angle $\delta_m$ which cannot be exceeded;
(b)   provided $\delta < \delta_m$, at any given value of $M_1$ there are two physically possible plane
       shock waves having different values of the shock angle $\beta$. The shock wave having
       the smaller $\beta$ is the weaker shock, i.e. it produces the smaller pressure ratio $p_2/p_1$
       and, although the flow velocity is reduced by the shock, the flow can still be super-
       sonic downstream of it.

Oblique shock waves are formed when a body is immersed in a supersonic stream.
If we consider the simple case of a wedge of semi-angle $\theta$, two types of flow are found
to occur as depicted in Fig. A13.

(i)    *When* $\theta \leqslant \delta_m$ a plane oblique shock wave is attached to the apex of the wedge. For
       $\theta < \delta_m$ it is the weak shock which is formed, while for $\theta = \delta_m$ there is no distinction
       between the two shocks and only one value of $\beta$ is possible.
(ii)   *When* $\theta > \delta_m$ a plane oblique shock wave is not possible, and it is replaced by a
       detached curved shock wave as in Fig. A13(b). (The same type of flow would be set

up by a round nosed body such as a compressor blade of aerofoil shape.) On the axis of the wedge the shock is normal, $\delta = 0$ and $M_2 < 1$. For a certain distance from the axis, $\delta < \delta_m$ and the curved shock can be regarded as being made up of a number of plane oblique shocks of elemental length and of successively increasing $\delta$. These are always the stronger of the two possible shocks, i.e. those having the larger $\beta$. Downstream of these shocks $M_2 < 1$, and thus the flow in the region near the body is subsonic. Further out in the mainstream, beyond the point where $\delta$ has increased to $\delta_m$, $\beta$ decreases until it is zero in the free stream (strictly speaking at an infinite distance from the body). In this region the curved shock can be regarded as a number of elemental plane shocks of the weak type (small $\beta$), and the flow remains supersonic with $M_2 \rightarrow M_1$ as the distance from the body increases.

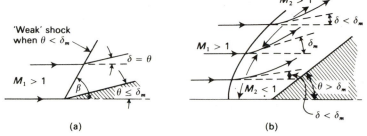

**FIG. A13**

Reverting to Fig. A12, consider the components of the velocities $C_1$ and $C_2$ in the tangential and normal directions relative to the plane shock. On either side of the shock there is no change of state along the shock, and with no force acting in this direction there can be no change in the tangential component of momentum or velocity across the shock. Thus

$$C_2 \cos (\beta - \delta) = C_1 \cos \beta \tag{31}$$

From the point of view of the normal component, the shock is a plane normal shock which we have already analysed. Thus equation (27) can be used but with $M_2 \sin (\beta - \delta)$ substituted for $M_2$ and $M_1 \sin \beta$ for $M_1$. Similarly, equations (28) and (29) for $p_2/p_1$ and $T_2/T_1$ will apply if $M_1 \sin \beta$ replaces $M_1$. To relate $\delta$, $\beta$ and $M_1$ we apply the continuity equation in the form

$$\rho_2 C_2 \sin (\beta - \delta) = \rho_1 C_1 \sin \beta$$

and substitute $C_1/C_2$ from (31) to give

$$\frac{\rho_2}{\rho_1} = \frac{\tan \beta}{\tan (\beta - \delta)}$$

But $\rho_2/\rho_1 = (p_2/p_1)(T_1/T_2) = f(M_1, \beta)$ from the modified (28) and (29), so that we have an equation for $\delta$ in terms of $M_1$ and $\beta$ which finally reduces to

$$\tan \delta = \cot \beta \frac{M_1^2 \sin^2 \beta - 1}{\frac{\gamma + 1}{2} M_1^2 - (M_1^2 \sin^2 \beta - 1)} \tag{32}$$

For given values of $\delta(< \delta_m)$ this equation has been shown to have two real positive roots, yielding the two values of $\beta$ corresponding to the weak and strong shocks. After differentiating, and some algebra, equation (32) also yields the maximum deflection $\delta_m$ in terms of $M_1$. The results are often plotted as in Fig. A14. They are also given in the oblique shock table of Refs. (1) and (2). Such a table gives the two possible values of $\beta_1$, $p_2/p_1$, $T_2/T_1$ and $M_2$, for each value of $M_1$ and a series of values of $\delta$. The curve which divides the weak and strong shock regions in Fig. A14 gives the maximum deflection $\delta_m$ which

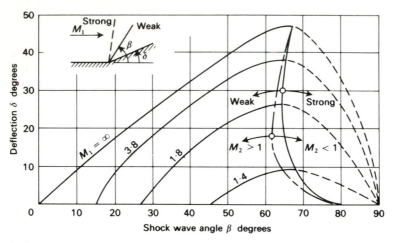

**FIG. A14**

the flow can undergo with a shock wave attached to the apex of the wedge. From the chain-dotted curve we see that although the Mach number downstream of a strong shock is always less than unity, it is not necessarily always greater than unity for a weak shock.

It is of interest to compare the efficiency of a compression by an oblique shock with that of the normal shock considered in section A.6. We took $M_1 = 2$, and found that a pressure ratio ($p_2/p_1$) of 4·5 was obtained with an efficiency of 78 per cent. From the oblique shock table of Ref. (1), with $M_1 = 2$ and an arbitrarily chosen value of $\delta = 10$ degrees, we obtain

|            | Weak   | Strong  |
|------------|--------|---------|
| $\beta$    | 39·3°  | 83·7°   |
| $M_2$      | 1·64   | 0·60    |
| $p_2/p_1$  | 1·71   | 4·45    |
| $T_2/T_1$  | 1·17   | 1·68    |
| $\eta$     | 0·98   | 0·79    |

The last line is calculated from

$$\left[ (p_2/p_1)^{(\gamma-1)/\gamma} - 1 \right] \Bigg/ \left[ \frac{T_2}{T_1} - 1 \right]$$

with $\gamma = 1·403$ as used in the tables.

Since a pressure ratio of 1·71 is obtainable from a weak plane oblique shock with an efficiency of 98 per cent, it would appear that a supersonic intake designed to make use of one or more such shocks might be more efficient than the simple type of 'pitot' intake referred to in section A.6; and this has been found to be the case. In the Oswatitsch intake, which makes use of a conical centre body as shown in Fig. A15, the flow is decelerated to a low supersonic velocity by several successive oblique shocks (two are shown) with the final transition to subsonic velocity taking place via a normal shock. The main problems in the design of such supersonic intakes are (a) the establishment of a stable shock pattern insensitive to minor changes in the flow downstream (e.g. in the engine), and (b) the maintenance of a good performance under off-design conditions.

**FIG. A15 Oswatitsch intake**

For aircraft engine intakes which have to operate over a complete speed range from $M_1 = 0$ to $M_1 \gg 1$, variable geometry is essential. This is accomplished by incorporating an adjustable centre body and/or cowl, and bleed slots. Figure A16 illustrates a variable geometry intake of the 'scoop' type, which is of rectangular cross-section: it may be slung under the wing or run along the side of the fuselage.

**FIG. A16 Intake with variable geometry**

## A.8 Isentropic two-dimensional supersonic expansion and compression

Perhaps the foregoing sections have left the reader with the impression that isentropic expansion and compression of a supersonic stream is impossible in principle. A moment's reflection on the existence of successful supersonic aeroplane wings, and efficient nozzles and diffusers for supersonic wind tunnels, should dispel this illusion. Processes which would be isentropic apart from the effect of viscous friction are possible if the duct walls, or immersed body as the case may be, are correctly shaped. Certainly it is possible to avoid the large loss of stagnation pressure due to breakdown of the flow which was illustrated in Fig. A4.

Consider supersonic flow that is initially parallel to a surface but which encounters a small change of direction of the surface: it may be a convex or concave deflection as in Fig. A17(a) and (b). If the change in direction is infinitesimal, the corner is the source

(a) Convex                    (b) Concave

**FIG. A17 Expansive and compressive Mach waves**

of an infinitesimal disturbance which is communicated to the main flow isentropically along a Mach wave. From section A.1 we know that this wave will make an angle $\mu = \sin^{-1}(1/M)$ with the direction of flow. And from the velocity triangles of Fig. A17 we see that the convex corner produces an increase in velocity, $dC$, which must be accompanied by a pressure drop $dp$; while the concave corner leads to a decrease in velocity implying a compression.

Any rounded convex corner giving a finite deflection can be regarded as the source of a series of infinitesimal deflections as illustrated in Fig. A18(a). The Mach waves, or *characteristic lines* as they are often called, diverge and do not interfere with each other, so permitting the flow to accelerate smoothly and isentropically to the downstream pressure. In the limit the same thing can occur at a sharp finite corner: in this case the flow expands smoothly through a fan of Mach waves as shown in Fig. A18(b). This is

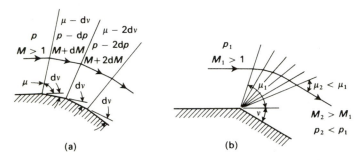

(a)                                        (b)

**FIG. A18 Isentropic expansion**

known as *Prandtl–Meyer flow*, and the evaluation of finite changes of pressure and Mach number associated with a finite deflection is possible with the aid of tabulations of Prandtl–Meyer 'expansion angles'. We shall proceed no further with this topic, but say merely that the analysis of supersonic flow patterns using a step-by-step method moving from one Mach line to the next, is referred to as the *method of characteristics*. When it is realized that in most situations the Mach waves will suffer reflections from neighbouring surfaces or jet boundaries, or intersect and interact with Mach waves from an opposing surface, it will be appreciated that the analysis is complex and beyond the scope of this Appendix.

To conclude, let us consider briefly what happens in a finite concave corner. If it is rounded as in Fig. A19(a), the Mach waves will appear as shown. They must converge because the Mach number is decreasing and hence the Mach angle $\mu$ is increasing. For

(a)                                        (b)

**FIG. A19 Compression (a) Isentropic (b) Non-isentropic**

a sharp concave corner there can be no compressive equivalent of an expansion fan, because this would involve successive Mach lines marching upstream of one another. What happens, since no corner is truly sharp, is that the Mach lines converge to a point in the stream near the surface and coalesce to form an oblique shock as in Fig. A19(b). The compression process is then no longer isentropic. In other words, an isentropic diffusion of a supersonic stream can only occur in a passage bounded by gradual concave curves.

# Appendix B
## Problems

The following problems are numbered by chapters. Unless otherwise stated the following values have been used:

For air: $c_p = 1.005 \text{ kJ/kg K}, \gamma = 1.40$
For combustion gas: $c_p = 1.148 \text{ kJ/kg K}, \gamma = 1.333$
For air and combustion gas: $R = 0.287 \text{ kJ/kg K}$

A manual of solutions can be purchased from: The Bookstore, Carleton University, Ottawa, Ontario K1S-5B6, Canada.

**2.1** In an ideal gas turbine cycle with reheat, air at state $(p_1, T_1)$ is compressed to pressure $rp_1$ and heated to $T_3$. The air is then expanded in two stages, each turbine having the same pressure ratio, with reheat to $T_3$ between the stages. Assuming the working fluid to be a perfect gas with constant specific heats, and that the compression and expansion are isentropic, show that the specific work output will be a maximum when $r$ is given by

$$r^{(\gamma-1)/\gamma} = (T_3/T_1)^{2/3}$$

**2.2** A simple gas turbine with heat-exchanger has a compressor and turbine having respective isentropic efficiencies $\eta_c$ and $\eta_t$. Show that the combined effect of *small* pressure drops $\Delta p_{hg}$ (in gas-side of heat-exchanger) and $\Delta p$ (total in combustion chamber and air-side of heat-exchanger) is to reduce the specific work output by an amount given by

$$\frac{\gamma-1}{\gamma} \times \frac{c_p T_3 \eta_t}{r^{(\gamma-1)/\gamma} p_1} \left[ \Delta p_{hg} + \frac{\Delta p}{r} \right]$$

where $T_3$ = turbine inlet temperature, $p_1$ = compressor inlet pressure and $r$ = compressor pressure ratio.
Assume that $c_p$ and $\gamma$ are constant throughout the cycle.

**2.3** Consider the ideal cycle for a gas turbine with heat-exchanger and separate LP power turbine. When the heat-exchanger is fitted in the normal way, using the exhaust from the LP turbine to heat the air after compression, the ideal cycle efficiency is

$$\eta = 1 - (c/t)$$

where $c = (p_2/p_1)^{(\gamma-1)/\gamma}$ and $t = T_3/T_1$. Another possibility is to pass the gas leaving the HP turbine through the heat-exchanger before it enters the LP turbine, and to let the LP turbine exhaust to atmosphere. Derive an expression for the ideal cycle efficiency of this scheme in terms of $c$ and $t$, and hence show that for all values of $c > 1$ the efficiency is higher than that of the ordinary scheme.

Finally, by referring to sketches of the cycles on the $T$–$s$ diagram, say in what respects the normal scheme might be superior in spite of the lower ideal cycle efficiency.

**2.4** In a gas turbine plant, air is compressed from state $(p_1, T_1)$ to a pressure $rp_1$ and heated to temperature $T_3$. The air is then expanded in two stages with reheat to $T_3$ between the turbines. The isentropic efficiencies of the compressor and each turbine are $\eta_c$ and $\eta_t$. If $xp_1$ is the intermediate pressure between the turbines, show that, for given values $p_1$, $T_1$, $T_3$, $\eta_c$, $\eta_t$ and $r$, the specific work output is a maximum when $x = \sqrt{r}$.

If this division of the expansion between the turbines is maintained, show that:

(a) when $r$ is varied, the specific work output is a maximum with $r$ given by

$$r^{3/2} = (\eta_c\eta_t T_3/T_1)^{\gamma/(\gamma-1)}$$

(b) When a perfect heat-exchanger is added the cycle efficiency is given by

$$\eta = 1 - \frac{T_1 r^{(\gamma-1)/2\gamma}\left[r^{(\gamma-1)/2\gamma}+1\right]}{2\eta_c\eta_t T_3}$$

Assume that the working fluid is a perfect gas with constant specific heats, and that pressure losses in the heater, reheater, and heat-exchanger are negligible.

**2.5** A gas turbine plant has a compressor in which air is compressed from atmospheric pressure and delivered to two turbines arranged in parallel, the combustion gases expanding to atmospheric pressure in each turbine. One of the turbines drives the compressor, to which it is mechanically coupled, while the other develops the power output of the plant. Each turbine has its own combustion chamber, the fuel supply to each being capable of control independently of the other.

(a) The power output of the plant is reduced by varying the fuel supply to the combustion chambers in such a way that the inlet temperature to the turbine driving the compressor is kept constant, while the inlet temperature to the power turbine is reduced. If the mass flow through a turbine is proportional to $p/\sqrt{T}$, where $p$ and $T$ refer to the turbine inlet conditions, show that the mass flow through the compressor is proportional to $r^{1/\gamma}$ and the power output of the plant is proportional to

$$r\left[r^{(\gamma-1)/\gamma}-1\right]/\left[1-kr^{(\gamma-1)/\gamma}\right]$$

where $r$ = pressure ratio
$k = T_1/(\eta_c\eta_t T_3)$
$T_1$ = compressor inlet temperature
$T_3$ = temperature at inlet to the turbine driving the compressor
$\eta_c$ = isentropic efficiency of compressor
$\eta_t$ = isentropic efficiency of turbine driving the compressor.

The isentropic efficiencies of the compressor and turbines may be assumed to remain constant under all conditions of the problem. Pressure losses, mechanical losses and the increase in the mass flow through the combustion chambers due to the mass of fuel burnt, may be neglected. $c_p$ and $\gamma$ may be assumed to be constant.

(b) Derive corresponding expressions for the alternative method of control in which the fuel supply to the combustion chambers is reduced in such a way that the inlet temperatures to both turbines are always equal.

(c) Calculate the percentage of full power developed under each of the foregoing methods of control when the pressure ratio of the compressor has fallen to 3·0, on a plant designed to give full power under the following conditions:

pressure ratio of the compressor     4·0
inlet temperature to the compressor    288 K
inlet temperature to both turbines     1100 K
$\eta_c = 0·85$, $\eta_t = 0·88$, $\gamma = 1·4$

[0·521, 0·594]

**2.6** A compressor has an isentropic efficiency of 0·85 at a pressure ratio of 4·0. Calculate the corresponding polytropic efficiency, and thence plot the variation of isentropic

efficiency over a range of pressure ratio from 2·0 to 10·0.
[0·876; 0·863 at 2·0 and 0·828 at 10·0]

**2.7** A peak load generator is to be powered by a simple gas turbine with free power turbine delivering 20 MW of shaft power. The following data are applicable:

| | |
|---|---|
| Compressor pressure ratio | 11·0 |
| Compressor isentropic efficiency | 0·82 |
| Combustion pressure loss | 0·4 bar |
| Combustion efficiency | 0·99 |
| Turbine inlet temperature | 1150 K |
| Gas generator turbine isentropic efficiency | 0·87 |
| Power turbine isentropic efficiency | 0·89 |
| Mechanical efficiency (each shaft) | 0·98 |
| Ambient conditions $p_a$, $T_a$ | 1 bar, 288 K |

Calculate the air mass flow required and the s.f.c.
[119·4 kg/s, 0·307 kg/kW h]

**2.8** A gas turbine is to consist of a compressor, combustion chamber, turbine and heat-exchanger. It is proposed to examine the advantage of bleeding off a fraction ($\Delta m/m$) of the air delivered by the compressor and using it to cool the turbine blades. By so doing the maximum permissible cycle temperature may be increased from $T$ to $(T + \Delta T)$. The gain in efficiency due to the increase of temperature will be offset by a loss due to the decrease in effective air flow through the turbine. Show that, on the following assumptions, there is no net gain in efficiency when

$$\frac{\Delta m}{m} = \frac{\Delta T/T}{(1 + \Delta T/T)}$$

and that this result is independent of the compressor and turbine efficiencies. Assumptions:
(1) No pressure loss in combustion chamber or heat-exchanger.
(2) The working fluid is air throughout and the specific heats are constant.
(3) The air bled for cooling purposes does no work in the turbine.
(4) The temperature of the air entering the combustion chamber is equal to that of the turbine exhaust.

A plant of this kind operates with an inlet temperature of 288 K, a pressure ratio of 6·0, a turbine isentropic efficiency of 90 per cent and a compressor isentropic efficiency of 87 per cent. Heat transfer calculations indicate that if 5 per cent of the compressor delivery is bled off for cooling purposes, the maximum temperature of the cycle can be raised from 1000 to 1250 K. Find the percentage increase in (a) efficiency, and (b) specific work output, which is achieved by the combined bleeding and cooling process. Make the same assumptions as before and take $\gamma = 1·4$ throughout.

What results would you expect if these calculations were repeated for the plant with the heat-exchanger omitted?
[25·1 per cent, 47·8 per cent]

**2.9** An auxiliary gas turbine for use on a large airliner uses a single-shaft configuration with air bled from the compressor discharge for aircraft services. The unit must provide 1·5 kg/s bleed air and a shaft power of 200 kW. Calculate (a) the total compressor air mass flow and (b) the power available with no bleed flow, assuming the following:

| | |
|---|---|
| Compressor pressure ratio | 3·80 |
| Compressor isentropic efficiency | 0·85 |
| Combustion pressure loss | 0·12 bar |
| Turbine inlet temperature | 1050 K |
| Turbine isentropic efficiency | 0·88 |

Mechanical efficiency (compressor rotor)   0·99
Mechanical efficiency (driven load)        0·98
Ambient conditions                         1 bar, 288 K
[4·78 kg/s, 633 kW]

**2.10**   A closed cycle gas turbine is to be used in conjunction with a gas cooled nuclear reactor. The working fluid is helium ($c_p = 5·19$ kJ/kg K and $\gamma = 1·66$).

The layout of the plant consists of two-stage compression with intercooling followed by a heat-exchanger; after leaving the cold side of the heat-exchanger the helium passes through the reactor channels and on to the turbine; from the turbine it passes through the hot-side of the heat-exchanger and then a pre-cooler before returning to the compressor inlet. The following data are applicable:

Compressor and turbine polytropic efficiencies        0·88
Temperature at LP compressor inlet                    310 K
Pressure at LP compressor inlet                       14·0 bar
Compressor pressure ratios (LP and HP)                2·0
Temperature at HP compressor inlet                    300 K
Mass flow of helium                                   180 kg/s
Reactor thermal output (heat input to gas turbine)    500 MW
Pressure loss in pre-cooler and intercooler (each)    0·34 bar
Pressure loss in heat-exchanger (each side)           0·27 bar
Pressure loss in reactor channels                     1·03 bar
Helium temperature at entry to reactor channels       700 K

Calculate the power output and thermal efficiency, and the heat-exchanger effectiveness implied by the data.
[214·5 MW, 0·429, 0·782]

**3.1**   A simple turbojet is operating with a compressor pressure ratio of 8·0, a turbine inlet temperature of 1200 K and a mass flow of 15 kg/s, when the aircraft is flying at 260 m/s at an altitude of 7000 m. Assuming the following component efficiencies, and I.S.A. conditions, calculate the propelling nozzle area required, the net thrust developed and the s.f.c.

Polytropic efficiencies of compressor and turbine   0·87
Isentropic efficiency of intake                     0·95
Isentropic efficiency of propelling nozzle          0·95
Mechanical efficiency                               0·99
Combustion chamber pressure loss                   6 per cent comp. deliv. press.
Combustion efficiency                              0·97
[0·0713 m², 7896 N, 0·126 kg/h N]

**3.2**   The gases in the jet pipe of the engine considered in problem **3.1** are reheated to 2000 K, and the combustion pressure loss incurred is 3 per cent of the pressure at outlet from the turbine. Calculate the percentage increase in nozzle area required if the mass flow is to be unchanged, and also the percentage increase in net thrust.
[48·3 per cent, 64·5 per cent]

**3.3**   A naval aircraft is powered by a turbojet engine, with provision for flap blowing. When landing at 55 m/s, 15 per cent of the compressor delivery air is bled off for flap blowing and it can be assumed to be discharged perpendicularly to the direction of flight. If a propelling nozzle area of 0·13 m² is used, calculate the net thrust during landing given that the engine operating conditions are as follows.

Compressor pressure ratio            9·0
Compressor isentropic efficiency     0·82

Turbine inlet temperature              1275 K
Turbine isentropic efficiency          0·87
Combustion pressure loss               0·45 bar
Nozzle isentropic efficiency           0·95
Mechanical efficiency                  0·98
Ambient conditions                     1 bar, 288 K

The ram pressure and temperature rise can be regarded as negligible.
[18·77 kN]

**3.4**   Under take-off conditions when the ambient pressure and temperature are 1·01
bar and 288 K, the stagnation pressure and temperature in the jet pipe of a turbojet
engine are 2·4 bar and 1000 K, and the mass flow is 23 kg/s. Assuming that the expansion
in the converging propelling nozzle is isentropic, calculate the exit area required and the
thrust produced.

   For a new version of the engine the thrust is to be increased by the addition of an aft
fan which provides a separate cold exhaust stream. The fan has a by-pass ratio of 2·0
and a pressure ratio of 1·75, the isentropic efficiencies of the fan and fan-turbine sections
being 0·88 and 0·90 respectively. Calculate the take-off thrust assuming that the expansion in the cold nozzle is also isentropic, and that the hot nozzle area is adjusted so that
the hot mass flow remains at 23 kg/s.
[0·0763 m², 15·35 kN; 24·9 kN]

**3.5**   Extending the example on the turbofan in Chapter 3, with the additional information that the combustion efficiency is 0·99, determine the s.f.c. Also, calculate the thrust
and s.f.c. when a combustion chamber is incorporated in the by-pass duct and the 'cold'
stream is heated to 1000 K. The combustion efficiency and pressure loss for this process
may be assumed to be 0·97 and 0·05 bar respectively.
[0·0429 kg/h N; 55·95 kN, 0·128 kg/h N]

**4.1**   The following data refer to the eye of a single-sided impeller.

Inner radius 6·5 cm          Mass flow            8 kg/s
Outer radius 15·0 cm         Ambient conditions  1·00 bar, 288 K
Speed 270 rev/s

Assuming no pre-whirl and no losses in the intake duct, calculate the blade inlet angle
at root and tip of the eye, and the Mach number at the tip of the eye.
[48°12′, 25°51′, 0·843]

**4.2**   An aircraft engine is fitted with a single-sided centrifugal compressor. The aircraft
flies with a speed of 230 m/s at an altitude where the pressure is 0·23 bar and the temperature 217 K. The intake duct of the impeller eye contains fixed vanes which give the air
pre-whirl of 25° at all radii. The inner and outer diameters of the eye are 18 and 33 cm
respectively, the diameter of the impeller periphery is 54 cm and the rotational speed
270 rev/s. Estimate the stagnation pressure at the compressor outlet when the mass
flow is 3·60 kg/s.

   Neglect losses in the inlet duct and fixed vanes, and assume that the isentropic
efficiency of the compressor is 0·80. Take the slip factor as 0·9 and the power input factor
as 1·04.
[1·75 bar]

**4.3**   The following results were obtained from a test on a small single-sided centrifugal
compressor:

Compressor delivery stagnation pressure         2·97 bar
Compressor delivery stagnation temperature      429 K
Static pressure at impeller tip                 1·92 bar

| Mass flow | 0·60 kg/s |
|---|---|
| Rotational speed | 766 rev/s |
| Ambient conditions | 0·99 bar and 288 K |

Calculate the overall isentropic efficiency of the compressor.

The diameter of the impeller is 16·5 cm, the axial depth of the vaneless diffuser is 1·0 cm and the number of impeller vanes ($n$) is 17. Making use of the Stanitz equation for slip factor, namely $\sigma = 1 - (0·63\pi/n)$, calculate the stagnation pressure at the impeller tip and hence find the fraction of the overall loss which occurs in the impeller. [0·75; 3·35 bar, 0·60]

**4.4** The following design data apply to a double-sided centrifugal compressor:

| | |
|---|---|
| Outer diameter of impeller | 50 cm |
| Speed | 270 rev/s |
| Mass flow | 16·0 kg/s |
| Inlet temperature | 288 K |
| Inlet pressure | 1·01 bar |
| Isentropic efficiency of impeller only | 0·90 |
| Radial gap of vaneless space | 4·0 cm |
| Axial depth of vaneless space | 5·0 cm |
| Slip factor | 0·9 |
| Power input factor | 1·04 |

(a) Calculate the stagnation pressure and temperature at the outlet of the impeller, assuming no pre-whirl.

(b) Show that the radial outlet velocity at the impeller tip is about 96 m/s and hence find the Mach number and air leaving angle at the impeller tip. (In calculating the circumferential area at the tip, the thickness of the impeller disc may be neglected.)

(c) Assuming isentropic diffusion in the vaneless space, find the correct angle of the leading edges of the diffuser vanes, and also find the Mach number at this radius.

[(a) 4·40 bar, 455 K (b) 1·01, 14°5' (c) 12°24', 0·842]

**4.5** A single-sided centrifugal compressor is to deliver 14 kg/s of air when operating at a pressure ratio of 4:1 and a speed of 200 rev/s. The inlet stagnation conditions may be taken as 288 K and 1·0 bar. Assuming a slip factor of 0·9, a power input factor of 1·04 and an overall isentropic efficiency of 0·80, estimate the overall diameter of the impeller.

If the Mach number is not to exceed unity at the impeller tip, and 50 per cent of the losses are assumed to occur in the impeller, find the minimum possible axial depth of the diffuser.

[68·9 cm, 5·26 cm]

**5.1** An axial flow compressor stage has blade root, mean and tip velocities of 150, 200 and 250 m/s. The stage is to be designed for a stagnation temperature rise of 20 K and an axial velocity of 150 m/s, both constant from root to tip. The work done factor is 0·93. Assuming 50 per cent reaction at mean radius calculate the stage air angles at root, mean and tip and the degree of reaction at root and tip for a free vortex design.

[$\alpha_1 = 17°4'$ ($= \beta_2$), $\beta_1 = 45°44'$ ($= \alpha_2$) at mean radius; $\alpha_1 = 13°46'$, $\beta_1 = 54°53'$, $\beta_2 = 40°14'$, $\alpha_2 = 39°26'$ at tip; $\alpha_1 = 22°15'$, $\beta_1 = 30°36'$, $\beta_2 = -20°15'$, $\alpha_2 = 53°51'$ at root; $\Lambda = 11·2$ per cent at root and 67·4 per cent at tip]

**5.2** Recalculate the stage air angles for the same data as in the previous question for a stage with 50 per cent reaction at all radii, and compare the results with those for the free vortex design.

[$\alpha_1 = 28°36'$ ($= \beta_2$), $\beta_1 = 48°16'$ ($= \alpha_2$) at tip; $\alpha_1 = 1°9'$ ($= \beta_2$), $\beta_1 = 44°25'$ ($= \alpha_2$) at root]

**5.3**   The first stage of an axial compressor is designed on free vortex principles, with no inlet guide vanes. The rotational speed is 6000 rev/min and the stagnation temperature rise is 20 K. The hub-tip ratio is 0·60, the work done factor is 0·93 and the isentropic efficiency of the stage is 0·89. Assuming an inlet velocity of 140 m/s and ambient conditions of 1·01 bar and 288 K, calculate:

(a)   The tip radius and corresponding rotor air angles $\beta_1$ and $\beta_2$, if the Mach number relative to the tip is limited to 0·95.
(b)   The mass flow entering the stage.
(c)   The stage stagnation pressure ratio and power required.
(d)   The rotor air angles at the root section.
[(a) 0·456 m, 63°57′ and 56°24′, (b) 65·5 kg/s, (c) 1·233, 1317 kW, (d) 50°50′ and 18°19′]

**5.4**   An axial flow compressor has an overall pressure ratio of 4·0 and mass flow of 3 kg/s. If the polytropic efficiency is 88 per cent and the stagnation temperature rise per stage must not exceed 25 K, calculate the number of stages required and the pressure ratio of the first and last stages. Assume equal temperature rise in all stages. If the absolute velocity approaching the last rotor is 165 m/s at an angle of 20° from the axial direction, the work done factor is 0·83, the velocity diagram is symmetrical, and the mean diameter of the last stage rotor is 18 cm, calculate the rotational speed and the length of the last stage rotor blade at inlet to the stage. Ambient conditions are 1·01 bar and 288 K.
[7, 1·273, 1·178; 414 rev/s, 1·325 cm]

**5.5**   A helicopter gas turbine requires an overall compressor pressure ratio of 10 : 1. This is to be obtained using a two-spool layout consisting of a four-stage axial compressor followed by a single-stage centrifugal compressor. The polytropic efficiency of the axial compressor is 92 per cent and that of the centrifugal is 83 per cent.

The axial compressor has a stage temperature rise of 30 K, using a 50 per cent reaction design with a stator outlet angle of 20°. If the mean diameter of each stage is 25·0 cm and each stage is identical, calculate the required rotational speed. Assume a work done factor of 0·86 and a constant axial velocity of 150 m/s.

Assuming an axial velocity at the eye of the impeller, an impeller tip diameter of 33·0 cm, a slip factor of 0·90 and a power input factor of 1·04, calculate the rotational speed required for the centrifugal compressor.

Ambient conditions are 1·01 bar and 288 K.
[Axial compressor 318 rev/s, centrifugal compressor 454 rev/s]

**6.1**   (Chapter 2 also refers) The reference hydrocarbon fuel for which the combustion temperature rise curves of Fig. 2.15 have been drawn contains 13·92 per cent hydrogen and 86·08 per cent carbon, and the relevant value of the enthalpy of reaction, $\Delta H_{298K}$, is −43 100 kJ/kg of fuel. An actual adiabatic steady flow combustion process employs this fuel, with fuel and air entering at 298 K in the ratio 0·0150 by weight. A chemical analysis of the products shows that 4·0 per cent of the carbon is burnt only to carbon monoxide, and the combustion temperature rise $\Delta T$ is found to be 583 K.

Using Fig. 2.15, calculate the combustion efficiency based on (a) ratio of actual to theoretical $f$ for the actual $\Delta T$, (b) ratio of actual to theoretical $\Delta T$ for the actual $f$. Compare these values with the efficiency based on the ratio of actual energy released to that theoretically obtainable, given that $\Delta H_{298K}$ for CO is −10 110 kJ/kg.
[0·980, 0·983, 0·981]

**6.2**   (Chapter 2 also refers) A gas turbine combustion chamber is supplied with liquid fuel at 325 K and air at 450 K. The fuel approximates to $C_{10}H_{12}$, and five times the quantity of air required for stoichiometric combustion is supplied. Calculate the fuel/air ratio, and estimate the fuel products temperature assuming the combustion to be adiabatic and complete.

In addition to the following data, use appropriate values of $c_p$ from p. 17 of the abridged

tables of Ref. (1) in Chapter 3: the combustion temperature rise curves of Fig. 2.15 may be used to obtain an initial approximate value of the products temperature.
Data:

$$C_{10}H_{12}(liq) + 13O_2 \rightarrow 10CO_2 + 6H_2O(liq); \quad \Delta H_{298K} = -42\,500 \text{ kJ/kg } C_{10}H_{12}$$

For water at 298 K, $h_{fg} = 2442$ kJ/kg.
For the liquid fuel, mean $c_p = 1.945$ kJ/kg K.
Composition of air by volume: $0.79 \text{ N}_2$, $0.21 \text{ O}_2$.
[0.0148, 984 K]

**6.3** The overall pressure loss factor of a combustion chamber may be assumed to vary with the temperature ratio according to the law

$$\frac{\Delta p_0}{m^2/2\rho_1 A_m^2} = K_1 + K_2 \left[ \frac{T_{02}}{T_{01}} - 1 \right]$$

For a particular chamber having an inlet area of $0.0389$ m$^2$ and a maximum cross-sectional area $A_m$ of $0.0975$ m$^2$, cold loss tests show that $K_1$ has the value 19.0. When tested under design conditions the following readings were obtained.

| | |
|---|---|
| Air mass flow $m$ | 9.0 kg/s |
| Inlet stagnation temperature $T_{01}$ | 475 K |
| Outlet stagnation temperature $T_{02}$ | 1023 K |
| Inlet static pressure $p_1$ | 4.47 bar |
| Stagnation pressure loss $\Delta p_0$ | 0.27 bar |

Estimate the pressure loss at a part load condition for which $m$ is 7.40 kg/s, $T_{01}$ is 439 K, $T_{02}$ is 900 K and $p_1$ is 3.52 bar.

Also, for these two operating conditions compare the values of (a) the velocity at inlet to the chamber and (b) the pressure loss as a fraction of the inlet stagnation (i.e. compressor delivery) pressure, and comment on the result.
[0.213 bar; 70.2 m/s, 67.7 m/s; 0.0593, 0.0597]

**7.1** A mean-diameter design of a turbine stage having equal inlet and outlet velocities leads to the following data.

| | |
|---|---|
| Mass flow $m$ | 20 kg/s |
| Inlet temperature $T_{01}$ | 1000 K |
| Inlet pressure $p_{01}$ | 4.0 bar |
| Axial velocity (constant through stage) $C_a$ | 260 m/s |
| Blade speed $U$ | 360 m/s |
| Nozzle efflux angle $\alpha_2$ | 65 degrees |
| Stage exit swirl $\alpha_3$ | 10 degrees |

Determine the rotor blade gas angles, degree of reaction, temperature drop coefficient $(2c_p \Delta T_{0s}/U^2)$ and power output.

Assuming a nozzle loss coefficient $\lambda_N$ of 0.05, calculate the nozzle throat area required (ignoring the effect of friction on the critical conditions).
[37°9', 57°22', 0.29, 3.35, 4340 kW, 0.0398 m$^2$]

**7.2** The following particulars relate to a single-stage turbine of free vortex design.

| | |
|---|---|
| Inlet temperature $T_{01}$ | 1050 K |
| Inlet pressure $p_{01}$ | 3.8 bar |
| Pressure ratio $p_{01}/p_{03}$ | 2.0 |
| Outlet velocity $C_3$ | 275 m/s |
| Blade speed at root radius, $U_r$ | 300 m/s |
| Turbine isentropic efficiency $\eta_t$ | 0.88 |

The turbine is designed for zero reaction ($\Lambda = 0$) at the root radius, and the velocities at inlet and outlet ($C_1$ and $C_3$) are both equal *and* axial. Calculate the nozzle efflux angle $\alpha_2$ and blade inlet gas angle $\beta_2$ at the root radius.

If the tip/root radius ratio of the annulus at exit from the nozzle blades is 1·4, determine the nozzle efflux angle and degree of reaction at the tip radius.

Assuming a nozzle blade loss coefficient $\lambda_N$ of 0·05 calculate the static pressure at inlet and outlet of the rotor blades at the root radius and thereby show that even at the root there is some expansion in the rotor blade passages under these conditions.
[61°9′, 40°14′; 52°15′, 0·46; 1·72 and 1·64 bar]

**7.3**   The following data apply to a single-stage turbine designed on free-vortex theory.

| | |
|---|---|
| Mass flow | 36 kg/s |
| Inlet temperature $T_{01}$ | 1200 K |
| Inlet pressure $p_{01}$ | 8·0 bar |
| Temperature drop $\Delta T_{013}$ | 150 K |
| Isentropic efficiency $\eta_t$ | 0·90 |
| Mean blade speed $U_m$ | 320 m/s |
| Rotational speed $N$ | 250 rev/s |
| Outlet velocity $C_3$ | 400 m/s |

The outlet velocity is axial. Calculate the blade height and radius ratio of the annulus from the outlet conditions.

The turbine is designed with a constant annulus area through the stage, i.e. with no flare. Assuming a nozzle loss coefficient $\lambda_N$ of 0·07, show that continuity is satisfied when the axial velocity at exit from the nozzles is 346 m/s. Thence calculate the inlet Mach number relative to the rotor blade at the root radius.
[0·0591 m, 1·34; 0·81]

**7.4**   The example of a free vortex turbine design considered in section 7.2 yielded the following results for the gas angles.

| | $\alpha_2$ | $\beta_2$ | $\alpha_3$ | $\beta_3$ |
|---|---|---|---|---|
| tip | 54°56′ | 0° | 8°31′ | 58°20′ |
| mean | 58°23′ | 20°29′ | 10° | 54°57′ |
| root | 62°9′ | 39°19′ | 12°7′ | 51°8′ |

The values of the radius ratios in plane 2 were $(r_m/r_r)_2 = 1·164$ and $(r_m/r_t)_2 = 0·877$.

Using the same mean diameter angles, calculate $\beta_2$ at tip and root for a constant nozzle angle design in which $\alpha_2$ and $C_{w2}r$ are constant over the annulus. Compare the two designs by sketching the velocity diagrams and commenting qualitatively on such aspects as the radial variation of degree of reaction and blade inlet Mach number.

To satisfy radial equilibrium with constant nozzle angle, the constant angular momentum condition should strictly be replaced by constant $C_w r^{\sin^2 \alpha}$. Show, by recalculating $\beta_2$ at tip and root, that this refinement has only a small effect on the required blade angle.
[$\beta_{2r}$ 35°20′, $\beta_{2t}$ 0°; $\beta_{2r}$ 33°30′. $\beta_{2t}$ 3°12′]

**7.5**   In certain designs the maximum mass flow passed by a turbine may be determined by choking in the turbine annulus at outlet instead of in the turbine nozzles. The maximum mass flow will then depend not merely on the inlet conditions as in the case of expansion in a nozzle, but also upon the work output per unit mass flow. Given fixed values of the inlet conditions $p_{01}$ and $T_{01}$, the temperature equivalent of the specific work output $\Delta T_{013}$, the annulus area $A$ at the turbine outlet and that there is no outlet swirl, show that for an isentropic expansion to a varying outlet static temperature $T_3$ the maximum flow can be expressed by

$$\frac{m\sqrt{T_{01}}}{p_{01}} = \left\{\frac{\gamma}{R}\left[\left(\frac{2}{\gamma+1}\right)\left(1-\frac{\Delta T_{013}}{T_{01}}\right)\right]^{(\gamma+1)/(\gamma-1)}\right\}^{\frac{1}{2}}$$

Comment on the effect this will have on the turbine mass flow versus pressure ratio characteristic.

**8.1**  The following data refer to a single-shaft gas turbine running at its design speed.

| Compressor characteristic | | | Turbine characteristic | | |
|---|---|---|---|---|---|
| $p_2/p_1$ | $m\sqrt{T_1}/p_1$ | $\eta_c$ | $p_3/p_4$ | $m\sqrt{T_3}/p_3$ | $\eta_t$ |
| 5·0 | 32·9 | 0·80 | 5·0 | 14·2 | 0·845 |
| 4·7 | 33·8 | 0·79 | 4·5 | 14·2 | 0·850 |
| 4·5 | 34·3 | 0·77 | 4·0 | 14·2 | 0·842 |

The combustion pressure loss is 5 per cent of the compressor delivery pressure and the ambient conditions are 1·01 bar and 288 K. Mechanical losses can be neglected. The 'non-dimensional' flows are based on $m$ in kg/s, $p$ in bar and $T$ in K, all pressures and temperatures being stagnation values.

Calculate the power output when operating at a turbine inlet temperature of 1100 K. Comment briefly on the variation in thermal efficiency as the load is reduced at constant speed.

[264 kW]

**8.2**  The following data refer to a gas turbine with a free power turbine, operating at design speed.

| Compressor characteristic | | | Gas generator turbine characteristic | | |
|---|---|---|---|---|---|
| $p_2/p_1$ | $m\sqrt{T_1}/p_1$ | $\eta_c$ | $p_3/p_4$ | $m\sqrt{T_3}/p_3$ | $\eta_t$ |
| 5·2 | 220 | 0·82 | 2·50 | 90·2 | 0·85 |
| 5·0 | 236 | 0·83 | 2·25 | 90·2 | 0·85 |
| 4·8 | 244 | 0·82 | 2·00 | 88·2 | 0·85 |

Assuming that the power turbine is choked, the value of $m\sqrt{T_4}/p_4$ being 188, determine the design values of compressor pressure ratio and turbine inlet temperature.

Neglect all pressure losses and assume the mechanical efficiency of the gas generator rotor to be 0·98 and take the ambient temperature as 288 K. The 'non-dimensional' flows quoted are based on $m$ in kg/s, $p$ in bar and $T$ in K, all pressures and temperatures being stagnation values. The suffixes 1, 2, 3 and 4 refer to the following locations:

    1—compressor inlet
    2—compressor delivery
    3—gas generator turbine inlet
    4—power turbine inlet.

[5·10, 1170 K]

**8.3**  When running at a low power condition a gas turbine with free power turbine operates at a compressor pressure ratio of 2·60. The combustion chamber pressure loss is 4 per cent of the compressor delivery pressure and the exhaust pressure loss can be ignored.

The turbine characteristics are given below:

| Gas generator turbine | | | Power turbine | |
|---|---|---|---|---|
| $p_3/p_4$ | $m\sqrt{T_3}/p_3$ | $\eta_t$ | $p_4/p_a$ | $m\sqrt{T_4}/p_4$ |
| 1·3 | 20·0 | 0·85 | 1·4 | 60·0 |
| 1·5 | 44·0 | 0·85 | 1·6 | 85·0 |
| 1·8 | 62·0 | 0·85 | 1·8 | 95·0 |

The 'non-dimensional' flows are based on $m$ in kg/s, $p$ in bar and $T$ in K, all pressures and temperatures being stagnation values.

Calculate the gas generator turbine pressure ratio at this condition. Assuming the compressor characteristic to be known, indicate briefly how you would calculate the turbine inlet temperature.
[1·61]

**8.4** A gas turbine with a free power turbine gives the following results when tested at ambient conditions of 1·0 bar and 288 K.

| $\dfrac{N\sqrt{T_1}}{(\%\ design)}$ | $m\sqrt{T_1}/p_1$ | $p_2/p_1$ | $\eta_c$ |
|---|---|---|---|
| 100 | 454·5 | 4·60 | 0·859 |
| 95 | 420·0 | 4·00 | 0·863 |
| 90 | 370·0 | 3·60 | 0·858 |

The 'non-dimensional' flows are based on $m$ in kg/s, $p$ in bar and $T$ in K, all pressures and temperatures being stagnation values.

The power turbine remains choked for all of these conditions. The gas generator turbine inlet temperature at 95 per cent design *mechanical* speed was found to be 1075 K. Assuming single line turbine flow characteristics, constant turbine efficiency, and constant mechanical efficiency of the gas generator rotor, calculate

(a) the gas generator turbine inlet temperature at design mechanical speed with the same ambient conditions;

(b) the compressor power absorbed when running at 95 per cent design mechanical speed with ambient conditions of 0·76 bar and 273 K.

Sketch the operating line on the compressor characteristic, and discuss the effect of ambient temperature on net power output for a fixed gas generator mechanical speed.
[1215 K, 3318 kW]

**8.5** A simple gas turbine is to be used as a source of compressed air. The turbine produces just enough work to drive the compressor which delivers an airflow, $m_c$, greater than that required for the turbine by an amount $m_b$. The design point operating conditions are:

$$\frac{m_c\sqrt{T_1}}{p_1} = 22\cdot8, \quad \frac{p_2}{p_1} = 4\cdot0$$

and at the design point we also have $T_3/T_1 = 3\cdot3$, $\eta_c = 0\cdot8$ and $\eta_t = 0\cdot85$.

A rough estimate of the equilibrium running line is required on the compressor characteristic for the conditions where $T_3/T_1$ is maintained constant and the amount of air bled off, $m_b$, is varied. For this purpose the efficiencies of the compressor and turbine can be assumed constant, the mechanical transmission loss and the combustion pressure loss can be neglected, and the turbine 'non-dimensional' mass flow can be assumed independent of speed and related to the pressure ratio $r$ by the expression

$$\frac{m\sqrt{T_3}}{p_3} = k \sqrt{\left(1 - \frac{1}{r^2}\right)}$$

where $m = m_c - m_b$ and $k$ is a constant.

As an example of the calculations required, find the operating point on the compressor characteristic for the condition where $m_b\sqrt{T_1}/p_1$ is reduced to three-quarters of its value at the design point. Take $\gamma = 1\cdot40$ throughout.

$[p_2/p_1 = 2\cdot1,\ m_c\sqrt{T_1}/p_1 = 13\cdot0]$

# Appendix C
# References

## Chapter 1

(1) WOOD, G. R. 'The Ruston Tornado: A 6 MW gas turbine for industrial application', *Am. Soc. Mech. Engrs* Paper 81-GT-171, 1981.
(2) MORRIS, R. E. 'The Pratt & Whitney PW 100 – evolution of the design concept', *Can. Aero. and Space Jnl*, **28**, 1982.

## Chapter 2

(1) HODGE, J. *Cycles and Performance Estimation*. Butterworths, 1955.
(2) ROGERS, G. F. C. and MAYHEW, Y. R. *Engineering Thermodynamics Work and Heat Transfer*. Longman, 1980.
(3) BANES, B., MCINTYRE, R. W. and SIMS, J. A. *Properties of Air and Combustion Products with Kerosene and Hydrogen Fuels*. AGARD, 1967.
(4) FIELDING, D. and TOPPS, J. E. C. *Thermodynamic Data for the Calculation of Gas Turbine Performance*. H.M.S.O., A.R.C., R. & M. No. 3099, 1959.
(5) YANG, O. *et al.* 'A study on the total energy system of the gas turbine', *Bull. JSME*, **14**, 1971, 234–47.

## Chapter 3

(1) MAYHEW, Y. R. and ROGERS, G. F. C. *Thermodynamic and Transport Properties of Fluids—SI units*. Blackwell, 1980.
(2) 'Relationships between some common intake parameters', *R. aeronaut. Soc., data sheet* 66028, 1966.
(3) HESSE, W. J. and MUMFORD, N. V. S. *Jet Propulsion for Aerospace Applications*. Pitman, 1964.
(4) ASHWOOD, P. F. 'A review of the performance of exhaust systems for gas-turbine aero-engines', *Proc. Instn mech. Engrs*, **171**, 1957, 129–58.
(5) ROGERS, G. F. C. and MAYHEW, Y. R. 'One-dimensional irreversible gas flow in nozzles', *Engineering, Lond.*, **175**, 1953, 355–8.
(6) FROST, T. H. 'Practical bypass mixing systems for fan jet aero engines', *Aeronaut. Q.*, **17**, 1966, 141–60.
(7) SARAVANAMUTTOO, H. I. H. 'Development of modern turboprop engines', AGARD Conference ProceedingsNo. 366, *Aerodynamics and Acoustics of Propellers*, 1984.

# Chapter 4

(1) WHITTLE, F. 'The early history of Whittle jet propulsion gas turbines', *Proc. Instn mech Engrs*, **152**, 1945, 419–35.
(2) CHESHIRE, L. J. 'The design and development of centrifugal compressors for aircraft gas turbines', *Proc. Instn mech. Engrs*, **153**, 1945, 426–40.
(3) WRONG, C. B. 'An introduction to the JT15D engine', *Trans. Am. Soc. mech. Engrs*, 69-GT-119, 1969.
(4) STANITZ, J. D. 'Some theoretical aerodynamic investigations of impellers in radial and mixed-flow centrifugal compressors', *Trans. Am. Soc. mech. Engrs*, **74**, 1952, 473–97.
(5) KENNY, D. P. 'A novel low-cost diffuser for high-performance centrifugal compressors', *Trans. Am. Soc. mech. Engrs*, Series A, **91**, 1969, 37–46.
(6) FERGUSON, T. B. *The Centrifugal Compressor Stage*. Butterworths, 1963.
(7) HANKINS, G. A. and COPE, W. F. Discussion on 'The flow of gases at sonic and supersonic speeds', *Proc. Instn mech. Engrs*, **155**, 1947, 401–16.
(8) ENGINEERING SCIENCES DATA UNIT: Fluid Mechanics—internal flow, 4 – Duct expansions and duct contractions, Data Sheets 73024, 74015, 76027.
(9) CAME, P. M. 'The development, application and experimental evaluation of a design procedure for centrifugal compressors', *Proc. Instn mech. Engrs*, **192**, No. 5, 1978, 49–67.
(10) HERBERT, M. V. 'A method of performance prediction for centrifugal compressors', *H.M.S.O., A.R.C., R. & M*. No. 3843, 1980.

# Chapter 5

(1) CONSTANT, H. 'The early history of the axial type of gas turbine engine', *Proc. Instn mech. Engrs*, **153**, W.E.P. No. 12, 1954.
(2) HOWELL, A. R. 'Fluid dynamics of axial compressors' and 'Design of axial compressors', *Proc. Instn mech. Engrs*, **153**, W.E.P. No. 12, 1954.
(3) HOWELL, A. R. 'The present basis of axial compressor design Part I – Cascade Theory', *Aeronaut. Res. Coun.*, R. & M. No. 2095, 1942.
(4) JOHNSEN, I. A. and BULLOCK, R. O. *Aerodynamic Design of Axial-flow Compressors*. NASA SP-36, 1965.
(5) HORLOCK, J. H. *Axial Flow Compressors*. Butterworth, 1958.
(6) LIEBLEIN, S. and JOHNSEN, I. A. 'Resume of transonic compressor research at NACA Lewis Laboratory', *Trans. Am. Soc mech. Engrs*, Journal of Engineering for Power, **83**, 1961, 219–34.
(7) TODD, K. W. 'Practical aspects of cascade wind tunnel research', *Proc. Instn mech. Engrs*, **157**, W.E.P. No. 36, 1947.
(8) GOSTELOW, J. P. *Cascade Aerodynamics*. Pergamon Press, 1984.
(9) CARTER, A. D. S. 'Blade profiles for axial flow fans, pumps, compressors, etc.', *Proc. Instn mech. Engrs*, **175**, No. 16, 1961, 775–88.
(10) MILLER, G. R., LEWIS, G. W. and HARTMAN, M. J. 'Shock losses in transonic blade rows', *Trans. Am. Soc. mech. Engrs*, Journal of Engineering for Power, **83**, 1961, 235–42.
(11) SCHWENK, F. C., LEWIS, G. W. and HARTMAN, M. J. 'A preliminary analysis of the magnitude of shock losses in transonic compressors', NACA RM E57A30, 1957.
(12) KERREBROCK, J. L. 'Flow in transonic compressors', *A.I.A.A. Journal*, **19**, 1981, 4–19.
(13) GREITZER, E. M. 'Review – axial compressor stall phenomena', *Trans. Am. Soc mech. Engrs*, Journal of Fluids Engineering, **102**, 1980, 134–151.

(14) STONE, A. 'Effects of stage characteristics and matching on axial-flow-compressor performance', *Trans. Am. Soc mech. Engrs*, **80**, 1958, 1273–93.
(15) CARCHEDI, F. and WOOD, G. R. 'Design and development of a 12:1 pressure ratio compressor for the Ruston 6 MW gas turbine', *Trans. Am. Soc mech. Engrs*, Journal of Engineering for Power, **104**, 1982, 823–31.
(16) DAVIS, W. R. and MILLAR, D. A. J. 'A comparison of the matrix and streamline curvature methods of axial flow turbomachinery analysis, from a user's point of view', *Trans. Am. Soc. mech. Engrs*, Journal of Engineering for Power, **97**, 1975, 549–60.
(17) DENTON, J. D. 'An improved time marching method for turbomachinery flow calculation', *Am. Soc. mech. Engrs*, Paper 82-GT-239, 1982.
(18) McNALLY, W. D. and SOCKOL, P. M. 'Review – Computational methods for internal flows with emphasis on turbomachinery', *Trans. Am. Soc mech. Engrs*, Journal of Fluids Engineering, **107**, 1985, 6–22.
(19) FREEMAN, C. and STOW, P. 'The application of computational fluid mechanics to aero gas turbine compressor design and development', *Instn. mech. Engrs*, Conference Publications 1984–3, C70/84.

# Chapter 6

(1) SMITH, I. E., ed. *Combustion in Advanced Gas Turbine Systems*. Pergamon, 1968.
(2) HAWTHORNE, W. R. and OLSON, W. T., eds. *Design and Performance of Gas Turbine Power Plants*. Oxford U.P., 1960.
(3) *Some Developments in Techniques for Temperature Measurement*, Instn mech. Engrs symposium, 1962.
(4) SPALDING, D. B. *Some Fundamentals of Combustion*. Butterworths Scientific Publications, 1955.
(5) *Technical Advances in Gas Turbine Design*, Instn mech. Engrs symposium, 1969.
(6) JONES, R. E. 'Gas turbine engine emissions – problems, progress and future', *Progress in Energy and Combustion Science*, **4**, 1978, 73–113.

# Chapter 7

(1) HAWTHORNE, W. R., ed. *Aerodynamics of Turbines and Compressors*. Oxford U.P., 1964.
(2) HORLOCK, J. H. *Axial Flow Turbines*. Butterworth, 1966.
(3) AINLEY, D. G. and MATHIESON, G. C. R. 'An Examination of the flow and pressure losses in blade rows of axial flow turbines', *H.M.S.O.*, *A.R.C.*, *R. & M.* 2891, 1955.
(4) JOHNSTON, I. H. and KNIGHT, L. R. 'Tests on a single-stage turbine comparing the performance of twisted with untwisted rotor blades', *H.M.S.O.*, *A.R.C.*, *R. & M.* 2927, 1953.
(5) ANDREWS, S. J. and OGDEN, H. 'A detailed experimental comparison of (compressor) blades for free vortex flow and equivalent untwisted constant section blades', *H.M.S.O.*, *A.R.C.*, *R. & M.* 2928, 1953.
(6) SAWYER, J. W., ed. *Gas Turbine Engineering Handbook*. Gas Turbine Publications, 1976.
(7) SMITH, D. J. L. 'Turbulent boundary layer theory and its application to blade profile design', *H.M.S.O.*, *A.R.C.*, *C.P.* 868, 1966.

(8) AINLEY, D. G. and MATHIESON, G. C. R. 'A method of performance estimation for axial-flow turbines', *H.M.S.O., A.R.C., R. & M.* 2974, 1957.
(9) AINLEY, D. G. 'Internal air-cooling for turbine blades – a general design survey', *H.M.S.O., A.R.C., R. & M.* 3013, 1957.
(10) HAWTHORNE, W. R. 'Thermodynamics of cooled turbines' Parts I and II. *Trans. Am. Soc. mech. Engrs*, **78**, 1956, 1765–81.
(11) BARNES, J. F. and FRAY, D. E. 'An experimental high-temperature turbine (No. 126)', *H.M.S.O., A.R.C., R. & M.* 3405, 1965.
(12) Technical Advances in Gas Turbine Design. Instn mech. Engrs symposium, 1969.
(13) DUNHAM, J. and CAME, P. M. 'Improvements to the Ainley–Mathieson method of turbine performance prediction', *Am. Soc. mech. Engrs* paper 70-GT-2, 1970.
(14) KACKER, S. C. and OKAPUU, U. 'A mean line prediction method for axial flow turbine efficiency', *Am. Soc. mech. Engrs* paper 81-GT-58, 1981.
(15) MOWILL, J. and STROM, S. 'An advanced radial-component industrial gas turbine', *Trans. ASME Jnl of Engng for Power*, **105**, 1983, 947–52.
(16) DIXON, S. L. *Fluid Mechanics, Thermodynamics of Turbomachinery.* Pergamon, 1975.
(17) *Aerodynamic Performance of Radial Inflow Turbines.* First (1963) and Second (1964) Reports, Motor Industry Research Association. (Alternatively, the work is summarized in: HIETT, G. F. and JOHNSTON, I. H. 'Experiments concerning the aerodynamic performance of inward radial flow turbines', *Proc. Instn mech. Engrs*, **178**, Part 3I(ii), 1964.)
(18) BENSON, R. S. 'A review of methods for assessing loss coefficients in radial gas turbines', *Int. J. mech. Sci.*, **12**, 1970, 905–32.
(19) BRIDLE, E. A. and BOULTER, R. A. 'A simple theory for the prediction of losses in the rotors of inward radial flow turbines', *Proc. Instn mech. Engrs*, **182**, Part 3H, 1968.
(20) BENSON, R. S. 'Prediction of performance of radial gas turbines in automotive turbochargers', *Am. Soc. mech. Engrs* paper 71-GT-66, 1971.

## Chapter 9

(1) MALLINSON, D. H. and LEWIS, W. G. E. 'The part-load performance of various gas-turbine engine schemes', *Proc. Instn mech. Engrs*, **159**, 1948, 198–219.
(2) TREWBY, G. F. A. 'British naval gas turbines', *Trans. Instn marine Engrs*, **66**, 1954, 125–67.
(3) SWATMAN, I. M. and MALOHN, D. A. 'An advanced automotive gas turbine concept', *Trans. S.A.E.* **69**, 1961, 219–27.
(4) BARBEAU, D. R. 'The performance of vehicle gas turbines', *Trans. S.A.E.* **76**[v], 1967, 90.
(5) KAYS, W. M. and LONDON, A. L. *Compact Heat Exchangers.* McGraw-Hill, 1964.
(6) RAHNKE, C. J. 'The variable-geometry power turbine', *Trans. S.A.E.*, **78**[i], 1969, 213–23.
(7) TURUNEN, W. A. and COLLMAN, J. S. 'The General Motors Research GT-309 gas turbine engine', *S.A.E. Trans.*, **74**[iv], 1965, 357–77.
(8) YOUNG, P. H. 'Propulsion controls on the Concorde', *Jnl. R.Ae.S.*, **70**, 1966, 863–81.
(9) SARAVANAMUTTOO, H. I. H. and FAWKE, A. J. 'Simulation of gas turbine dynamic performance', *Am. Soc. Mech Engrs* paper 70-GT-23, 1970.
(10) FAWKE, A. J. and SARAVANAMUTTOO, H. I. H. 'Experimental investigation of methods for improving the dynamic response of a twin-spool turbojet engine', *Trans. Am. Soc. mech. Engrs*, **93**, series A, 1971, 418–24.
(11) FAWKE, A. J. and SARAVANAMUTTOO, H. I. H. 'Digital computer methods for prediction of gas turbine dynamic response', *Trans. S.A.E.*, 80 [iii], 1971, 1805–13.

(12) SARAVANAMUTTOO, H. I. H. and MacISAAC, B. D. 'An overview of engine dynamic response and mathematical modelling concepts', AGARD Conference Proceedings No. 324, 'Engine Handling', 1982.
(13) SARAVANAMUTTOO, H. I. H. and MacISAAC, B. D. 'Thermodynamic models for pipeline gas turbine diagnostics', *Trans. Am. Soc mech. Engrs*, **105**, Series A, 1983, pp. 875–84.

## Appendix A

(1) HOUGHTON, E. L. and BROCK, A. E. *Tables for the Compressible Flow of Dry Air.* Arnold, 1970.
(2) KEENAN, J. H. and KAYE, J. *Gas Tables*, Wiley, 1948.
(3) SHAPIRO, A. H. *The Dynamics and Thermodynamics of Compressible Flow.* Ronald. 1954.
(4) LIEPMANN, H. W. and ROSHKO, A. *Elements of Gas Dynamics.* Wiley, 1957.

## Supplementary texts

(1) BATHIE, W. W. *Fundamentals of Gas Turbines.* John Wiley & Sons, 1984.
(2) DIXON, S. L. *Fluid Mechanics, Thermodynamics of Turbomachinery*, 3rd Edition. Pergamon, 1978.
(3) HARTMAN, R. T. C. *Gas Turbine Engineering.* Macmillan, 1981.
(4) LEFEBVRE, A. H. *Gas Turbine Combustion.* McGraw-Hill, 1983.
(5) OATES, G. C. *Aerothermodynamics of Gas Turbine and Rocket Propulsion.* AIAA Education Series, 1984.
(6) SAWYER, J. W. ed. *Gas Turbine Handbook.* Gas Turbine Publications, 3rd Edition, 1980.
(7) WILSON, D. G. *Design of High Efficiency Turbomachinery and Gas Turbines.* MIT Press, 1984.

# Index